"Using his nimble narrative gifts, Mr. Ingrassia turns the creation stories behind the Prius and other cars into gripping accounts of how visionary design, corporate competition and inventive engineering combined to produce automobiles that would come to represent an era or a mind-set."

—Michiko Kakutani, *The New York Times*

"Highly entertaining . . . lucid . . . *Engines of Change* informed and charmed me . . ."

—Joseph Epstein, *The Wall Street Journal*

"You will never look at a car the same way after reading *Engines of Change*—as I strongly recommend to anyone who relishes great storytelling that combines biography, social and political history, science, and romance. Having driven and virtually lived in a 1953 Plymouth on a year's journey across Eisenhower's America, and having followed that up many driving years later by writing on the innovations of Henry Ford, I thought I knew something of the history of cars. I was all the more surprised—and vastly entertained—by the riches in Ingrassia's stories of fifteen vehicles embodying the American dream from the Model T to the Beetle, the Corvair, the Corvette, and the Mustang to the pickups and the Prius (driven by the Pious). Even readers who cannot tell a camshaft from a camiknicker will find fascination in a gallery of characters depicted by Ingrassia with vivacity and wit."

—Sir Harold Evans

"Entertaining and instructive . . ."

—George Will, *The Washington Post*

"In *Engines of Change*, Mr. Ingrassia arguably does for cars and culture what David Halberstam did for a decade in *The Fifties*. History well researched, made alive, relevant and eminently readable."

—John Lamm, *The New York Times*

"Sure, cars suck up gas, and they promote suburban sprawl, but they also help drive the economy, and drive families from home to school to soccer field. And, of course, cars fire our imaginations. Paul Ingrassia, who won a Pulitzer Prize for reporting from Detroit for *The Wall Street Journal,* has written a book about cars that may not all be cherished classics or engineering marvels, but have earned a place in America's scrapbook."

—Scott Simon, National Public Radio

"Paul Ingrassia knows where the bodies are buried, or maybe where the keys to the American car business got lost. With a swift, sure scalpel honed by years as the industry reporter, he anatomizes Detroit in all its glory and inglorious decline. A thoughtful, propulsive assay of the machine that changed a nation, a world."

—Dan Neil, car critic, *The Wall Street Journal*

"In this new book, Ingrassia traces the history of some iconic cars and how those models reflected shifts in politics, culture, and technology. He also takes readers inside the industry, skillfully navigating among the soaring tail fins, egomaniacal visionaries, and corporate intrigue that surrounded the creation of these vehicles."

—*Boston Globe*

"The prose is lapidary, the tone informed by humor. Paul Ingrassia has written an automobile book that goes beyond the genre; it's for anyone interested in modernity and what led us to where we are."

—Miles Collier, The Revs Institute for Automotive Research

"Ingrassia takes great pleasure in historical irony, and the unpredictable conclusion of each car's story is so fascinating even those who prefer their MetroCard to the BQE will appreciate the inherent paradoxes of the vehicle's road to glory."

—*New York Daily News*

"Ingrassia succeeds in fashioning well-researched, swift-paced narratives around each of these 15 select automobiles. Using colorful detail, he effectively recasts these significant driving machines in their respective cultural contexts and brings to life the eras they influenced."

— *Kirkus Reviews*

"Paul Ingrassia . . . is probably the best broadsheet reporter ever to cover the car business. . . . Picking 15 vehicles as tent poles for this sprawling canvas was a good idea, and Ingrassia chose well. . . . Any book on a topic so overwhelming as the car in America has to be more of a goad to, than a proof of, argument. And here Ingrassia has succeeded."

— *Weekly Standard*

"Paul Ingrassia's *Engines of Change: A History of the American Dream in Fifteen Cars* ranges as widely and quirkily as the title suggests among the people, passions and foibles of the automotive industry. As a journalist for the *Wall Street Journal,* Ingrassia shared a 1993 Pulitzer Prize for writing on General Motors Co. In this book he lets out the journalistic stays, enjoying the freedom to openly needle an industry and admire its pioneers without any loss of the good reporter's delight in detail and a fine tale."

— Jeffrey Burke, *Bloomberg BusinessWeek*

"The whole country in fifteen cars—that's crowded! And *Engines of Change* is indeed packed from rocker panels to sunroof with good stories and salient facts about the automobiles that shaped America, from the oddity of the Model T to the oddballs driving the Prius."

— P.J. O'Rourke

ALSO BY PAUL INGRASSIA

Crash Course: The American Automobile Industry's Road to Bankruptcy and Bailout—and Beyond

Comeback: The Fall & Rise of the American Automobile Industry (with Joseph B. White)

Engines of Change

A HISTORY OF THE AMERICAN DREAM
IN FIFTEEN CARS

PAUL INGRASSIA

SIMON & SCHUSTER PAPERBACKS

NEW YORK   LONDON   TORONTO   SYDNEY   NEW DELHI

For Charlie

Simon & Schuster
1230 Avenue of the Americas
New York, NY 10020

First Simon & Schuster trade paperback edition May 2013

SIMON & SCHUSTER PAPERBACKS and colophon are registered
trademarks of Simon & Schuster, Inc.

For information about special discounts for bulk purchases,
please contact Simon & Schuster Special Sales at
1-866-506-1949 or business@simonandschuster.com

The Simon & Schuster Speakers Bureau can bring authors to
your live event. For more information or to book an event,
contact the Simon & Schuster Speakers Bureau at
1-866-248-3049 or visit our website at www.simonspeakers.com.

Designed by Ruth Lee-Mui

Manufactured in the United States of America

1 3 5 7 9 10 8 6 4 2

The Library of Congress has cataloged the hardcover edition as follows:

Ingrassia, Paul.
Engines of change : a history of the American dream
in fifteen cars / Paul Ingrassia.
p. cm.
1. Automobiles—United States—History. 2. Automobiles—
Social aspects—United States—History. I. Title.
TL23.I54 2012
629.2220973—dc23
2012002303

ISBN 978-1-4516-4063-2
ISBN 978-1-4516-4064-9 (pbk)
ISBN 978-1-4516-4065-6 (ebook)

CONTENTS

Introduction   xi

1. When Henry Met Sallie: Car Wars and Culture Clashes
   at the Dawn of America's Automotive Age   1
2. Zora, Zora, Zora: A Bolshevik Boy Escapes the Nazis
   and Saves the Great American Sports Car   31
3. The 1959 Cadillacs: Style, Status, and the Race for
   the Biggest Tail Fins Ever   57
4. Volkswagen's Beetle and Microbus: The Long and Winding
   Road from Hitler to the Hippies   81
5. The Chevy Corvair Makes Ralph Nader Famous,
   Lawyers Ubiquitous, and (Eventually) George W. Bush
   President of the United States   111
6. Turning a "Librarian" into a "Sexpot": The Youth
   Boom, the Sixties, and the Making of the Mustang   141
7. The Brief but Glorious Reign of John Z. DeLorean
   and the Pontiac GTO   163
8. Ohio Gozaimasu: Godzilla, Mr. Thunder, and
   How a Little Japanese Car Became America's
   Big Ichiban   191

CONTENTS

9. The Chrysler Minivans: Baby Boomers Become
   Soccer Moms and a, um, Driving Force in
   American Politics   219
10. The BMW 3 Series: The Rise of the Yuppies and
    the Road to Arugula   241
11. The Jeep: From War to Suburbia, or How to Look
    Like You're Going Rock Climbing When You're
    Really Going to Nordstrom   265
12. The Ford F-Series: Cowboys, Country Music, and
    Red-Meat Wheels for Red-State Americans   289
13. An Innovative Car (the Prius), Its Insufferable Drivers
    (the Pious), and the Advent of a New Era   313

Afterword   341
Acknowledgments   347
Notes   349
Selected Bibliography   373
Index   377

Engines of Change

INTRODUCTION

In the 1990s Chrysler executives used to joke that they could tell a Jeep driver from a minivan driver just by looking at his or her watch. Timex watches were for practical people, those utterly unconcerned about putting up appearances and thus unworried about driving a mommy-mobile. Even if they were mommies.

But Rolexes signaled people whose self-image couldn't cope with a minivan and who wanted to flaunt their rugged, outdoor lifestyle. Even if ruggedness only meant hitting potholes en route to the mall in their Jeep Grand Cherokee Orvis Edition. With a skinny Venti Latte in the cup holder, of course.

For decades, the connection between cars and self-image has been understood and appreciated by prominent philosophers. Consider the Beach Boys. Their song "Fun, Fun, Fun" (1964) wasn't so much about the Ford Thunderbird as about the free-spirited teenaged girl who drove one.

Other songs celebrated drivers whose personas defied the stereotypes of their cars. "The Little Old Lady from Pasadena" (Jan

& Dean, 1964) was about a granny who flew around the freeways in a "brand new shiny red Super Stock Dodge." In researching this book I learned about a real-life grandmother who terrorized other drivers just like the one in the song, except that she drove a Mustang. She lived in Oconomowoc, Wisconsin, which probably explains why she didn't wind up in a song. "The Little Old Lady from Oconomowoc" would have left a whole generation of Americans tongue-tied.

A handful of cars in American history, however, rose above merely defining the people who drove them. Instead, like some movies (e.g., *The Big Chill*) and some books (*The Catcher in the Rye,* etc.), they defined large swaths of American culture, helped to shape their era, and uniquely reflected the spirit of their age. These cars, and the cultural trends that they helped define, are the subject of this book.

The underlying premise here is that modern American culture is basically a big tug-of-war. It's a yin-versus-yang contest between the practical and the pretentious, the frugal versus the flamboyant, haute cuisine versus hot wings, uptown versus downtown, big-is-better versus small-is-beautiful, and Saturday night versus Sunday morning. The elemental conflict between these two sets of values is amply evident in the contrast between the first two cars in the book, the Ford Model T and the General Motors LaSalle. The inexpensive Model T was the pinnacle of practicality, the first people's car.

It didn't have much style. But during its remarkable twenty-year run, from 1908 to 1927, the Model T gave farm families unprecedented mobility and a taste of the city lights. Even if they were only the lights of Muncie.

The LaSalle, in contrast, was the first mass-market designer car, an early yuppie-mobile that was intended for getting attention as well as for getting around. It debuted in 1927, the very year that the Model T died. The two cars were perfect bookends. They're the only pre–World War II cars in this book, because American cultural evolution hit a roadblock in the 1930s and 1940s.

Those decades, of course, were dominated by Depression and war. Production of civilian cars was suspended during World War II, and Detroit's factories were converted to produce planes and tanks. American cultural upheaval, at least overtly, took a time-out, too. During the Thirties and Forties Americans were mostly focused on finding food and work, and on staying alive.

But twenty-five straight years of virtually nonstop Depression and war came to an end with the armistice in Korea in 1953. By then Americans were ready to let loose, which made the timing perfect for the Chevrolet Corvette, the first modern American sports car. The pivotal figure in Corvette history was a Chevrolet engineer who had been raised as a Bolshevik boy in Russia before coming to America and winding up at General Motors.

The defining design statement of postwar America's sky's-the-limit ethos was tail fins. Chrysler actually sold them as safety devices, and the company's top designer was honored by the Harvard Business School (really). Chrysler almost won Detroit's great tail fin war, but General Motors struck back in 1959 with Cadillacs that had the biggest tail fins ever.

The man who designed them, Chuck Jordan, died in late 2010, at age eighty-three. I was fortunate to have interviewed him before he passed away. Nearly half a century on, Jordan told the tale of the fins with relish.

Corvettes and tail fins were all about pretension, but an import from Germany, America's erstwhile wartime enemy, pulled the pendulum back toward practicality. The little Volkswagen Beetle, which was the ultimate anti-Cadillac, debuted in the 1930s as Adolf

Hitler's "people's car." Decades later it became the unofficial car of American hippies, completing an automotive and cultural journey of epic sweep.

Volkswagen marketed the Beetle with clever, self-deprecating advertising. One ad told how a farm couple in the Ozarks, living in a log cabin, bought a Beetle to replace their dearly departed mule.

GM's belated response to the Beetle was the practical but problematic Chevrolet Corvair, launched in late 1959, which fared less well. It inspired an unknown young lawyer named Ralph Nader to write a book called *Unsafe at Any Speed*.

The Corvair's lasting legacy has been America's greatest growth industry: lawsuits. The car sentenced millions of Americans to watch personal-injury television commercials from law firms, unless they're really quick with the remote control button.

Every vehicle in this book represents either practicality or pretension, although a couple of them straddle the great divide. Pickup trucks started out as down-and-dirty work tools until Detroit discovered it could make billions by selling lavish designer trucks. The seats on some of them have more leather than most cows.

For many people, especially those under thirty-five, cars aren't nearly as important as iPads, iPods, cell phones, apps, personal computers, and BlackBerries. The modern fascination with electronic devices might make the idea of writing about the social significance of the Ford Mustang seem quaint, a relic of an era when "laptop" wasn't a high-tech term.

But cars continue to provide unique personal freedom and mobility. They spawn powerful emotions, experiences, and memories, of family road trips, one's first car, or one's first sexual adventure.

Hardly anyone keeps the purchase papers for their first computer, or gives the device a name. But some Americans do both for their first cars. The Beach Boys sang a song about a drag race called

"Shut Down," but nobody has yet recorded one called "Download."

This book reflects my fascination with cars and car culture, which grew slowly, over many decades. As a boy in 1950s Laurel, Mississippi (where our family was known as the EYE-talians), I learned about the early explorers through our cars: Hudsons and DeSotos. It took me years to realize that not all cars were station wagons.

In 1960, when I was ten, we moved to suburban Chicago and became a two-car family for the first time, just like millions of other Americans. Our station-wagon treks to visit my two grandmothers back East created a family tradition: the annual breakdown on the Pennsylvania Turnpike. I cherish the memories . . . sort of.

I wasn't a hot-rodder in high school. In fact, I wasn't a hot anything. But automobiles were taking on important dimensions in my adolescent psyche.

In 1964, the year that the Mustang and the Pontiac GTO debuted, I found myself in Mr. McGowan's freshman math class at St. Francis High School in Wheaton, Illinois. I hated math, but Mr. McGowan, bless him, gave us boys an occasional break from dreary decimals and unfathomable fractions to talk about cars.

We discussed whether Pontiac or Chevy had the better lines that year (it was Pontiac, as I recall), and whether anybody's dad had bought a Mustang V8 (yes, but not my dad). Alas, I learned to drive on Dad's pedestrian six-cylinder Chevy Bel Air with a three-speed stick shift on the steering column, the proverbial "three on the tree."

I didn't buy my first car until college at the University of Illinois. It was a 1969 Chevy Nova with an anemic six-cylinder engine that made passing on the two-lane roads around Champaign a death-defying adventure. That Nova cost around $2,200, about the price of a new fender today.

The hot car in college, a '69 Pontiac GTO, belonged to my friend Dale Sachtleben. Once we took it on a road trip out East and Dale got a speeding ticket in Delaware for going ninety miles an hour, which was his slowest speed of the entire trip.

More than thirty years later, when Pontiac launched an updated version of the long-dead GTO, I reconnected with Dale for a nostalgia-trip test-drive in the new model. He had gray hair, I was a cancer survivor, and we were both (gulp) Republicans.

We drove the new GTO to tiny Greenview, Illinois, his hometown just north of Springfield, best described as sitting at the corner of corn and soybeans. There we raced up and down the little farm road that once served as the local drag strip, and for a while we were kids again. Cars can do that.

My first job out of college was in 1973 in Decatur, Illinois, where our next-door neighbors, the Whitneys, owned a 1960 Thunderbird. By then Ford had added a backseat to the 'Bird, which had been a taut two-seater when it debuted in 1955, but the low-slung, boulevard-cruiser styling remained. The Whitneys' son Clay, then a boy and now a business owner, still keeps the classic car today.

The first car in this book that I owned was way more pedestrian and practical: a 1984 Chrysler minivan. My wife and I were in Cleveland, where I worked for the *Wall Street Journal*, and we had three boys under age six.

The minivan's interior was so spacious that it seemed to have designed-in demilitarized zones that kept the kids from killing each other, and from driving us nuts on road trips. For a few years, before they became mommy-mobiles, minivans actually were cool. (No kidding.)

In 1985 my boss, *Journal* managing editor Norm Pearlstine, transferred me to Detroit. It wasn't clear what I had done to be sentenced to Detroit after living in Cleveland, but the truth was I loved it.

I felt like an anthropologist living among exotic natives who worshipped strange gods called "Multivalve Engine," "Zero to Sixty," and "Pound-Feet of Torque." Not to mention the twin deities "Intercooler" and "Supercharger," devices that boosted the power of internal combustion engines, and thus made them worthy of worship at the automotive altar.

In Detroit I discovered the nuts and bolts, pardon the pun, of the car business, from balance sheets to balance shafts. Covering the auto industry was a journey of discovery that led to a Pulitzer Prize. Our son Charlie, then in grade school, told his teachers his dad had just won the "Pulitzer Surprise." In my case he was about right.

In 1994 I left Detroit to become an executive with the *Journal*'s parent company, Dow Jones. But my fascination with the automobile industry and its culture continued, partly because I wrote occasional car reviews for one of the company's magazines, *Smart-Money*.

One vehicle I tested was the Ford Excursion, launched in 2000 as the largest SUV ever. The Excursion was so big that Ford held the press preview in Montana, about the only place the vehicle would fit.

I skipped that event and, instead, drove the Excursion through the somewhat tighter confines of Greenwich Village, where I managed to parallel park it on a street. Of course the two curbside wheels climbed onto the sidewalk, but the Excursion was so obscenely heavy (about four tons) that I didn't even feel the bump.

The Nissan Titan pickup truck, which I reviewed in late 2003, was almost equally massive. I attended that press preview, held in Napa Valley, as much for the wine as for the roads.

Around that time I started thinking about this book. Automobiles are ubiquitous. But there's little appreciation of how certain cars have reflected the way we think and live, becoming shapers and symbols of their eras. I found the stories of the cars and of the people behind them to be full of surprising twists and turns, even though I had been writing about the auto industry for nearly twenty-five years.

The research, which I began in 2007, took me all around America. I drove Priuses in Michigan, Jeeps through Colorado, and pickup trucks around the Texas Hill Country and Midtown Manhattan, feeling right at home in the former and like an alien invader in the latter. At least the pickup I drove in New York didn't have a Confederate flag decal or a gun rack.

I attended a slew of car conventions and shows, including the centennial celebration for the Model T Ford in Indiana in the summer of 2008. One man there had driven his Model T from "UCLA," which he explained meant the "Upper Corner of Lower Alabama."

His hometown there happened to be Monroeville, also the home of the reclusive Harper Lee, author of *To Kill a Mockingbird*. It was a book that, like the Model T Ford, had reshaped American life and thought.

At one car show a man displaying his AMC Gremlin told me his favorite story: about a woman who confided she had been conceived in the backseat of a Gremlin. Not a good start in life. Later comedian Jon Stewart told me his first car was a 1975 Gremlin, and that his cat peed in the backseat on the day of his high school graduation. No wonder he can laugh at anything.

The annual Bloomington Gold Corvette exhibition in Illinois and the National Corvette Museum in Kentucky provided memorable visits. The museum, complete with relics and records from the car's history, should be called the Corvette Cathedral.

The epitome of car shows is the annual Concours d'Elegance held every August at Pebble Beach, California. There's nothing like gathering at 6 a.m. to feel the cold mist rolling in from the ocean and watch the priceless Hispano-Suizas and Delage De Villars Roadsters—names that shouldn't be pronounced without an affected accent—rolling onto the 18th fairway for the show. Sacred cars on sacred ground.

I also tackled the more mundane but ultimately rewarding work of delving into the depths of automotive archives. They included the Benson Ford Research Center in Dearborn, Michigan, the Collier Museum and Library in Naples, Florida, various branches of the New York Public Library, and the National Automotive History Collection at the Detroit Public Library.

My research was hitting high gear when, suddenly, I took a detour. In the fall of 2008 Detroit's car companies careened into a crisis that resulted in the bankruptcies of two of them: General Motors and Chrysler. These were historic, albeit tragic, events, and I was pulled into writing about them.

The result was *Crash Course*, my 2010 book about the bailouts and bankruptcies, which was really a book about human behavior. So is, in a different sense, *Engines of Change*.

When I returned to this book, after my hiatus, I travelled to the National Steinbeck Center in Salinas, California, a shrine to native son and Nobel Prize–winning author John Steinbeck, who often put cars in his books. *Cannery Row* says the Model T revolutionized sex and marriage. *The Grapes of Wrath* describes a bitter trek westward in a makeshift pickup truck. The Steinbeck Center still displays the dark-green GMC pickup that the author drove around America to write his 1962 travelogue, *Travels with Charley*.

The research also took me to Japan and to a less likely overseas destination: Copenhagen. There I participated in a weekend road rally with the Cadillac Club of Denmark, an organization whose very existence caught me by surprise. The club's members cruise past medieval castles in tail-finned cars that seem almost equally medieval, even though they were "alive" during my lifetime.

My research was both a literary and an automotive journey. I reread *On the Road*, Jack Kerouac's automotive journey of self-discovery. And two wonderful books, also from the 1950s, that lampooned the automotive excesses of the era: *The Insolent Chariots* and *The Hidden Persuaders*. P. J. O'Rourke's *Republican Party Reptile* captured the close connection between pickup trucks and beer.

This road of research also included film and television, including *The Roy Rogers Show* of my boyhood, *Route 66* of my adolescence, and *Curb Your Enthusiasm* of my, um, maturity. Automobiles played prominent roles in all three, as well as many other TV shows and movies.

In the afterword I'll mention some cars that might have been included in this book but weren't. My reasoning probably won't sit well with fans of cars that didn't make the cut, but this isn't a book about the best cars or the worst cars. Instead it's about the cars that helped to shape how we think and live, which is a very select group. Put another way, like a play in four acts, this is a history of modern American culture in fifteen cars.

Whether the cars shaped the culture or the culture shaped the cars is just another version of whether the chicken came before the egg, or vice versa. Let's just say it's both. But I will answer straightaway a question I suspect some readers will want to ask me: What kind of car do you drive?

If you really need to know . . . it's a red one.

1

WHEN HENRY MET SALLIE: CAR WARS AND CULTURE CLASHES AT THE DAWN OF AMERICA'S AUTOMOTIVE AGE

Someone should write an erudite essay on the moral, physical, and esthetic effect of the Model T Ford on the American nation. Two generations of Americans knew more about the Ford coil than the clitoris, about the planetary system of gears than the solar system of stars.

—John Steinbeck, *Cannery Row*[1]

Just north of downtown Detroit on a small street called Piquette sits an inner-city storefront church called the Abundant Faith Cathedral. By the looks of the surrounding weed-choked lots and empty factories, abundant faith is exactly what's needed, not to mention plenty of hope. The neighborhood is a postindustrial ghetto, although right across the street from the church is a functioning business called the General Linen & Uniform Service. It occupies the first floor of an old building where, as unlikely as it seems, modern America began.

It was here, in the early autumn of 1908, that Henry Ford

started producing a car called the Model T, so named because it followed previous Ford cars called the Models N, R, and S. But the Model T was so radically different that Henry Ford had to fight to build it, even within the company that bore his name.

Instead of being built and priced for America's emerging industrial elite like most other cars of its day, the Model T was simple, practical, and affordable. "I will build a motor car for the great multitude," Henry Ford said. "No man making a good salary will be unable to own one—and enjoy with his family the blessing of hours of pleasure in God's great open spaces."[2]

The key to hours of pleasure as opposed to days in the repair shop was a reliable and light design. The Model T's chassis flexed with the road, often uncomfortably so, but it could be driven or pushed out of places from which bigger, heavier cars wouldn't budge. Henry Ford's favorite joke was about the farmer who wanted to be buried in his Model T, because it had gotten him out of every hole he had ever been in. The car was available in body styles ranging from a racy open-air "speedster" to a five-passenger sedan. It even could be adapted to fit campers equipped with water tanks and sleeping beds, presaging the Volkswagen Microbus half a century later.

The "Tin Lizzie," as the Model T was nicknamed for reasons now obscure, provided unprecedented mobility, easing the isolation of farm life and ending rural peasantry in America. It initially cost $850, compared to upward of $1,000 for comparable cars, but by 1924 the Model T's price would be just $260. The drastic price cuts were made possible by another innovation, the moving assembly line, which Ford developed after outgrowing the Piquette Street plant and moving to a larger factory a few miles away. He followed the moving assembly line with the $5 day, more than double the average factory wage at the time.

The Model T and mass production sparked a chain reaction that created a new world. To paraphrase the story of the original creation: "Yea, Henry begat the Model T which begat mass production which begat the $5 day. And verily, those begat the middle class, the suburbs, shopping malls, McDonald's, Taco Bell, drive-through banking, and other things beloved of the modern-day philistines." That might not have been the erudite essay Steinbeck had in mind, but it does describe what Ford wrought.

The Model T promoted social networking one hundred years before Facebook and fostered a sexual revolution a half century before the pill. "Most of the babies of the period were conceived in Model T Fords and not a few were born in them," wrote Steinbeck. "The theory of the Anglo Saxon home became so warped that it never quite recovered."[3] And it wasn't just Anglo-Saxon homes. Eventually the Model T would be made in nineteen countries from Australia to Argentina, and many places in between.

This rich legacy caused some 13,000 Model T collectors to gather in America's heartland in July 2008 to celebrate the centennial of the revolutionary car. "The Ford Model T has never been just a car—in many ways, it's *the* car," a representative from the Ford Motor Company told the crowd. "The car that started it all. The car that put the world on wheels. The car that changed everything."[4] Looking strikingly like his great-great-grandfather, the speaker was a twenty-eight-year-old purchasing manager named Henry Ford III.

The Model T ruled the roads for twenty years, from 1908 to 1927, until suddenly its day was done. It was hit by the meteor of the Roaring Twenties, when cars became vehicles for personal expression as well as for transportation. The automobiles that epitomized this change debuted in 1927, the Model T's last year, as the first mass-market designer cars—"inherently smart, individual and racy," declared a sales leaflet, "the epitome of this zestful age."[5] They were LaSalles, junior versions of Cadillacs, the most prestigious marque in the General Motors hierarchy of brands.

Billed as a "companion brand" to Cadillac and sold by the same dealers, LaSalles were smaller, lighter, and sportier. They were aimed at the "smart set," the 1920s term for "yuppies." While the Model T was relentlessly practical, the ultimate household appliance, the LaSalle was exuberantly stylish. Harley Earl, the man who designed it, was the polar opposite of Henry Ford.

While Ford was the consummate country mouse, raised in then-rural southeast Michigan in the years following the Civil War, Earl was a sophisticated city mouse. He hailed from a well-to-do family in Los Angeles. He graduated from Hollywood High in 1912, attended Southern Cal and Stanford, and launched a career building custom car bodies. One of his clients was a silver-screen cowboy named Tom Mix, for whom Earl designed a car with a saddle on the hood. (It was, to be sure, a one-off job.)

Through an unlikely chain of connections Earl was invited to Detroit in 1926 by the bigwigs at General Motors to try his hand at design. He soon realized that Detroit, just like Hollywood, could manufacture dreams. A year later Earl was hired by GM to form the auto industry's first design department, which he would run for the next thirty-one years, from the Jazz Age to the Space Age. "People like . . . visual entertainment," Harley Earl would say, and that's what he would give them.[6]

During its fourteen-year life the LaSalle did for upward mobility what the Model T had done for personal mobility. The two cars epitomized different philosophies about cars, society, and people. One was country, the other was country club. One was dutiful and self-reliant; the other beautiful and self-indulgent. The Model T might have been used for sex, but the LaSalle was designed for sex appeal. The yin-yang contrast between the two cars would reflect different philosophies that helped shape American culture and would echo in future cars that, like the Model T and the LaSalle, defined the ethos of their day.

In the early 1920s, the story goes, a farmwoman was asked by a social scientist why her family had a Model T Ford but not indoor plumbing. She replied: "You can't go to town in a bathtub."[7] Apocryphal or not, the story says everything about the Model T's appeal. But Henry Ford's road to success, like the Model T's famously bumpy ride, was anything but smooth.

It was fortunate that Ford lived a long life, beginning during America's Civil War and lasting until after the Second World War. He was the proverbial late bloomer, born on July 30, 1863, less than a month after the Battle of Gettysburg, in Dearborn, Michigan, then a farming community ten miles west of downtown Detroit. The oldest of six children, young Henry hated farm chores but loved to tinker with machinery. At age sixteen he left home to work in the machine shops of Detroit.

By 1893, at age thirty, Ford had become chief engineer for the city's Edison Electric Illuminating Company, but his mind was restless. Three years later, using a design he had seen in a magazine, Ford built his first car in a shed behind his house, much like Steve Jobs and Steve Wozniak would build the first Apple computer in a garage some eighty years later.

Ford's car was basically a platform steered by a rudder, powered by a two-cylinder engine, and borne on four bicycle tires. Thus it was named the Quadricycle. Ford had been so focused on the mechanics of his creation that he hadn't noticed it was too big to fit through the door of his shed. But he was so eager to see it run that he tore through a wall with an axe and, around 4 a.m. on June

4, 1896, the first Ford car rolled onto the streets of Detroit.[8] After running a few blocks it stalled out, which prompted jeers from the street transients who witnessed the spectacle. But Henry Ford was jubilant that he had created a working automobile.[9]

Three years later Ford quit his job and found backing from local investors to start the Detroit Automobile Company. But that company went broke after less than two years. Like dozens of other early automotive entrepreneurs, Henry Ford had failed, and it wouldn't be his last time.

He kept tinkering with cars, and in October 1901 he entered a race on a dirt track in Grosse Pointe, just east of Detroit. The race would pay a $1,000 prize to the winner, and luckily for Ford only one other car was running. The other entrant happened to be Alexander Winton, a mogul from Cleveland who owned a car company bearing his name. Henry Ford, in contrast, was a little-known Winton wannabe.

At first Winton's car led handily but midway through the race a smoking engine forced him to slow down. Ford, meanwhile, struggled to keep his car from tipping over. But he had deployed a local mechanic to hang on the side of his car during the race to provide balance, like someone hiking out over the side of a sailboat. The tactic looked comical but it worked, allowing Ford to cruise to victory and claim the $1,000 purse.

Ford's enhanced reputation attracted investors for another car company, the Henry Ford Company, with Henry as chief engineer and one-sixth owner. It seemed promising but just four months later the single-minded and cantankerous Ford quit in a dispute with his backers. They wanted to build luxury cars aimed at upscale buyers, while Ford wanted to focus on racing cars to capitalize on his recent success.

With Ford gone the financiers found a new chief engineer

named Henry Leland and changed the company's name to Cadillac. It would later be bought by General Motors and its brand would reign for decades as America's premier automotive status symbol. By age thirty-eight Henry Ford had helped form two car companies and had lost them both. He vowed that his days as an employee were over, or as he put it, "never again to put myself under orders."[10]

The odds against the young inventor were long, however. He was not only a two-time loser but America already had lots of car companies. There were around sixty at the time, including such long-forgotten names as Maxwell, Thomas, Holsman, White, and the Babcock Electric Carriage Company of Buffalo, which made an electric car a century ahead of its time.

Nonetheless, Ford had a third chance courtesy of local businessmen, who deployed a young bookkeeper named James Couzens to round up investors for his latest venture. Couzens had the financial and administrative aptitude that Ford lacked. In less than fifteen years the two men would revolutionize the world as much as two of their contemporaries, Lenin and Trotsky, minus all the mayhem.

The Ford Motor Company formed on June 16, 1903, and a month later sold its first car, a four-passenger Model A, to a dentist in Chicago. Within ten months Ford Motor had sold 657 more Model As, booking nearly $100,000 in profit, equivalent to about $2.4 million today. In January 1904 Ford set a new speed record of 91.37 miles an hour with a car running on the frozen ice of nearby Lake St. Clair. Less than two years after his second company had failed, Henry Ford was on a roll.

While his luck had changed, Ford's cocksure abrasiveness hadn't. He started squabbling with his investors over the familiar issue of what kinds of cars to develop. Most of them favored big, heavy cars such as the Ford Model K, with a price tag of upward of $2,000. But Ford's view of the fledgling car market had evolved. He wanted to pursue the direction set by the new Ford Model N, a lighter and cheaper car launched in 1906 priced initially at only $500 and aimed at downscale buyers who couldn't afford cars like the K. It was a classic big-car-versus-small-car conflict of the sort that would rage in the auto industry for the next hundred years.

As the argument heated, Ford and his allies squeezed their troublesome investors financially. They formed an entity called the Ford Manufacturing Company to supply components for Ford cars, and sold stock in the new company only to Henry's allies. Ford Manufacturing charged Ford Motor high prices for each component, reaping handsome profits while throwing Ford Motor itself into crisis.[11]

Had Ford and Couzens tried this financial finagling a century later they would have violated laws against corporate self-dealing and found themselves making license plates in a federal prison instead of making cars. But their brazen tactics worked and the dissident investors capitulated, selling all their stock in Ford Motor and leaving Henry with a 58 percent stake of the company. Couzens became the second-largest shareholder with 11 percent. Not long afterward, Ford Manufacturing, a company that had served its purpose, sold all its stock to Ford Motor, and was dissolved. Henry Ford had gained control of his own company and could now develop the car he wanted.

Ford described his dream in a 1906 letter to a magazine called *The Automobile*. The "greatest need today," he wrote, "is a light, low priced car with an up-to-date engine with ample horsepower . . . powerful enough for American roads and capable of carrying its passengers anywhere that a horse-drawn vehicle will go."[12]

Ford walled off the northeast corner on the third floor of the Piquette Street plant, space that is still preserved today, padlocked the door, and set to work with a small team of employees. They started with the Ford Model N and two slightly improved versions, Models R and S. The three had been modestly successful even though they were small, slow, and less than durable on the rutted roads of the day. Henry didn't trust draftsmen's drawings, only actual prototype parts that he could test and evaluate. The process was thus time-consuming, involving lots of trial and error. The first running models of the new car were ready by October 1907, but developing the final version would take another year.

In September 1908 Ford himself led the car's final shakedown cruise, a 1,357-mile trip around Lake Michigan. The group drove up the spine of Michigan's Lower Peninsula, crossed the Straits of Mackinac by ferry, proceeded through the remote wilderness of Michigan's Upper Peninsula, and headed down the west side of the lake. They then drove through Chicago's urban wilderness before returning to Detroit in a car that looked like it had taken a mud bath. But the Model T had made it, suffering only a punctured tire along the way. Ford deemed his creation ready for sale.

When the Model T debuted on October 1, 1908, it wasn't the cheapest car on the market. The Brush "Everyman's Car" Runabout cost just $500, which was $350 less than the Model T. But the Brush had a one-cylinder engine and its chassis, axles, and wheels were all made of wood. The Brush's detractors sniped: "Wooden frame, wooden axles and wouldn' run."[13]

The Model T's key components, in contrast, were crafted from a new type of steel, vanadium, that was lighter but stronger than traditional carbon steel. "We defy any man to break a Ford Vanadium steel shaft or spring or axle . . . ," the company's sales literature boasted.[14] The car's four-cylinder engine also had a one-piece block and a detachable head, an unusual design for its day. This made it easy to get inside the engine, which in turn made it simpler to manufacture and repair than most other engines.

Most critically, the Model T was the first car with fully interchangeable parts. If one part failed or was damaged, it could be quickly and cheaply replaced. At 1,200 pounds, the five-passenger touring model weighed 400 to 600 pounds less than comparable cars. Instead of relying on a heavy chassis to withstand primitive roads, the Model T used a light "three-point" chassis and a suspension that flexed with the road, a blessing with certain drawbacks. One joke of the day described a man who named his Model T the Teddy Roosevelt because, he explained, it was the "Rough Rider."[15]

The Model T could go up to 40 miles an hour and got nearly 20 miles on a gallon of gas. The driving controls were quirky, but effective. There were two forward speeds and three floor pedals: one for reverse gear, one for the brake, and the third for the clutch. The accelerator was a stalk mounted on the steering column, like a modern turn signal. The car's construction had no gas gauge, no shock absorbers, and no fuel pump. The carburetor drew in gasoline by gravity. Thus a Model T low on fuel couldn't climb steep hills because the car's angle prevented gas from flowing into the engine. The solution was to back up the hill in reverse.

Despite those drawbacks, as word of Henry Ford's new creation spread, the public reacted with the enthusiasm reserved a century later for iPhones and iPads. Advance orders for 15,000 Model Ts—nearly twice the company's total sales the previous year—flooded Ford. In May 1909 Ford stopped accepting Model T orders for two months because the backlog was so big. A month later a Model T finished first in a highly publicized cross-country race from New York to Seattle, taking twenty-two days and averaging 7.75 miles an hour. The victory later lost its luster when the car was disqualified because the engine had been replaced during the race, which was a blatant violation of the rules. But by then the publicity bonanza paid off.

The combination of affordability and versatility made the Model T a sensation, bringing car ownership to thousands of people who previously had deemed it beyond their means. A whole new group of companies soon were launched to produce accessories for Ford's "flivver," an idiom of the day for a small and inexpensive car. One device had spiked steel wheels that would let a farmer drive his car into his fields to haul a mechanical reaper, and others harnessed the Model T's engine to saw wood or to pump water. Dozens of smaller attachments came to include one that converted the car's engine manifold into a cooking grill.

In 1912 the company built nearly 70,000 Model Ts, and the price of the basic two-seat "Torpedo Runabout" model had been cut to $590. A year later, in 1913, Henry Ford unveiled another innovation: the moving assembly line. His engineers had been inspired, in part, by the *disassembly* lines of the stockyards of Chicago, where each worker performed a distinct task in cutting up the carcass of a cow. Ford first tried the assembly line concept on the subassembly of components, and found that productivity for those parts immediately surged some 40 percent.[16]

Ford spread the concept to other subassembly areas: dashboards, engines, and the chassis. He then created a main assembly line for the full car. To simplify production and boost productivity further, the Model T would no longer be available in red, green, gray, and dark blue, as it had been for years. Instead, Ford declared, customers could have the car in "any color they want, as long as it's black." Ironically, the exact shade was "Japan black enamel." Had

Henry known what Japanese competitors would do to Detroit decades later, he might have picked a different color.

With sales surging and profits booming, the company next transformed not just the auto industry, but all of America. On January 5, 1914, Couzens summoned reporters to his office and read a statement. Ford Motor, he said, would "reduce the [daily] hours of labor from nine to eight, and add to every man's pay a share of the profits of the house. The smallest amount to be received by any man 22 years old and upwards will be $5.00 per day."[17] Initially the new policy didn't include women workers. They weren't deemed to be supporting families, though that policy was changed a couple years later. The same reasoning excluded men under twenty-two, though Couzens announced that "Young men who are supporting families, widowed mothers, younger brothers and sisters, will be treated like those over 22."[18]

Couzens, the no-nonsense finance man, championed the $5 day instead of Henry himself, many historians say. Either way, it's clear that commercial considerations were as important as idealistic ones. Alienation was growing among Ford workers, many of whom struggled to support families on the average Ford wage of $2.34 a day. Turnover at Ford's Highland Park factory approached 400 percent a year, and the constant cost of training new employees was high. Perhaps most important, Couzens, whose duties included managing sales and distribution, argued that a $5 day would be a masterstroke of marketing.

Other industrialists condemned the move. The *Wall Street Journal* editorialized that Ford "has in his social endeavor committed economic blunders, if not crimes. They may return to plague him and the industry he represents, as well as organized society."[19] For all its progressiveness, the new pay policy came with paternalistic strings attached. Staffers in the company's Sociological

Department, established just before the $5 day, visited employees' homes regularly to inspect them for order and cleanliness. Ford and Couzens wanted to be sure workers weren't squandering their prosperity on liquor, prostitutes, or other dissolute living. Offenders were counseled to mend their ways, and could be fired if they persistently refused.

Couzens proved spot-on about the move's marketing impact. Even as Henry Ford was on his way to becoming the world's richest man, the $5 day made him a working-class hero. Letters arrived from grateful workers, thanking Ford because they no longer had to indenture their children as servants to make ends meet. A new boomlet of immigration began. Nearly a century later many elderly Detroiters would describe how their grandparents had left Europe for the lure of the $5 day. Job applicants swamped Ford, and other companies were forced to match Ford's wages. Ford sold 300,000 Model Ts in 1914, more than four times the number of just three years earlier. Within two years sales more than doubled again, topping 700,000 cars.

The $5 day capped a breathtaking six years during which Henry Ford unleashed enough creativity for two or three lifetimes. During that brief span he created a people's car, invented mass production, and started paying workers enough to create mass prosperity. The Model T became such a staple of American life that jokes about its foibles spread from coast to coast. The *Original Ford Joke Book*, published in 1915, included "The Twenty-Third (Ford) Psalm."

> The Ford is my auto, I shall not want another.
> It maketh me to lie beneath it.
> It soureth my soul.
> It leadeth me in the paths of ridicule for its name sake.

Yea, though I ride through the valleys, I am towed up the hills.

And I fear much evil for thy rods and thy engines discomforteth
  me.

I anoint thy tire with patches. Thy radiator runneth over.

I prepare for blow-outs in the presence of mine enemies.

Surely if this thing follow me all the days of my life,

I will dwell in the bug-house forever.[20]

By 1915 Henry Ford's success convinced him that he could do most anything, and his willful eccentricity began taking a bizarre and destructive turn. In December of that year he chartered a ship and sailed with other prominent Americans to Norway in a quixotic effort to mediate Europe's Great War. The "Peace Ship" mission was as comic as it was controversial. It flopped. The mission also prompted a falling-out with Couzens, who resigned in 1915 and launched a political career that eventually made him a U.S. senator from Michigan.

In 1918 Ford himself ran for the Senate in Michigan, though in a bizarre fashion. He entered both the Republican and Democratic primaries, lost the former but won the latter, and then ran in the general election as a Democrat, even though he had long been a Republican. Despite his enormous wealth he spent almost no money on the campaign. He almost won the election anyway, but was so shocked at his narrow loss that he hired a small army of private investigators to probe possible vote fraud. Nothing came of it. Next Ford turned from politics to publishing, with equally destructive results. He bought his hometown newspaper, the *Dearborn Independent*. It soon started spouting Ford's nativist anti-Semitism, including his theory that the Jews had started World War I so the gentiles would kill each other.

Meanwhile, in late 1918 Ford declared he would resign as

president of Ford Motor, and turn the reins over to his only child, Edsel, who was just twenty-five years old. It was a shocking announcement, but five months later Edsel followed with his own stunner. He also would quit Ford Motor and, with his father, start another, separate car company.

The announcements were a brazen effort to scare Ford's minority shareholders into selling their shares to Henry Ford. They worked. Henry consolidated his control over Ford Motor by paying $105.8 million to buy all the non-family shares. Couzens got more than $29 million, most of which he later donated to charity. Henry rid himself of nettlesome outside investors and became Ford's de facto dictator, even though Edsel held the title of president.

As Henry believed the Model T represented the ultimate in automotive evolution, he allowed improvements only slowly and reluctantly. The Model T didn't get an electric starter to replace its cumbersome hand crank until 1919, seven years after Cadillac introduced the device, and even then Ford made it an extra-cost option. Throughout the early 1920s, more and more of Ford's managers grew convinced that the Model T was becoming obsolete. Few dared to say that openly. Once, when Henry and his wife took an extended vacation in Europe, some enterprising Ford engineers developed the prototype for a new car and decided to surprise the boss when he returned. When Henry saw the new car he flew into a rage and started tearing it apart, piece by piece, with his bare hands. So much for employee initiative.

For a while it appeared Ford was right to resist change. In 1921, Ford Motor's share of the U.S. market topped 60 percent, a new record. More than 1.9 million Model Ts were sold in 1923, another record, thanks in part to its amazing versatility. The Model T's chassis could be adapted to create a host of vehicular variants.

Among them was the Lamsteed Kampkar, a $535 bolt-on body that converted Ford's car into a home on wheels, complete with fold-out beds enclosed with canvas walls, a tank for running water, and a stove for cooking. The Kampkar attachment was manufactured, improbably, by Anheuser-Busch, the brewer, which had entered new businesses after Prohibition became law in 1919.

All the while, Henry Ford continued his basic marketing strategy of making his manufacturing process more efficient, and passing on the benefits to consumers by continually cutting prices. For three straight years, 1924 to 1926, a basic Model T "runabout" with a hand crank instead of an electric starter could be bought for as little as $260, or about $3,500 in today's dollars.

But after 1923, the peak year, sales of the Model T began a steady decline. Even a styling face-lift in 1926, along with the belated addition of an electric starter as standard equipment to replace the outmoded, dangerous hand crank, couldn't halt the trend. In dozens of ways, big and small, the car that put America on wheels fell further behind the competition.

Edsel Ford was convinced of the need for a more modern and stylish car, but his father wouldn't listen. Ironically, Henry Ford wasn't comfortable with the affectations of America's new prosperity, which he himself had done so much to create. By 1926 Ford's market share had dropped below 50 percent. The big winner was GM's Chevrolet, which was emphasizing comfort, convenience, and prestige as opposed to Ford's single-minded focus on a low price. Many Americans had come to want status and style, which more and more of them could now afford.

The Twenties weren't called Roaring for nothing. The bankrupt Duesenberg Automobile & Motors Company was revived under new ownership with a top-of-the-line model costing upward of $20,000—equivalent to $245,000 today. The phrase "It's a

Duesie" entered the language as an expression of admiration, even though Duesenberg would fold during the Depression. The Dow Jones Industrial Average, which had begun the Twenties around 90, closed at 171.75 on May 20, 1927, the day that a young aviator named Charles Lindbergh departed on a solo flight to Paris.

On May 25, five days after Lindbergh's flight into history, Ford also made history by announcing it would discontinue the Model T. During the prior two decades, more than 15 million Model Ts had been built, a record for any single model that would stand for forty-five years, until another people's car, made in Germany, would surpass it. Even remote corners of the country mourned the Model T's passing as sad but inevitable. In North Dakota, the *Bismarck Tribune* wrote: "For years the homely flivver sold as fast as Ford could make them. Then about three years ago there came a change. Slowly at first, then more rapidly, people passed up the flivver for more ornamental machines."[21] One such machine in particular pointed the way to the future.

When the television sitcom *All in the Family* debuted in 1971, Archie and Edith Bunker introduced each episode with the show's theme song, "Those Were the Days." Many people were mystified, however, by the lyrics of the fourth line: "Gee, our old LaSalle ran great." It referred to a long-dead and mostly forgotten brand.

LaSalle debuted on March 5, 1927, just eleven weeks before Ford announced the demise of the Model T. The cars weren't direct competitors—the cheapest LaSalles cost nearly seven times as much as a two-seat Model T runabout—but the two events represented a passing of the guard. As America transformed from a rural to an urban nation, the dull was giving way to the stylish, the practical to the pretentious, and the old-fashioned to the modern.

Other car companies tried to cash in on the trend. In 1927 Studebaker introduced an upscale model called the Dictator, a name intended to suggest that the car would dictate the standard for all other cars to follow. By the mid-1930s, however, it suggested instead images of Mussolini and Hitler, prompting Studebaker to drop the name.

But LaSalles were special. In September of 1927 a newspaper in central Ohio reported on "the escapades of a band of youthful miscreants" from the little town of Delaware who had stolen twenty-five cars over the previous five months and taken them out for joyrides. One was a LaSalle, "which the boys told the sheriff was the 'spiffiest' car they had stolen. They told Sheriff Lambert that they intended to steal the LaSalle again for another ride because the car ran so well," the newspaper reported.[22]

The young Ohioans lacked common sense, but they had great taste in cars. Thanks partly to LaSalle, General Motors was supplanting the dominance of Ford and would be the world's largest car company for the next eight decades. And the homespun values of Henry Ford were giving way to the sophisticated sensibilities of a man named Harley Earl.

Harley J. Earl was born in 1893 in circumstances that couldn't have been more different from the rural roots of Henry Ford. He and his four siblings grew up in a three-story house in Hollywood. One of the family's neighbors was director Cecil B. DeMille. Young Harley was tall (six foot four), handsome, and athletic, more interested in football, rugby, and track than in academics.

He participated in sports at the University of Southern California and at Stanford, where he sudied prelaw, before dropping out of both schools to work for his father. J. W. Earl owned the Earl Automobile Works, which hand-crafted custom coaches, as automobile bodies were then called. In 1919 J. W. sold his business to a local Cadillac dealer. One of the conditions was that the talented young Harley would stay with the company.

Earl's designs sported sleek, stylish lines that contrasted with the upright, boxy dimensions of most cars of the day. When some of his creations were displayed at automotive exhibitions in Chicago and New York, they won admiring reviews from several General Motors executives, including the powerful and influential Fisher brothers. The seven brothers had transformed their father's Ohio blacksmith shop into a company that mass-produced automotive bodies, and in 1919 they sold a 60 percent interest to General Motors. They would sell the remaining 40 percent to GM in 1926, becoming for a time the company's second-largest shareholders. To this day there is a Fisher Freeway, as well as a Ford Freeway, in Detroit.

In December 1925, a few days before Christmas, Larry Fisher, the head of GM's Cadillac division, phoned Earl in Los Angeles. Cadillac wanted to develop sportier, more youthful luxury cars, he explained, to compete with rival Packard, which was luring young, well-heeled buyers away from Cadillac. But GM's engineers seemed incurably wedded to the dull and stolid styling that resembled, well, many Cadillac owners themselves. Would Earl become a consultant for Cadillac, Fisher asked, and try his hand at designing something different?

On January 6, 1926, thirty-two-year-old Harley Earl boarded a train in Los Angeles for Detroit. For the next three months he buried himself in the design and development shops of the Cadillac plant on the city's southwest side. That spring he presented his work to GM's Executive Committee, including CEO Alfred P. Sloan Jr. Instead of simply displaying his drawings, however, Earl staged a showing worthy of a Hollywood impresario.

He unveiled four full-sized models of his designs, sculpted

from clay applied over wooden frames. Painted with black enamel and finished with a shiny coat of clear varnish, they had a wet and luscious look. The GM bosses circled around the models again and again. "Get Earl over here so everyone can meet him," Sloan said.

With the young designer summoned, Sloan announced: "Earl, I thought that you'd like to know that your design has been accepted!" He then suggested to Larry Fisher that Cadillac reward Earl with a trip to the upcoming Paris Auto Show. The beaming Fisher replied: "Mr. Sloan, I already have his ticket!"[23]

Paris was a fitting reward, because Earl had drawn inspiration from French cars, particularly the Hispano-Suiza. Like the "Hisso," Earl's models featured prominent vertical front grilles and louvered vents along each side of the hood, creating a longer, more horizontal stance. From every angle the clay models had interesting lines that invited a closer look. They appeared elegant without seeming overbearing and compact without looking cramped. The designs provided just what the GM bigwigs wanted: a youthful and sporty flavor to broaden the prestigious but stuffy Cadillac line. Earl's designs were so different that, while GM decided to sell them in Cadillac dealerships, the company gave them their own marque: LaSalle. The names fit. Both Cadillac and La Salle were early French explorers in America. Cadillac was the founder of Detroit while La Salle claimed the territory of Louisiana, before being murdered by his own men. That fit, too, as Harley Earl's management style would prove.

Alfred Sloan found Harley Earl at just the right time. Sloan had taken the helm at GM in 1923, just a few years after the company had survived a brush with bankruptcy under the reckless leadership of its founder, William "Billy" Durant. Sloan concluded that he couldn't beat Henry Ford at his own game of constantly cutting costs. He decided to change the game instead. He outlined his strategy in a letter to shareholders in GM's 1924 annual report. General Motors, he wrote, would "build a car for every purse and purpose." Instead of adhering to Henry Ford's "once size fits all" philosophy, GM would create a hierarchy of brands. Most buyers would start with the practical and inexpensive Chevrolet and hopefully graduate to Pontiacs, Oldsmobiles, and Buicks, and eventually those who climbed America's ladder of social and financial success would buy Cadillacs.

What Henry Ford had done for mass manufacturing Alfred Sloan would do for mass marketing. Ironically, Sloan's personality lacked any hint of marketing flamboyance; he was as stiff and cerebral as the starched high white collars on the shirts that he wore every day. It took a strait-laced, no-nonsense businessman to spot the enormous potential in selling to people's pretensions. Sloan and Earl were a perfect match.

A year after Sloan approved Earl's designs, the first LaSalles were unveiled to the public at the Copley Plaza Hotel in Boston. The guests followed white-coated musicians into the ballroom, where the daughter of a Boston Cadillac dealer launched the new car like it was a ship, breaking a bottle of champagne on its radiator. "Lovely creation of many minds and hearts and hands, go forth into the highways and byways of the world," she intoned. "I christen thee, LaSalle."[24]

There were six body styles, ranging from the two-passenger roadster with a rumble seat priced at $2,495 to a five-passenger sedan priced at $2,685—competitive with Packard, and some $500 less than the least expensive Cadillac. The sedan's styling was traditional, prompting the *New Yorker* to sniff that it was "squarish, ample, redolent of the suburban family . . . a concession to the

Rotarian market." The magazine was more impressed, however, with the LaSalle coupes and convertibles, adding that "the line, as a whole, is as refreshing as a Paris frock in a Des Moines, Iowa, ballroom."[25] GM would use the magazine's praise in LaSalle advertising, without the reference to Des Moines.

The LaSalles had more than just pretty grilles. Their technical capabilities were impressive. Every LaSalle came with a 75-horsepower V8 engine at a time when V8s connoted luxury on American roads. In May 1927 a GM mechanic piloted a LaSalle around the company's Michigan Proving Ground while averaging just over 95 miles an hour for nearly 952 miles, qualifying the LaSalle to be the pace car in that year's Indianapolis 500. It was a remarkable feat, considering that a 160-horsepower Duesenberg won the race that year averaging only two miles an hour faster over half the distance.

LaSalles proved especially popular with the small but growing ranks of women drivers. "LaSalle is a graceful car, with a great deal of charm for the female eye on account of its smart lines," *Vogue* reported, describing women driving LaSalles.[26] In Los Angeles, actress Clara Bow, sex symbol of the silent-film era, was regularly photographed in her 1927 LaSalle roadster.

Women found LaSalles easy to handle because their wheelbase was seven inches shorter than that of the smallest Cadillac. Also, the gear ratios allowed LaSalles to turn many corners without downshifting, minimizing an aspect of driving that many women hated. "The LaSalle obeys the feminine hand instantly, with only the slightest effort," declared one advertisement aimed at women, which was unusual for its day.[27] Other ads carried French headlines, such as *"La Nouvelle Arrivee"* or *"Bon Voyage."* GM sold nearly 27,000 LaSalles the first year, a success by any measure. General Motors had a new brand and, before long, a new executive: Harley Earl.

• • •

From the 1930s through the 1950s, designers at General Motors would recite a little ditty that went:

> *Our father, who art in styling,*
> *Harley be thy name.*

It was about the only time they called the boss Harley instead of Mr. Earl, which they invariably pronounced as "Misterl," as if one word. But Harley Earl wasn't merely *in* styling. He *was* styling.

So impressed was Sloan with Earl's work on the LaSalle that in the summer of 1927, just a few months after the launch of the new line, the CEO brought a high-level personnel matter before GM's Executive Committee. Harley Earl, Sloan proposed, should be hired full-time as head of a small but important new corporate staff, the Art and Colour Section (with the British spelling of "colour"). There was nothing else like it in America's adolescent automobile industry.

Earl and his designers would make design an integral part of the process of car development at General Motors, and thus give the company an edge over its more pedestrian competitors. Earl had to move from balmy Los Angeles to a place that seemed to have two seasons: winter, and winter-is-coming. At age thirty-three he transported his young family to Detroit, settling in the highbrow suburb of Grosse Pointe.

Settling into General Motors proved more difficult. Earl had no experience working in large organizations filled with fiefdoms, executive intrigue, and office politics. Fisher Body, the division that stamped out the sheet metal for the cars Earl designed, was a semiautonomous satrapy, and neither the Fisher brothers nor their underlings brooked much interference. Sometimes they summarily altered Earl's designs, claiming that his specifications would compromise the structural integrity of the cars or would increase production costs.

In 1929 GM launched a new Buick sedan, designed by Earl, with styling that flared outward at the "belt line," the imaginary line just below the car's windows. Walter Chrysler, a former GM executive turned competitor, described the car as a "Pregnant Buick." The name stuck. Buick's sales plunged more than 25 percent that year, even before the stock market crash.

The fiasco sparked a finger-pointing war. The Fishers claimed they had executed Earl's design faithfully. Earl retorted that the Buick's belly bulge was unauthorized and took to calling the Fishers, short men who wore old-fashioned homburg hats, "the Seven Dwarfs."[28] Meanwhile, he insisted on approving any design changes, and hired a couple of engineers to work on his own staff so he wouldn't be hoodwinked by the metal-benders at Fisher Body.

Earl also invoked his personal relationship with Sloan to settle arguments. "Let's see what Alfred thinks about this," he would declare, but only rarely did he place the call. It was clear that Earl's power emanated from Sloan. In 1937, after a decade at GM, Earl was promoted to vice president, and the Art and Colour Section became the General Motors Styling Department. With his status enhanced, Earl formalized the structure within his domain.

He created separate, locked styling studios for each GM division—Chevrolet, Pontiac, Oldsmobile, Buick, and Cadillac/LaSalle—reserving access to all of them for himself alone. He wanted his designers to compete with one another. All the while, styling took on added importance as the practice of putting reshaped sheet metal on the same old mechanical underpinnings became the annual model change. Henry Ford disdained the practice, but GM and Earl led the way. Planned obsolescence brought people into showrooms, and showroom traffic sold cars.

In 1938 Earl designed the first futuristic "dream car," a vehicle intended not for production but simply for making a styling statement and creating buzz. It was the Buick Y-Job, a name inspired by experimental aircraft. It featured a sleek, low body; hideaway headlamps; prominent chrome bumpers; and a horizontal front grille that contrasted with the typical vertical grilles of the day.

The Y-Job served both as Earl's personal car—he drove it to and from work for years—and his personal statement. "My primary purpose," he would say, "has been to lengthen and lower the American automobile, at times in reality and always at least in appearance."[29] He also had a more pithy expression of his design philosophy: "If you drive by a schoolyard and the kids don't whistle, go back to the drawing board."[30]

Harley Earl had left Hollywood, but Hollywood hadn't left Harley Earl. His various offices at GM were studio sets, in effect, with Earl himself cast as the star. Dark paneling lined offices crowned with beamed ceilings and floored with plush Oriental carpets. The boss's desk was set on a raised dais, so he was always talking down to those who sat before him.

Most GM executives dressed with all the flair of an asphalt parking lot—dark suits with thin, dark ties and white shirts. But not Harley Earl. He wore bespoke suits of tan, gray, or even white, brightly colored silk shirts and ties, a matching pocket hankie, and matching shoes of suede or soft leather. He kept identical sets of clothes in his office closets so he could change at noon, always looking pressed and fresh even as others wilted in Michigan's midsummer, pre-air-conditioning humidity. With his height, broad shoulders, pale blue eyes, and fair skin that often sported a just-right Florida tan, Earl cut an imposing figure.

He would stalk the halls of his studios until 10 p.m. or midnight, assessing the work that would make or break men's careers. He often judged the drawings of his designers by having them stand around him in a semicircle while he sat cross-legged in front of their work, pointing the toe of his butter-soft shoe at features he liked. Focus groups and marketing surveys hadn't been invented yet (though when they were, their benefits would be dubious). Earl relied on instinct. After he designed the LaSalle he rarely, if ever, sketched new cars himself, always leaving that task to underlings.

"You were always a little bit in terror of Mr. Earl," a longtime Cadillac designer, Dave Holls, wrote in his memoirs. "I think he loved it." Once, early in his career, Holls stood nearby while Earl discussed a new design with Cadillac's studio chief. Earl asked: "Why don't we ask some of the young fellows, and see what they like?" Holls was asked for his opinion, spoke his mind, and then learned that Earl had a different opinion. "If I want the young fellow to say anything," Earl snapped, "I'll ask him." For the next two weeks, Holls feared he would be fired.[31]

So did another young designer who, in the midst of Earl's long career, spotted his boss one day striding toward the executive parking garage with two big packages under his arm. "How are you, Misterl?" the designer blurted out. "I see you just cashed your check!" Earl stopped and glared hard before striding away, leaving the gulping man relieved that he hadn't been fired on the spot.[32]

Others weren't so lucky. Bill Mitchell, Earl's longtime second-in-command and eventual successor, tried once to intercede with the boss on behalf of two colleagues whose spirits crumbled under Earl's constant criticism. "A couple of fellas, you just scared them to death," Mitchell confided to Earl one evening after work. "They're going to psychiatrists." Earl listened with apparent sympathy and said, "I'm glad you told me that." But a couple days later he summoned Mitchell and barked: "Goddam son of a bitch, [if] they don't like the work, throw their asses out of here." The two stylists were fired.[33]

While Harley Earl was cementing his ascendancy at General Motors, his LaSalle fared less well. The Great Depression caused people who previously could afford luxury cars to settle for something less. By 1933 LaSalle advertisements, in a nod to the nation's prevailing frugality, touted the car's durability as well as its prestige. Still, LaSalle sales plunged that year to just 3,500 cars even though the price had been cut to $2,235, 10 percent below the price of 1927.

Rumors spread that the LaSalle line would be axed. But when GM's top brass gathered to review the designs of the new 1934 models, Earl addressed the issue head-on. "Gentlemen, if you decide to discontinue the LaSalle," he said, "this is the car that you are not going to build."[34] The curtains drew back to reveal a stunning new design.

The car's new lines were longer and more rounded than before. The grille had been narrowed to look like a tall tower. Each side of the hood had five round portholes. A stylish abstract ornament, suggesting a bird taking flight, sprouted atop the elongated hood. In a bow to commercial reality, the 1934 LaSalles borrowed engines and other key components from GM's lower-priced Oldsmobile division to reduce production costs. The tactics worked. GM sold 7,200 LaSalles in 1934, more than double the year before. Three years later, in 1937, LaSalle sales hit a record 32,000 cars. That year LaSalles got Cadillac engines again, and louverlike stainless-steel strips to replace the hood portholes. To fuel sales, LaSalle prices were cut to as low as $1,000, depending on the body style. Nineteen thirty-seven marked LaSalle's apogee, but the success didn't last.

LaSalle sales dropped sharply in 1938. By that time, LaSalles had grown big enough to serve as less costly substitutes for Cadillacs. In Buffalo, New York, for example, the Paske family—father, mother, and five sons—would pile into their 1939 LaSalle sedan and motor across the state to Lake George for family holidays. Raymond, the middle son, had to sit on the floor of the backseat, dodging his brothers' feet and inhaling exhaust fumes. But all seven Paskes and their vacation luggage fit snugly into the LaSalle.[35] "It was becoming increasingly clear to GM that the LaSalle and Cadillac had become practically the same thing," *Automobile Quarterly* magazine later observed. "One of them had to go."[36] The 1940 LaSalles would be the last in the line.

By then the world was at war, even though it would be nearly two years before America joined the fray. During World War II civilian car production would cease. America became the "Arsenal of Democracy," Detroit's factories converted to produce airplanes, tanks, troop haulers, and a quirky military vehicle called the jeep. Automobiles stopped signaling trends in American culture until the dawn of a new decade, the 1950s.

But the two cultural strains—the simple versus the stylish, the practical versus the pretentious—established by the Model T and by the LaSalles would endure and evolve. Henry Ford had sold to the head; Harley Earl sold to the heart. The two men's cars captured opposing values that would define fault lines in American society for decades. They would gain expression in future generations of vehicles as the pace of cultural change in America accelerated in the Fifties, the Sixties, and beyond.

"I have in my office a scale model of the first sedan I ever designed for the company, a 1927 LaSalle V-8," Earl wrote in 1954.[37]

"I have a great affection for the old crock, but I must admit it is slab-sided, top-heavy and stiff-shouldered." Indeed, by that time automobile styling had evolved. And Harley Earl was creating some of the most powerful totems of America's affluent and optimistic postwar era: the Corvette, chrome-slathered cars, and tail fins.

2

ZORA, ZORA, ZORA:
A BOLSHEVIK BOY ESCAPES THE NAZIS AND SAVES THE
GREAT AMERICAN SPORTS CAR[1]

Every one of us is born into solitary confinement. And we spend
the rest of our lives sending out a small SOS we hope someone will
hear.

—George Maharis, as Buz Murdock, in *Route 66*[2]

When *Route 66* debuted on television in October 1960, critics
hailed it as sophisticated drama unlike anything else in TV's bland
Wonder bread wasteland. The show told the story of "two itinerant young Easterners trekking westward in a sports car," wrote *TV
Guide* in a complimentary critique. "[Martin] Milner plays a young
and recently orphaned Yale man, George Maharis a near-hood off
the streets of New York. The sports car, felicitously, is played by
the sponsor's product."[3] The sponsor was the Chevrolet division
of General Motors. The product was the Corvette.

In the show's initial episode, the Corvette breaks down in
a small Mississippi town. The two young men run afoul of the
venal local sheriff and are on the verge of being lynched, until the

sheriff's long-suffering son defies his father and intervenes to save them. The plot bears an eerie, though coincidental, resemblance to the story of the Corvette itself, a car nearly killed in its early days, only to be saved by an unlikely hero.[4]

The Corvette debuted in December 1953, nearly a year after being exhibited as a prototype sports car called the EX (for "experimental") 122. At the outset its curvy-sleek body was an eye-catcher, but under the hood sat an anemic six-cylinder engine that made the car more of a go-cart for grown-ups than a true sports car. "The Austin-Healy will eat it alive and so will the Jaguar," wrote *Mechanix Illustrated*'s Tom McCahill, then the dean of American car critics. "If you want an American-made 'sporty' car . . . to impress the hillbillies, the Corvette has a lot to recommend it."[5] The faint praise was damning indeed, and the Corvette's early sales were sluggish. Before long Chevrolet considered discontinuing the car, which was losing money despite a hefty price that approached $4,000, more than triple the average car of its day.

But 1953 was a year of transition in America. Dwight Eisenhower moved into the White House, and the *Saturday Evening Post* profiled the freshman senator from Massachusetts in an article headlined: "Jack Kennedy—The Senate's Gay Young Bachelor."[6] Only in the Fifties sense, of course.

On July 27, an uneasy armistice ended three years of bitter and bloody fighting in Korea. The cease-fire meant that a quarter century of nearly nonstop Depression and war had come to an end. A generation of Americans who had come of age knowing self-denial and personal privation wanted to let loose. A booming postwar economy was starting to make that possible.

Just as a new cast of political leaders came to the fore, a few unlikely young men started to challenge long-held cultural norms. One was a poor white boy from rural northeastern Mississippi who could sing like a black man and could swivel his hips like a gyroscope. Elvis Presley launched his career in music in 1953.

Another young rebel was Hugh Hefner, who had a Chicago middle-class upbringing and went to the University of Illinois, where he gained his first publishing experience drawing cartoons for the student paper, the *Daily Illini*. In December 1953, the same month the Corvette first appeared, Hefner launched a new magazine called *Playboy*. Highbrow articles provided cover, as it were, for photos of naked women. The inaugural issue carried a stunning nude photograph of actress Marilyn Monroe, the star of one of the year's hit movies, *Gentlemen Prefer Blondes*. *Playboy* became a quick success. Presley's background was southern rural, while Hefner's northern urban. But the two men sparked enough condemnations from Catholic and Baptist pulpits alike that, it might be said, they did more to foster ecumenical unity than any two men of their time.

A lesser-known young rebel of the day was Zora Arkus-Duntov, whose early years, though also humble, were far more dramatic than Hefner's or Presley's. The son of Russian-Jewish parents, Duntov grew up during the Russian Revolution in St. Petersburg. Later he witnessed the rise of Nazism as a young man in Berlin, where like young men everywhere he tinkered with cars and chased girls, both with particular passion. He came to the United States in 1940, fleeing France just ahead of the victorious Germans, when he was thirty years old.

In America, as he had in Europe, Duntov embraced a playboy lifestyle that would have made Hefner proud. He made his interest in cars into a livelihood, developing high-performance engine components. Then he finagled his way into a job at GM because he wanted to work on just one car, the Corvette. So Duntov was chagrined, understandably, at the prospect of the Corvette's early demise. He pleaded for the car's life with memos that combined the charms of immigrant English with the dreariness of corporate speak.

"If the value of a car consists of practical values and emotional appeal," he wrote in one, "the sports car has very little of the first and consequently has to have an exaggerated amount of the second. If a passenger car must have an appeal, nothing short of a mating call will extract $4,000 for a small two-seater."[7] A primal, hormonal appeal is exactly what Duntov's work would bestow upon the Corvette. The eventual result would be the rise of youth marketing and, for better or worse, nonstop commercials for pimple cream, designer jeans, and MTV.

Zora Arkus-Duntov didn't create the Corvette. He did save it from doom and transform it into a real sports car. Had his Bolshevik boyhood been known to Senator Joe McCarthy, the anticommunist crusader from Wisconsin, the Corvette might have been deemed a commie plot instead of a symbol of the American way of life. As it was, however, the Corvette would evolve from a car into a cult, and would spawn its own mini-industry created by a disparate cast of very capitalist entrepreneurs.

After World War II a hot-rod subculture began to take root in and around Los Angeles. Kids there took to "hopping up" their engines to produce extra horsepower and racing the cars on dry lake beds. In 1948 a local publicist named Robert "Pete" Petersen, who started his career as a messenger boy at MGM, launched a magazine called *Hot Rod*, which wrote with adulation about these backyard custom jobs. Petersen peddled the magazine at local racetracks for twenty-five cents a copy. A year later, *Hot Rod* was so successful that Petersen launched another title, *Motor Trend*, which focused on new cars.

The hot-rodding boomlet and Petersen's success attracted the attention of Harley Earl. Attending the races at Le Mans after the war, Earl witnessed firsthand the popularity of two-seat sports cars, notably the Jaguar XK120 roadster, which had been introduced in 1948 as Jaguar's first postwar car. It had a six-cylinder engine, a long and curvaceous front end, and short, muscular rear haunches that made the car seem ready to pounce, just like its namesake. The XK120 and other European sports cars were also fan favorites on the track at Watkins Glen in Upstate New York, where in 1951 Earl himself drove the pace car. He started spotting roadsters on American college campuses while visiting his two sons.

Inspired by these cars, Earl directed his designers to develop a two-seat, sporty roadster with clean, simple lines that he hoped would upstage the XK120. In the spring of 1952, Earl unveiled his new roadster to two key people: Harlow Curtice, GM's president, and Edward N. Cole, Chevrolet's new chief engineer, a hard-charging man whose directive was to add some pizzazz to the lackluster Chevrolet lineup. The three men decided to gauge public reaction to the car by displaying a prototype the following January at a GM exhibition, dubbed the Motorama, at the Waldorf-Astoria hotel in New York. GM created Motoramas as venues for showing its new cars to the press and public and creating buzz. The company had held two such shows, one in New York and the other in Boston, in 1949 and 1950, but then dropped the idea for a couple of years.

The Motorama that opened at the Waldorf on January 17, 1953, however, made up for the lapse. Unlike the first two events, the 1953 show featured futuristic vehicles and prototypes instead of just current models. The EX 122 prototype stole the show. GM's financial report for the first quarter of 1953 showed a photo of well-dressed men and women crowding around the car, mesmerized.[8] Actress Dinah Shore dropped by to see the car, and was photographed sitting in the driver's seat. The crowd around the car also included a middle-aged immigrant engineer living in New York named Zora Arkus-Duntov.

Duntov had been born Zachary Arkus on Christmas Day 1909, in Brussels, where his Russian parents, Jacques and Rachel Arkus, were students. Being born outside of Russia would prove convenient decades later, at the height of the Cold War, when GM's publicists said Duntov's ancestry was Belgian. His parents returned with their infant son to St. Petersburg, where his mother became active in Bolshevik politics. Young Zora, as he was called, witnessed the 1917–18 Bolshevik Revolution first hand. Because his mother took a position with the new Soviet government the family had food, though sometimes barely enough. As a boy Zora would eat so many onion sandwiches that, in his adult years, he never ate onion again. Violence was never far away. When ten-year-old Zora was assigned to pick up the family's ration of bread, he took a Smith & Wesson .45 caliber revolver to guard it. Once he brandished the pistol to threaten a doctor who refused to come treat his ailing mother. The doctor changed his mind.[9]

Zora was less effective at other tasks. Assigned to enroll his younger brother, Yura, in school because his parents were busy, Zora got there late. For a month he covered up his negligence by hiding five-year-old Yura in a cabinet during school hours. One school deemed Zora incorrigible, and expelled him. His mother called him stupid "so often that I believed it," he would recall decades later.[10]

The chaos of Zora's boyhood continued when his parents divorced. With housing scarce in the Bolshevik workers' paradise, his father continued to live with the family even while his mother's new man, Josef Duntov, moved in with them. Eventually Zora and Yura added the name Duntov to their own. "We took the English way out," Duntov would later explain. "We hyphenated."[11] Over time Zora, despite his hyphenated surname, became known simply as "Duntov."

In 1927 the family moved to Berlin, where the Soviet government posted his stepfather as a trade representative. Zora bought a motorcycle and started racing it, much to his parents' chagrin. In 1934 he received an engineering degree from Berlin's prestigious Charlottenburg Institute, only to bounce from one tedious job to another, including one where the future Corvette guru designed the engine for a chain saw. As a young man, Duntov pursued women and risk taking with equal ardor. He took up amateur boxing in Berlin, dislocated his shoulder in one match, and then broke his other arm in an auto accident while driving home. His boxing career was over.

In 1935, two things happened that would change Duntov's life. Along with friends, he tried to build his own sports car, called the Arkus, on the chassis of an old Talbot racer. The car ran into one mechanical problem after another and never made it into a race, but Duntov found the effort exhilarating. He also met a blond, blue-eyed teenager, Elfi Wolff, a petite amateur dancer and acrobat. Her father owned one of the largest greeting card companies in Germany.

The Wolffs, like the Arkus-Duntov clan, were Jewish, and as the young couple's courtship unfolded, it became clear that Germany was an increasingly dangerous place. In 1937 the couple landed in Paris, where Elfi got a job with the Folies Bergère dance troupe. Duntov, meanwhile, earned money by smuggling gold into Belgium, hiding it in a special hollow tube he installed on the chassis of an old Ford V8. On February 11, 1939, Elfi and Zora married. Seven months later, war broke out in Europe.

Duntov enlisted in the French air force, where he was trained, to his chagrin, as a tail gunner instead of a pilot. It hardly mattered. In May 1940, before he saw any action, the Germans routed the French army and occupied Paris. Elfi fled just ahead of them, in her MG roadster, heading for the village of Mérignac in the Bordeaux wine country, where Duntov was stationed. After four days of flight— during which she begged for gasoline and billeted with French and Scottish soldiers—she made it. Together with Yura, also stationed in Mérignac, the couple looked for a way to escape from France.

In Marseilles the threesome hid out in a bordello, deeming it safe because the owner had bribed the authorities to steer clear of the place. The local Spanish consul granted them visas, possibly to get the rakish Duntov brothers away from his sister. The threesome sold Elfi's MG, booked train passage to Madrid, and went on to Lisbon, where they met the Duntovs' mother and stepfather. The group sailed from Lisbon on a creaky freighter named the *Nyassa*, and on December 4, 1940, arrived in Hoboken, New Jersey. After enough excitement for two or three lifetimes, Zora Arkus-Duntov had arrived in the United States.

Duntov's next dozen years, while comparatively calm, were eventful. Through fellow refugees he found steady work as a consultant for companies that made engines for airplanes, ships, and trains. In 1942, with America in the war, he borrowed money to start Ardun (blending "Arkus" and "Duntov") Mechanical Corporation, which made machine tools for producing war matériel. The company prospered, allowing Zora and Elfi to move into a luxury apartment on Manhattan's Riverside Drive. Due to Duntov's continual philandering the couple separated for a while, and Elfi resumed her dancing career. They reunited after the war, at least officially, though they often spent long periods apart while they pursued their own interests.

For Duntov, this separation meant developing engines and driving race cars. He designed the Ardun cylinder head for the Ford "Flathead" V8 engine that provided a more potent air-fuel mixture to the combustion chambers, thus boosting horsepower by an astonishing 60 percent. He raced cars in the United States and later in Europe, but without success. In the fall of 1952 he began writing letters to American car companies, seeking a job.

Duntov was rebuffed, in succession, by Studebaker, Chrysler, and Ford. Then in January 1953, along with thousands of other New Yorkers, he bought a ticket to the GM Motorama show at the Waldorf-Astoria. There he laid eyes on a white fiberglass Chevrolet sports car, the EX 122. *Sports Illustrated* later would write that the car's toothy grille made it look, from the front, like "an albino toadfish with ill-fitting false teeth." [12] Duntov was enthralled.

On January 28 he wrote Maurice Olley, a senior engineer at Chevy. "Dear Mr. Olly," he began, misspelling the recipient's name. "I went to Motorama and found the Chevrolet sports car breathtacking [sic]." Duntov went on: "I think this is the turning point from which european [sic] body designer can look for inspiration to Detroit." [13] In mid-April Olley, apparently not fussy about spelling or grammar, offered Duntov a job as an assistant staff engineer. Duntov started on May 1, 1953, at $14,000 a year, plus bonus. He and Elfi moved to Detroit.

In many ways GM was more of a country than a company. It had a university (General Motors Institute in Flint, Michigan), an air force (though the corporate planes were unarmed), and its own social security system, in the form of salaries, benefits, and pensions that assured its managers lifetime comfort. The Duntovs were a striking couple: Elfi with her dancer's dexterity and svelte figure, and Zora with his blue eyes, prematurely gray hair, and an uncanny resemblance to actor Paul Newman (who made his Broadway debut that same year). But it wasn't only their looks that set them apart.

The Duntovs were worldly and foreign while Detroit in general, and GM in particular, were provincial. General Motors executives tended to be WASPs with small-town Michigan or Ohio boyhoods. Their inner sanctum was the Bloomfield Hills Country Club, where even a Ford was regarded as a foreign car. "Picasso to me is a queer," Bill Mitchell, who would succeed Harley Earl as GM's design chief, once proudly declared in a speech.[14] It was a sign of the insularity that eventually would lead to GM's demise.

The Duntovs, as foreign-born Jews who didn't care for golf, bridge, or country club culture, sometimes lamented their new atmosphere. GM executives, Duntov once confided to a friend, believed "the world is bordered on the east by Lake Huron and on the west by Lake Michigan."[15] But Duntov cared passionately about racing and about sports cars. Living in Detroit and adapting to GM, more or less, was the price he would have to pay for turning that passion into a livelihood.

A month after arriving at GM, Duntov casually informed Olley that he was heading over to France to drive in the races at Le Mans, which he had done once before, for the Allard car company. It was a prior commitment, Duntov claimed. Olley was incensed that his new hire would be so bold. Ever the rebel, Duntov appealed to Ed Cole, Olley's boss and a rebel himself. As it happened, Allard would be using a Cadillac engine in the race, which provided the excuse for Cole to overrule Olley. Duntov could go.

Just weeks into his job, Duntov began what would become a familiar pattern: challenging and irritating his bosses. His daring had consequences. Immediately after Duntov returned from France, Olley exiled him to GM's corporate Proving Ground in Milford, Michigan, to work on school buses. The assignment was

GM's version of Siberia, an analogy that Duntov, as a Russian, could appreciate.

But that September, thanks to his Le Mans experience, the prestigious Society of Automotive Engineers invited him to speak about the outlook for sports cars in America. The thought of giving a speech scared him—English was his fourth language, after Russian, German, and French—but Duntov agreed. "The interests and volume sales of sports cars can exist only in the economics of prosperity," he told the engineers, explaining why the market was ripe.

"In our age where the average person is a cog wheel who gets pushed in the subways, elevators, department stores, cafeterias, lives in the same house as the next fellow, has the same style of furniture, [and] wears the same clothing, . . . the ownership of a different car provides the means to ascertain his individuality to himself and everybody around," he added.[16] In his own awkward way, Duntov had expressed the potential of the Corvette.

Three months later on December 16, 1953, an emboldened Duntov wrote the manifesto that would define his life. It was a memo to Chevrolet management titled "Thoughts Pertaining to Youth, Hot Rodders and Chevrolet." His language was stilted, but there was no mistaking his message.

"The Hot Rod movement and interest in things connected with hop-up and speed is still growing. As an indication, the publications devoted to hot rodding and hop-upping, of which some half dozen have a very large circulation and are distributed nationally, did not exist some six years ago."[17] But the unfortunate problem, Duntov informed his bosses, was that these publications were filled with articles about enemy cars, *Ford* cars. Chevrolet had nothing to offer young hot-rodders. "This is not surprising then that the majority of hot rodders are eating, sleeping, and dreaming modified

Fords . . . ," he went on. "It is reasonable to assume that when hot rodders or hot rod-influenced persons buy transportation, they buy Fords. As they progress in age and income, they graduate from jalopies to second-hand Fords, then to new Fords. Should we consider that it would be desirable to make these youths Chevrolet-minded?"[18]

Duntov outlined an idea that would come to dominate American marketing for the next half century: selling to youth. The guiding principle behind youth marketing was to capture customers during their formative years and then, hopefully, keep them forever. The chief corollary was that products designed for young people would attract older customers to a brand.

Duntov's logic, if not always his language, was coherent and clear. Impressed by the memo, Cole recalled Duntov from school-bus exile to develop a new fuel injection system. Fuel injection, an alternative to a traditional carburetor, would boost a car's performance by spraying a precise amount of gasoline directly into an engine's combustion chamber. It wasn't strictly a Corvette job, but it was close enough.

The same month that Duntov wrote his memo, Chevrolet started selling the Corvette, a name it took from a class of warship designed for speed and agility. The company made only 300 of the 1953 models, all in "Polo White" with black tops and red interiors. Instead of using normal dealer channels, Chevrolet offered the cars to key celebrities and VIPs, including Dinah Shore and John Wayne. Chevrolet wanted to create cachet and stoke demand for 1954, when Corvette production would be ramped up to more than 3,600 cars. All would have bodies of fiberglass plastic, which originally was intended just for the prototype car, but was proving to be less expensive and more durable than GM had expected. The company's timing seemed propitious.

• • •

Rising prosperity and mounting hysteria aren't natural companions, but they coexisted in 1954, the height of the Red Scare in midcentury America. It was the year President Eisenhower coined the term "domino theory" to describe the threat of countries falling in succession to the Communists. Domestic threats to national security, whether real or imagined, also loomed large.

In mid-June the fear boiled over at the Chevrolet assembly plant in Flint, Michigan, where the Corvette was built. Workers staged wildcat strikes to protest working alongside alleged communists and beat up one of their fellow employees, twenty-seven-year-old James Zarichny, demanding that he be fired and "go to Moscow." Zarichny, a former GI, had been labeled a communist before the House Committee on Un-American Activities by one Mrs. Beatrice Churchill, identified by the *Detroit Times* as (no kidding) a "Flint grandmother and FBI spy." [19]

Amid such fright, however, Americans were making babies and money in unprecedented amounts, and signs of the new national prosperity were everywhere. On April 26, 1954, *Time* carried a cover story headlined "Road Racer Briggs Cunningham: Horsepower, Endurance, Sportsmanship." Cunningham was a millionaire American sportsman who was making his mark on the European auto racing scene.

The article described how, thanks in part to Cunningham, interest in sports cars was moving across the Atlantic. "In the modern era, it is the Europeans who have done most to define the sports car," the magazine wrote, referring to the Ferraris, MGs, and Jaguars that ruled the Continent's racetracks. "Detroit is obviously perking up and taking notice. The Chevrolet Corvette and Ford Thunderbird, though probably not sporty enough for European purists, are efforts to meet the conditions of the U.S. highway network and the tastes and pocketbooks of a potentially good-sized U.S. market."[20] *Time,* of course, wasn't some hot-rod sub-culture rag but the voice of the Eastern Establishment, chronicling, and indeed legitimizing, new trends.

But in truth the Corvette wasn't sporty enough for Americans, either. The first 'Vette was an embarrassment. It was powered by the Chevrolet "Blue Flame six," which, despite its marketing moniker, was a lackluster six-cylinder engine. Instead of a proper manual transmission that a driving enthusiast could shift with precision, the Corvette had a performance-sapping two-speed automatic called "Powerglide." The car took 11 seconds to accelerate from a standing start to 60 miles an hour, not much faster than contemporary sedans, and slower than most family cars of today.

It lacked exterior door handles. Drivers and passengers had to reach inside to open their doors. Windows wouldn't get in the way because there weren't any. All the Corvettes were convertibles. Instead of real windows they had snap-in plastic "side curtains." Despite those shortcomings, the 1954 Corvettes carried a sticker price of $3,254, far more than what Harley Earl had envisioned, and twice as much as Chevy's most expensive family sedan. The heater and radio (AM only in those days) cost extra, and brought the real price to well over $3,500.

Worse still, the Corvette's ill-fitting convertible tops leaked when it rained. Some owners, including GM executives, drilled holes in the floor of their Corvettes so water that leaked in could drain out. By the fall more than 1,000 unsold 1954 Corvettes, one-third of those made, sat on the lots of Chevrolet dealers like little orphan puppies, waiting for an owner. Chevrolet's management prepared to kill the car, with the blessing of the General Motors corporate brass.

Hallway chatter about the Corvette's likely demise reached

the ears of Duntov. The Corvette had inspired him to join General Motors, even though that meant leaving New York and moving to provincial Detroit. So on October 15, 1954, he wrote another memo to Cole and Olley, beginning with an uncharacteristic nod of deference to his bosses.

"In this note I am speaking out of turn . . . ," he wrote. "I realize this but am still offering my thoughts for what they are. In order to make the content clear and short, I will not use the polite, apologetic phrasing. . . ."[21] That said, he explained why killing the Corvette would be shortsighted.

"By the looks of it, Corvette is on its way out," the memo continued. "Dropping the car now will have adverse effect, internally and externally. It is admission of failure. Failure of aggressive thinking in the eyes of the organization, failure to develop a saleable product in the eyes of the outside world. Above-said can be dismissed as sentimentality. Let's see if it can hurt the cash register. I think it can."

He raised the specter of Ford's imminent introduction of its own stylish two-seater. "Ford enters the field with the Thunderbird, a car of the same class as the Corvette," Duntov wrote. "If Ford makes success where we failed, it may hurt. With aggressiveness of Ford publicity, they may turn the fact to their advantage. I don't mean in terms of Thunderbird sales, but in terms of promotion of theirs and depreciation of our general lines. We will leave an opening in which they can hit at will. 'Ford out-engineered, outsold, or ran Chevrolet's pride and joy off the market.' Maybe the idea is far-fetched. I can only gauge in terms of my own reactions or actions. In the bare-fisted fight we are in now, I would hit at any opening I could find and the situation where Ford enters and where Chevrolet retreats, it is not an opening, it is a hole!

"Now if they can hurt us, then we can hurt them! We are one

year ahead and we possibly learned some lessons which Ford has yet to learn. Is the effort worthwhile? This, I am in no positon to say. Obviously, in terms of direct sales a car for the discriminating low volume market is hardly an efficient investment of efforts. The value must be gauged by effects it may have on an overall picture." [22] He was arguing that the Corvette's greatest value to General Motors would lie not in its actual sales, but in its ability to help transform the lackluster image of Chevrolet.

It was a brash memo, even for a man who had survived the Russian Revolution and escaped the Nazis, but Duntov's timing was fortuitous. Earlier that year, with management's approval, he had raced again at Le Mans—this time in a Porsche—and won in his class, enhancing his reputation. What's more, Ford had outsold Chevy for the first six months of 1954. The threat of further embarrassment by the Thunderbird couldn't be dismissed.

Best of all, Chevrolet's new V8 engine, developed by Cole himself, would be available for the 1955 models, including the Corvette. To Duntov, the new V8 offered the chance to transform the Corvette from a pretty toy into a European-type sports car with an American flavor. The Europeans, with their narrow, winding roads, valued agility and precise handling above all else. Americans, blessed with big, straight streets, wanted power.

Zora Arkus-Duntov was greedy. He wanted both.

Duntov's passionate memo stayed the Corvette's execution, but the car's future was far from secure. In 1955 the archrival Thunderbird, which Ford billed as a "personal luxury car" instead of a sports car, racked up impressive sales of 14,190 cars. Chevrolet, still working off the hangover of unsold Corvettes, built only 700 of them the entire year. But the leaks got fixed, and the car got Cole's new V8, which was lighter but just as powerful as the Thunderbird's engine.

The Corvette also got a manual transmission, albeit only a three-speed.

Duntov improved the 1956 models further by adding better suspension hardware that allowed the Corvette to take curves at higher speeds. He also engineered a new camshaft that boosted the car's horsepower by adjusting the opening and closing of the engine valves, which made the air-fuel intake more potent. It came to be called the "Duntov camshaft." In January that year Duntov made headlines by driving a Corvette 150.583 miles an hour on the sands at Daytona Beach during a week of racing events. Corvette sales jumped fivefold that year, exceeding 3,400 cars.

The real breakthrough came in 1957 with the "fuelies." A new fuel injection system, developed by Duntov, boosted the Corvette to 283 horsepower, nearly 100 more than two years earlier. The Corvette also got an optional four-speed manual transmission. Corvette buyers also could opt to buy a performance package called RPO 684.

The letters stood for "regular production option," and the package included enhanced suspension, tougher shock absorbers, tighter steering, and heavy-duty brakes. These were just the sort of features hot-rod enthusiasts craved. In March, Duntov brought a hastily built special-version Corvette to the track at Sebring, Florida, to race against some of the world's fastest cars. The Maserati team won, but not before the Corvette gave the Italians a scare. "The seeds of a storybook tale were sown," gushed *Sports Illustrated.* "A Detroit sports car, of all things, traveling easily at lap speeds that would make it one of the major contenders in a world championship race."[23] An exultant Duntov started planning to drive a Corvette at Le Mans.

The storybook tale was cut short, however. In June of 1957, Detroit's car companies agreed to stop participating in automobile racing. Congressional critics sniffed that racing set a bad example for young people. GM, Ford, and Chrysler, fearing increased regulation of their business, wanted to appease them. "Zees people, zey try to keel the Corvette," a stunned Duntov moaned.[24] To him, racing did more than burnish the image of the Corvette, and therefore all of Chevrolet. It also let engineers learn how to improve ordinary cars by testing key components under rigorous conditions. But while Chevrolet couldn't officially sponsor racing teams, nothing could prevent teams sponsored by *other* organizations from racing Corvettes. And if Duntov and his colleagues provided some discreet help, well, what GM's bosses didn't know wouldn't hurt them.

So began a years-long cat-and-mouse game with the corporate brass. Duntov used elaborate and ingenious tactics to hide his activities. He set up a dummy corporation in Florida to place orders for Corvette engines, met secretly with drivers at obscure racetracks in New Jersey, and took business trips that brought him to major racing events. All by coincidence, of course.

When Briggs Cunningham raced three Corvettes at Le Mans in 1960, Duntov went along as an "advisor." (The cars didn't win, but they made their mark.) In December 1963 Duntov and several colleagues "vacationed" in Nassau during the annual Speed Week races, where their specially built, but privately owned, Corvette Grand Sport happened to be racing. "Best I think vee not tell management," Duntov would caution his colleagues. But due to GM's potent corporate rumor mill, time and again Ed Cole was summoned to the executive suite on the fourteenth floor of GM headquarters, where he pleaded to save Duntov's job.

Duntov fought battles on other fronts, too. In 1958 Ford added a backseat to the Thunderbird, converting it to a four-passenger car. Sales surged to more than 48,000 cars. By 1962, with Ford selling more than 74,000 Thunderbirds and Corvette sales stuck in low gear at just 14,000, pressure was building for Chevrolet to add a backseat to the Corvette. But to Duntov, a four-passenger Corvette would have been as ungainly as a sprinter saddled with a backpack. The reprieve came only when GM president John Gordon hopped into the backseat of a four-passenger Corvette prototype, and had to be pulled out because the seat was so small. The four-passenger Corvette was dead.

Duntov also waged war with Bill Mitchell, who succeeded Harley Earl as GM's chief of design in 1959. Mitchell had terrific talent, a wicked sense of humor, and a combative personality. He also was a skirt-chasing, hard-drinking bigot. He derided designs he disliked as "jewfish," which didn't exactly endear him to Duntov.[25] Once, when asked to define good taste, he untastefully explained: "Picture two guys in black tie. One fellow has a dash of red in the kerchief in his pocket. The other is standing there with his fly open and his cock hanging out. One is crude and the other is elegant."[26]

After a drinking binge in New York one night, Mitchell heisted a horse-drawn carriage near Central Park and tried to drive it into a hotel lobby—only to be foiled when the horse got stuck in the door.[27] He and Duntov both had expansive egos. It didn't help their relationship when *Car and Driver* wrote, in 1962, that "Zora Arkus-Duntov is so firmly identified with Corvettes they could bear his name."[28]

The biggest Duntov-Mitchell dispute came over the 1963 Corvette Sting Ray. It would be among the most successful Corvettes ever, featuring the car's first independent rear suspension, engineered by Duntov, and curvaceous new styling, personally overseen by Mitchell. But the two men clashed over the car's rear window, of all things. Mitchell insisted on a split rear window with a metal bar down the center. It would be distinctive, he argued, while Duntov countered that it would impair the driver's rear vision.

Mitchell won, but only for a year. In 1964 the split window disappeared, mainly because it cost more to make. Eventually the split window would make the '63 one of the most distinctive collectible Corvettes, fetching prices above $150,000. Meanwhile, though, it was boys-in-the-schoolyard time. Mitchell took to calling Zora "Zorro," and Duntov complained that Mitchell was a "red-faced baboon."[29] But as the Split Window War was reaching its comic zenith, the Corvette was crossing from car to icon.

The Thunderbird's addition of a backseat left the Corvette as the only American sports car available to a generation of baby boomers coming of age. Being a Chevrolet as opposed to, say, a Cadillac, the Corvette had an everyman appeal. While it wasn't cheap, it cost less than half the price of the pedigreed sports cars from Europe. And television's *Route 66* enhanced the car's common-man image.

The show blended Hollywood hip with Jack Kerouac, recounting the heroes' road trips in search of personal identity, the meaning of life, and, sometimes, cheap beer. It was shot entirely on location, a first for a TV series, though sometimes the locales strayed far from the actual Route 66 highway between Chicago and Los Angeles. The show's "engaging young actors promise to make Buzz and Tod as much a part of the American myth as Hemingway's Nick Adams," a critic wrote in *Television Quarterly,* describing the characters as "figures of the Sixties searching for human sustenance in a world more complex than Hemingway ever fathomed."[30]

It was hyperbole, perhaps, but the show portrayed life's choices as more subtle than black-or-white, even though each episode's plot followed a similar outline. Buz and Tod cruise into town in their Corvette and get tangled up in something—romance, a local injustice, or the dilemma of some unfortunate soul—before fighting or talking their way out of it and moving on. A typical episode was "Trap at Cordova," in which Tod is arrested on a trumped-up traffic charge in a tiny New Mexico town. But the townspeople, it turns out, commit the injustice against the young itinerant for a noble purpose. The town school lacks a teacher, and Tod is sentenced to one year of teaching at the local school. Tod beats the rap, but not before giving an impassioned speech about the importance of education for youngsters everywhere.[31]

In 1964, *Route 66*'s last year on the air, the Corvette got another cultural lift. Pop singers Jan and Dean recorded "Dead Man's Curve," a song about a race between a Corvette and a Jaguar XKE through Los Angeles:

> *I was cruisin' in my Sting Ray late one night*
> *When an XKE pulled up on the right*

The song ended with a fatal crash—of the Jaguar, of course—on Dead Man's Curve, a hairpin turn on Sunset Boulevard. Two years later, ironically, the duo's Jan Berry was injured in a real crash in his Sting Ray, not far from the site celebrated in the song. He would be partially paralyzed for life. The tragedy only added to the Corvette's status.

It gained even more stature from another "regular production option" called L88. It was a superhigh-performance package cooked up by Duntov and his colleagues with a monstrous engine topping 500 horsepower. Chevrolet built the L88 Corvettes from 1967 to 1969, and the car's surging, screaming engine mirrored the times: loud, rebellious, and barely in control. Only 216 L88s were produced, but forty years on, aging men would walk around

Corvette conventions wearing T-shirts that said simply: "L88."

By 1969 even basic Corvettes carried 350-horsepower engines, and buyers could upgrade to the massive 427-cubic-inch V8 with 435 horsepower. The Corvette was in full flower. "You don't have to beware of substitutes," boasted one Corvette ad. "There aren't any."

Duntov was in full bloom, too. He would return from two-martini lunches, some of them peacemaking sessions with designers, sneak up on a secretary from behind, cup his hands on her breasts, and say, "Guess who zeees ees!"[32] He provided great copy for journalists, whom he charmed. He would cruise into press previews, proudly driving the next year's Corvette, waving to the reporters with cigarette in hand before hopping out and fielding questions while they crowded around.

Duntov was the prime suspect, usually with good reason, whenever Corvette news leaked. He would invariably answer: "Leek must be design staff." It was on the list of well-known "Zora-isms," which also included: "Let me theeenk to talk" and "Put zee 427 in zee car; then it veel go."[33]

In December 1967 Duntov graced the cover of *Hot Rod*, seated grinning among four massive new GM engines like a general displaying his tanks. In 1968 GM gave him the title that fit the actual role he had held for nearly fifteen years: Corvette Chief Engineer. In 1972, *Sports Illustrated* wrote about the man and his car in an article headlined: "The Marque of Zora." "Although he was assigned to work on advance projects for the whole Chevrolet division," the magazine wrote, "in a style that Rasputin would have admired Duntov slowly appropriated the Corvette."[34]

Duntov didn't always get his way. He dreamed of building a Corvette with the engine behind the driver's seat, a "midengine" design in car parlance, that he believed would enhance the car's balance and driving dynamics. But his pleading and plotting never produced enough corporate support for the project.

By the mid-1970s Duntov was nearing retirement. The Corvette was touching its nadir. The Clean Air Act of 1970 brought the advent of low-lead and no-lead gasoline. Automotive engineers at all companies, including GM, didn't yet know how to make high-horsepower engines compatible with lead-free fuel. The 1973 oil crisis put a premium on fuel economy, and 1974 brought the abomination, as Duntov saw it, of a national 55-miles-an-hour speed limit.

The Corvette's horsepower, like the length of women's skirts, proved a reliable indicator of America's economic and psychic strength. In the years between 1969 and 1975, when America endured the Vietnam War and Watergate, came the emasculation of the Corvette. The car's basic Corvette engine shrank from 350 to 165 horsepower, barely more than the Corvette's original Blue Flame six, and about the same as four-cylinder compact cars today.

It was fitting, then, that at the end of 1974, Duntov retired at sixty-five, the mandatory retirement age for GM executives. He bought his own silver blue 1974 Corvette, personalized with the initials ZAD on the driver's door, to replace the company-owned Corvettes he had driven for two decades. Not long after his departure David McLellan, Duntov's successor as the Corvette's chief engineer, tried to put the best face on the car's humbler horsepower by explaining: "A person who drives a Corvette doesn't drive it for its high speed potential but because he likes driving a finely tuned machine."[35] They were words, it is safe to say, that Duntov never would have uttered.

Though he retired from GM, Duntov never retired from the Corvette. He would drop by the office to urge GM executives to build a midengine Corvette, but he never could convince them. He became a regular at Corvette events, attending conventions and ceremonies big and small, and reporters still sought his perspective.

In 1980, in the wake of America's second oil crisis, a reporter asked Duntov about rumors that the new Corvette would have a six-cylinder engine for the first time in more than twenty-five years. "I do not like to pronounce on this," he replied, with evident distaste.[36] (The rumor proved false.) He took to flying his own airplane, and flew around Detroit-area freeways in his Corvette at speeds well above the ambient traffic. He bought a Honda motorcycle and, despite being a septuagenarian, would pop wheelies in the driveway of his Grosse Pointe home.

The Corvette, for its part, recovered from the 1970s. In the 1980s and beyond Chevrolet restored its horsepower and enhanced it far beyond what it had been in the 1960s. In 1989 Chevy chief Jim Perkins led journalists on an extended test drive through the French Alps. In one village an admiring young couple persuaded the local priest to marry them in a Corvette on the spot.

For all the passion it inspired, the Corvette remained a low-volume car, with sales above 30,000, about 0.2 percent of the overall American market, in a good year. This sales level was just as Duntov had envisioned, but it meant that the Corvette was never safe from corporate cost-cutting campaigns.

In 1991, with GM in dire financial straits, GM's top management almost killed the car. An appalled Perkins, like Duntov before him, defied his bosses, diverting nearly $2 million from various other projects to keep the Corvette going. After corporate accountants discovered the money-shuffling, GM's internal auditors grilled Perkins and top management rebuked him. At one point he offered to resign.[37] But the Corvette and Perkins both survived, just as Duntov always had.

In June 1992 Duntov appeared at the Corvette assembly plant

in Bowling Green, Kentucky, to drive the millionth Corvette off the assembly line. A few months later, sporting his old driver's suit and a yellow helmet, he drove the bulldozer at the groundbreaking for the National Corvette Museum, to be built near the assembly plant. And on Labor Day weekend of 1994, when the Corvette Museum opened, 120,000 Corvette fans cheered him.

In March 1996 he appeared at a much smaller venue, Corvette Night at a Chevrolet dealership in suburban Detroit. It was to be his last Corvette event. Six weeks later, at age eighty-six, Duntov died of complications from cancer. If "you do not mourn his passing," wrote columnist George Will, "you are not a good American." [38]

Zora Arkus-Duntov made his final trip to the Corvette Museum in the urn that contains his ashes. They remain enshrined there today. The museum also features a life-sized plaster statue of Duntov in one of his own suits, white poplin with light-blue stripes. His original 1953 memo, "Thoughts Pertaining to Youth, Hot Rodders and Chevrolet," hangs nearby.

Corvette buyers can take delivery of their new car at the museum instead of at their local dealership. The event will be recorded on webcam in a special area of the museum called "the nursery," because it delivers people's babies, as it were. There are dozens of gatherings of Corvette enthusiasts around America each year, ranging from small regional conclaves to the annual Bloomington Gold extravaganza in Illinois. There, cars vie to be designated a Bloomington Gold Survivor, not a new TV reality show, but a car so well preserved that it needs little to no restoration. At the 2006 Bloomington Gold auction, a buyer paid $367,500 for a 1964 pink Corvette, with pink-wall tires, that had been specially built for a GM executive's wife.

The Corvette has also inspired a mini-industry. In 1972 a Georgia mailman launched a Corvette magazine on a mimeograph machine in his basement, and became a millionaire. A young fan who sold Corvette stickers from the trunk of his Oldsmobile built a thriving business called Mid America Motorworks, selling everything from Corvette parts to Corvette bathrobes, his and hers.

For many years, a 1990 Corvette arrived regularly at the U.S. Supreme Court on Capitol Hill, driven by its owner, Justice Clarence Thomas. The car's license plate read "RES IPSA," lawyerly Latin for "It speaks for itself."[39] Other Corvette drivers would launch into verbal overdrive to explain the car's appeal. "The touch of that kind of speed and power imparts a sense of endless possibilities," explained Shelby Coffey, a newspaper editor, who got a new Corvette in 1991 as an anniversary present from his wife. He got his first-ever speeding ticket two weeks later. "It's a sense that I am more powerful than I thought, that I can leave much of the ordinary world behind."[40]

In late 2008 General Motors launched the new Corvette ZR1, with 638 horsepower, enough to accelerate from 0 to 60 miles an hour in under 4 seconds. The ZR1, which cost $105,000, was over-the-top by any standard, all the more so with gas costing $4 a gallon at the time. But it sold out within weeks.

Such passion provided testament to the lasting power of Duntov's vision. It wasn't as noble as all men being created equal. But it furthered the pursuit of happiness at very high speeds. And the Corvette would endure far longer, mercifully, than the other automotive expression of America's we-are-the-world exuberance in the 1950s: tail fins.

3

THE 1959 CADILLACS:
STYLE, STATUS, AND THE RACE FOR
THE BIGGEST TAIL FINS EVER

Perry Lane was a typical 1950s bohemia. Everybody sat around shaking their heads over America's tailfin, housing-development civilization, and Christ, in Europe, so what if the plumbing didn't work, they had mastered the art of living.

—Tom Wolfe, *The Electric Kool-Aid Acid Test*

As the parade of cars wound through the town, the car that drew the most attention was as long as a small boat. Its enormous length was accented by sweeping tail fins, the tallest and biggest fins ever appended to the rear end of a vehicle that didn't fly. Hundreds of onlookers waved their admiring approval to the driver, who in turn smiled and waved back from his 1959 Cadillac Eldorado Biarritz convertible.

It might have been a scene from midcentury America but instead it took place in the early twenty-first century, nearly fifty years after the '59 Cadillacs appeared. And the parade had other oddities. Most people watching it were blond. The road signs pointed to towns with tongue-tying names such as Sigerslevvester and Svestrup. And standing in the background was a castle, not some plastic-fantastic Walt Disney tourist trap but a real-life castle that, as it happened, had been immortalized by a playwright named Shakespeare.

Helsingør, Denmark, is home to Hamlet's castle, and a hamlet on the route chosen by the Amerikaner Biltraef Koebenhavn (the American Car Club of Copenhagen) for a parade of classic cars from the 1950s and 1960s. The Copenhagen club is a local chapter of the Federation of Danish American Car Clubs, along with the Peggy Sue club, the California Dreamers, the Cadillac Club of Denmark, and dozens of similarly named chapters. Their members cruise around in Corvette Sting Rays, early Mustangs, and a plethora of 1960s high-horsepower muscle cars: Pontiac GTOs, Dodge Coronet Super Bees, and Oldsmobile 442s. But the cars that usually bring the widest smiles and the biggest cheers from onlookers are the tail-finned, pastel-colored Cadillacs from the late 1950s and early 1960s.

America's postwar prosperity was entering full flower when those cars were built. In the seesaw struggle between the practical versus the pretentious, the latter ruled the roads. There never have been any cars anywhere like the tail-finned land yachts of the day. And it's safe to say there never will be any such cars again.

The recent popularity of American cars from the 1950s and 1960s is surprising in the state of Denmark, and in Sweden and Norway, too. Cars familiar to middle-aged Americans are celebrated at Scandinavian road rallies, conventions, and in specialized magazines. AMCAR magazine, published by the American Car Club of Norway, has a merchandise store that offers a "cold weather cruise jacket," handy for, say, cruising through Trondheim in a Cadillac convertible. Even with the top up.

Going to Scandinavia to study how cars defined American culture might seem a waste. Then again, Margaret Mead did go to Samoa to gain insights into the sexual mores of American teens, and Samoa is a lot farther away. But Mead was onto something. Often the best way to look at yourself is in a mirror, even though it can be discomforting.

Most Americans look back on tail fins with foot-shuffling embarrassment, asking the question: "What *were* we thinking?" But fifty-plus years after the bohemian protohippies on Perry Lane in Palo Alto sniffed at tail fins as proof of America's cultural inferiority, many middle-aged Scandinavians see them as symbols of their childhood image of America: big, benevolent, and bountiful. Americans could be just who they wanted to be, even if their freedom led to the bombastic and the bizarre. The religion of self-expression was praised in the hymns of Elvis, whose music, to this day, often wafts in the background whenever Danish car enthusiasts convene. Their stereotype of 1950s America is quite the opposite of quaint and quiet Denmark, where the prevailing sensibility, both in the Fifties and now, is one of *janteloven*. That's the Danish word for aversion to putting on airs, trying to appear better than anyone else, or, God forbid, showing off.

Showing off, of course, was the very purpose of tail fins. They started out small, like the tails on tiny tadpoles, in 1948. But they soon grew. In 1957 Chrysler bedecked its cars with enormous fins that pointed to the sky, and touted them with advertisements that declared: "Suddenly, it's 1960!"[1] The company also claimed that its outsized tail fins were safety devices: "directional stabilizers," in its words, that kept its cars going straighter than ordinary, unfinned cars.

It was ludicrously outrageous, of course, even for an industry where hyperbole is as ambient as horsepower. But the mandarins at the Harvard Business School swallowed the claim anyway, lock, stock, and dipstick, and honored the Chrysler executive who fathered the big fins. Meanwhile, the styling gurus at Cadillac worked themselves into a near panic at the prospect of being out-finned by Chrysler. They launched a crash program to develop bigger and more ostentatious fins of their own. As America found itself locked in the arms race and the space race with the Soviet Union, Detroit's great tail fin race was on.

The most exuberant symbols of America's midcentury peacetime prosperity were inspired by a war machine. Sometime during World War II, Harley Earl visited the Selfridge Air Base northeast of Detroit. He wanted to see the Lockheed P-38 Lightning fighter. While Earl and his aides weren't allowed within thirty feet of the plane for security reasons, that was close enough to appreciate the design.

The Lockheed P-38 had a cockpit flanked by two large fuselages that housed the engines and fuel tanks. Extending from each fuselage, and giving the plane an unusual, catamaran-like look, were twin tails, both of which supported a vertical fin that thrust straight upward. To Earl, the fins would look as good on cars as on airplanes. At his direction, modest tail fins—looking more like bumps than fins—debuted on the Cadillacs of 1948.

The fins weren't an instant hit. GM executives worried about the public's reaction, and not without reason. "As soon as the '48 model appeared, the protests began to roll in," *Fortune* magazine reported. "It has long been rumored that in desperation, the company hurried up designs on a finless rear fender. The rumor is true; the company did. But somewhere along the line opinion began to change. The more fins that appeared on the road, the more people got used to them, and finally they began to like them."[2] Harley Earl turned out to be right, just as he knew he would be.

By 1947 Earl had been at General Motors for two decades and for almost all of that time he had remained, amazingly, the only design executive in Detroit. Henry Ford always disdained design, and efforts at Ford Motor to embrace styling were curtailed with the untimely death of his son, Edsel, in 1943. It wasn't until 1946, the year before old Henry died, that his grandson and successor, Henry Ford II, started to rely on a designer. His name was George William Walker.

A former semipro football player who carried 222 pounds on his five-foot-ten-inch frame, Walker owned forty pairs of shoes and seventy suits, and doused himself with so much Fabergé cologne that colleagues could tell when he had been in a room, long after he had left. He once gave a reporter a detailed description of his own sartorial splendor during one of his vacations in Florida. "I was terrific," Walker said. "There I was in my white [Lincoln] Continental, and I was wearing a pure silk, pure white, embroidered cowboy shirt and black gabardine trousers. Beside me in the car was my jet black Great Dane, imported from Europe, named Dana von Krupp. You just can't do any better than that."[3]

God help anyone who'd even try.

Chrysler was the last of Detroit's Big Three to embrace styling. And in comparison to Harley Earl and George Walker, its design director, Virgil Exner, was downright normal—"a hell of a nice guy," as a designer at rival American Motors put it. "He was an absolute, perfect example of a car-loving designer."[4]

Exner was born in Ann Arbor, Michigan, in 1909 as Virgil Anderson to an unwed Norwegian-American girl who couldn't support her baby. George and Iva Exner, a machinist and his wife, adopted the boy. They lived in Buchanan, Michigan, a tiny town in the southwestern part of the state, closer to Chicago than Detroit. As a schoolboy Virgil loved to draw, sketching cars for fun and doing illustrations for various publications at his high school. The first car he drew was a Model T Ford, but Exner wasn't satisfied with its no-frills appearance. So he put a Duesenberg emblem on the flivver's hood and gold pinstriping on its sides.

After high school he enrolled at Notre Dame University in South Bend, Indiana, just fifteen miles from Buchanan, to study art and design. After two years he was restless, and left school to work for Advertising Artists Inc., a studio in South Bend, as a messenger boy for $12 a week. Exner wangled his way into doing marketing illustrations for the local car company, Studebaker, but he wanted to design cars instead of sales brochures.

When he got the chance to do just that at General Motors, which was then expanding its design department, Exner jumped. His work had caught the eye of Earl, who hired "Ex," as the young man liked to be called, and put him in charge of the Pontiac design studio in 1934. Exner was just twenty-five years old, slim and distinguished-looking even at a young age. His future with the world's largest car company looked bright.

But four years later Exner left GM. Raymond Loewy, the most famed industrial designer of his day, lured him away. Loewy's independent New York studio had designed products ranging from International Harvester tractors to the package for Lucky Strike cigarettes. Earl tried to keep Exner by telling him he had a good shot at becoming GM's next design chief after Earl himself retired, but that wasn't due to happen for another couple of decades.

Exner departed Detroit, and moved his young family to Long Island. At Loewy's, he drew designs for Studebaker, this time for cars instead of for leaflets. He split his time between New York and South Bend, until Loewy transferred him to South Bend in 1941 to be closer to Studebaker. As it happened, Exner got closer to the client than Loewy had intended.

In May of 1945, Studebaker chose Exner's drawings over Loewy's own designs for its new cars of the postwar era. An incensed and jealous Loewy boarded a train from New York to South Bend and upon arrival, he marched into a meeting with Exner and Studebaker's vice president for engineering, Roy Cole (no relation to GM's Ed Cole). "You are immediately fired," Loewy told Exner. But Cole, who was Exner's biggest fan at the company, issued a quick retort: "You are immediately hired, Mr. Exner, by Studebaker." It was a classic comeuppance for Loewy and sweet vindication for Exner.[5]

But Exner's good fortune was too good to last. Loewy's prestige, which would later land him on the cover of *Time*,[6] provided him considerable clout at Studebaker. Roy Cole was nearing retirement age. After Cole's departure, Exner's standing at Studebaker would be uncertain. With Cole's blessing, Exner started scouting for other jobs.

Ford offered him its top design post, or so Exner thought, until he got a call saying George William Walker would handle the company's design duties instead. Exner's search eventually took him to K. T. Keller, the president of Chrysler, a stocky, ruddy man and an engineer by training. At Chrysler, the engineers disdained designers, not entirely without reason. The most daring design in Chrysler's history, the streamlined 1934 Airflow sedan with an elongated hood and virtually no backside, had been a flop (a forgotten fact, as collectors now prize the car). Airflow production halted after only three years.

In 1949, though, Chrysler was lagging in America's postwar car-buying boom, and Keller was concerned. He hired Exner, for a salary of $25,000, as head of Chrysler's advance styling studio, an assignment that would keep him on a tight leash. While Exner would have latitude to experiment with future styling concepts, the engineers would decide which cars got produced.

So Chrysler kept losing market share. Its cars had an old-fashioned "three-box" design—boxy hood, boxy cabin, and boxy trunk—that looked upright and uptight compared to the rounded, chrome-laden cars from Ford and GM. Cadillac was particularly daring. In late 1952 it adorned its new cars with chrome cones protruding from the front bumpers. Supposedly intended to evoke torpedos, to most men they evoked a more intimate shape. The chrome protrusions were nicknamed "Dagmars," after a blond starlet of the day famed for her ample bosom and, in her television persona, less than ample brains.

At about the same time, Keller showed Exner the designs for Chrysler's 1955 models and asked his opinion. Exner replied that the designs were lousy.[7] By this time Keller had seen enough disappointing sales numbers, so he offered Exner the chance to redesign the company's entire lineup: Plymouths, Dodges, DeSotos, and Chryslers. To be sure, there were a couple of caveats. The new designs would have to be done quickly, because the 1955 cars would be launched in October 1954. Production would begin in just over eighteen months. Because of the tight time frame, Exner would have to use the underlying chassis already in place for the cars.

Exner accepted the challenge. In the summer of 1953 he got a new title: director of styling, a first for Chrysler. At age forty-four, the boy born in the humblest of circumstances, who never earned an art degree and who didn't finish college, was in the right place at the right time. He was about to reshape Chrysler's cars, put Cadillac in a state of panic, and make his mark on American life.

• • •

Dissent wasn't unknown in 1950s Red Scare America. The hapless worker who had been beaten up at the Chevrolet Flint factory for his alleged communist sympathies could attest to that. But conformist normality ruled the day. It found fertile soil in Levittown, Long Island, created in 1947, and other far-flung suburbs where growth boomed after the first Interstate highways appeared in 1955.

A trio of television situation comedies—*The Adventures of Ozzie and Harriet*, *Leave It to Beaver*, and *The Donna Reed Show*—debuted between 1952 and 1958, and celebrated the ideal suburban family. All three featured workaday white-collar dads (whose actual jobs remained unclear), stay-at-home, cookie-baking moms, and mannerly children who groused about their homework but pretty much did what they were told.

Television parents weren't the only authority figures afforded respect. Government officials, corporate executives, and clergymen commanded pedestals their successors in the late 1960s would regard with envy. The 1954 hit movie *Executive Suite* celebrated "the executive as hero," as *Fortune* put it, "the man who, despite obstacles and consequences, seeks the prize of high management and wins it."[8] The lead character, played by William Holden, is McDonald Walling, vice president of design at a fictional furniture company, Tredway Corp. When Tredway's president dies unexpectedly, Walling makes a pitch for the top job, pledging to restore Tredway's traditional values. In the climactic scene, before Tredway's board of directors, Walling's passion and idealism prevail.

In January 1955 *Fortune* carried a first-person article by a real-life executive hero, David Sarnoff, the founder and chairman of RCA. He predicted America was entering an age of unprecedented prosperity born of new technology. "Small atomic generators, installed in homes and industrial plants, will provide power for years," Sarnoff wrote.[9] The prospect of a nuclear reactor in every home, however, wasn't as compelling as a chicken in every pot.

In January 1956 *Time* named the CEO of General Motors, Harlow Curtice, 1955 Man of the Year. As leader of the world's largest manufacturer, the magazine wrote, Curtice was "first among scores of equals whose skill, daring and foresight are forever opening new frontiers for the expanding American economy . . ."[10] It could have been a GM press release, but even the company's flacks might have blushed while writing it.

Four months later General Motors dedicated its new sixty-acre Technical Center on a campuslike setting north of Detroit. President Eisenhower came to speak and invited President Sukarno of Indonesia as his guest, to put America's prowess on display. Another speaker was Lawrence R. Hafstad, vice president of GM's research laboratories, who declared: "We in the United States have a new feature, a strictly American invention—the provision of continually increasing purchasing power for the consumer."[11]

Just twenty years after the Great Depression, boundless optimism and faith in the future filled America. Conformity and respect for authority were natural companions for this happy certitude. Americans were willing, by and large, to have their tastes and aspirations defined by advertising gurus, the original Mad Men. Bigger was always better to most people in midcentury America, and better was sure to come.

As the 1950s marched on, *Playboy* popularized pictures of Dagmars made of flesh instead of chrome. Hefner, Elvis, and the other cultural rebels of the day stood for bigger parties, not for social justice or conservation or anything like that. By happy coincidence, self-indulgence offered Detroit a lot more profit potential than self-denial. Cars that were bigger and flashier cost more than cars that were plain and simple. So there was an odd, if often wary, alignment between America's cultural rebels and its blue-suited corporate establishment. It was a perfect atmosphere for automotive extravagance.

Chrysler's lackluster styling was a key reason why its market share plunged to 13 percent in 1954, five or six points below its normal average. That summer rumors began to circulate that the company's 1955 models, the first cars designed by Exner, would look radically different. On the stock market, Chrysler's shares shot up five points. "Exner sits on the hottest spot in the auto industry today," wrote *Look* magazine.[12] Indeed, it was far from clear whether the market's faith in Chrysler, and Chrysler's faith in Exner, would be rewarded.

GM's 1955 Chevrolets, developed by chief engineer Ed Cole, promised stiff competition. The new Chevrolets stood three to six inches lower than their predecessors, depending on the model. The new 265-cubic-inch V8 engine was the most powerful yet in a Chevrolet. Dull colors such as dark blue and deep green gave way to bright two-tone schemes, including the art deco combination of charcoal and coral.

Ford's stylish new Thunderbird lacked major engineering innovations, but George William Walker, by now the company's vice president for design, wasn't fazed. "Beauty is what sells the American car," he would declare, explaining that developing cars that appealed to women would help Ford sell them to men.[13]

But even among those cars, Chrysler's new styling was eye-catching. Despite having little money to work with and less time, Exner had transformed Chrysler's designs. The plain-Jane fronts gave way to "Frenched" headlights that protruded forward like bushy eyebrows. The three-box look disappeared. The new cars' wedge-shaped bodies sloped upward, front to back, and reached a rear-end crescendo with pronounced tail fins. Chrysler's fins contained taillights that were, themselves, vertical wedges, with bright chrome bezels surrounding the lights. Chrysler called the new styling "The Forward Look."

Raymond Loewy, doubtless still smarting from his treatment at Studebaker, spoke out against his former protégé's work. The American car was becoming a "jukebox on wheels," he wrote in the *Atlantic*, adding: "Nothing about the . . . 1955 automobiles offsets the impression that Americans must be wasteful, swaggering, insensitive people."[14] Take that, Virgil Exner.

Exner's cars were indeed the gaudiest of them all, but they racked up the biggest sales gains. Chrysler's market share jumped back up to 18 percent in the first five months of 1955, making up almost all of the ground lost in the disastrous prior year. Even more impressive, the company's earnings for the first two months of 1955 exceeded its profits for all of 1954. Thus emboldened, Exner and Chrysler introduced more daring designs in 1956, with even bigger fins. The company touted it as "Flight-Sweep" styling and boasted: "From jutting headlight to crisply upswept tail fin, it displays a single clean line that says ACTION!"[15] Hyperbole, like tail fins, was also reaching new heights.

Nonetheless, critics responded with rave reviews. "Styling of the 1956 Plymouth is bright as a new silver dollar," wrote *Motor Life* magazine. "Where conservative design made it a wall-flower in 1952–54, for 1956 it is a year—and maybe even more—ahead of its rival makes, who were clearly beaten to the punch in the tail-fin department."[16]

Taller tail fins weren't Chrysler's only coup for 1956. The company also introduced push-button automatic transmission, dubbed PowerFlite, which offered the advantage of "completely avoiding any connotation of shifting," the company explained.[17]

Push-button cars seemed modern, high-tech, and uniquely suited for the dawn of the space age. In 1956, Exner went to Chicago to accept the Esquire Magazine Fashion Award. He was riding high and aiming higher.

But Exner was an intense, chain-smoking workaholic who consumed too much coffee and alcohol. He also suffered from the stress of constant bureaucratic battles with Chrysler's engineering and manufacturing executives, who often complained that his designs were too expensive or too difficult to build. On July 24, 1956—with production of Chrysler's 1957 models about to begin—Exner suffered a major heart attack. Rushed into open-heart surgery, he pulled through.

He got more bad news during his recuperation. A Chrysler experimental car called the Norseman, built in Italy by Ghia at the then astronomical cost of $100,000, was sitting on the bottom of the Atlantic Ocean. The Norseman was one of Exner's pet projects; its very name was a nod to his Norwegian heritage. His demanding specifications for the car, including a power-operated rear window and ultrathin steel rods supporting a lightweight roof, had caused a three-week delay in shipping it from Italy to the United States. Passage was rebooked on the first available boat, which happened to be the *Andrea Doria.* On July 26 the Italian luxury liner collided with a Swedish ship off the coast of Nantucket and sank, killing nearly 50 of the 1,700 people on board. It would be the last great transatlantic shipwreck, and Virgil Exner's Norseman forever would be a footnote to history.

Late that summer, as Exner's recuperation progressed, a young man named Chuck Jordan took another, much shorter journey that also would become a footnote to history. Jordan, a Cadillac designer, took a lunch-hour drive northward along Mound Road, a major suburban thoroughfare northeast of Detroit. Such drives were Jordan's way of relaxing from the inevitable stresses of the design studio. This day he had another purpose in mind.

He had heard rumors that Chrysler's 1957 cars, which were still under wraps, had shapes even more radical than the company's acclaimed designs of the previous two years. As Jordan passed a Chrysler facility he spotted some cars parked in a field behind a fence. He decided to drive in and look. The tall grass around the cars obscured his view, but Jordan saw enough to confirm his worst fears.

"I couldn't believe my eyes," Jordan would recall more than fifty years later. "The cars were beautiful, with thin roofs and long, lean lines. Our '57s weren't anything like that. They had thick roofs, heavy body shapes, bumpers down to the ground and lots of chrome. They were Harley Earl designs. And Harley was a tough boss; he got what he wanted."[18]

Jordan tore back to the GM Technical Center and ran into Bill Mitchell's office. "You've got to see what I just saw," sputtered Jordan. "You won't believe it." So Mitchell and Jordan grabbed a colleague, designer David Holls, and drove straight back up Mound Road. "Those fins just shot up out of the grass," as Holls would recall the scene. "It was just absolutely unbelievable. We all said, 'My, God, they just blew us out of the tub.'"[19]

Indeed, the sleek, high-finned new Chryslers couldn't have been more different from the chrome-bedecked blobs that GM was about to introduce for 1957. One rival stylist said, only half in jest, that it would have been easier just to cover GM's 1957 cars completely with chrome, and then mask off a few spots to be painted.[20]

The discovery put Mitchell in a dilemma. It was too late to change GM's 1957 styling, and even too late for all but minor tweaks to the 1958 cars. He had time to tear up the early drawings of GM's 1959 models and have them redone. But that would incur the wrath of Earl, who was taking a leisurely tour of Europe. Earl was just a year or so from retirement. Mitchell was his heir apparent. He had to do something, but he couldn't overtly challenge his boss.

So Mitchell hatched a clever plan. Instead of scrapping the existing designs and defying the boss, he ordered work begun on *alternative* designs for every major GM model. When Earl returned from Europe some weeks later, he cruised into the styling studios and saw the sleek new designs sitting alongside the squat, turtlelike shapes he had decreed for the 1959 models.

The man who had dictated GM's designs for three decades remained silent and retreated into his private office. His silence was hellish for Mitchell, Jordan, and their co-conspirators. Would Misterl have the alternative drawings publicly shredded, as an example to others who might dare to defy him? Would he fire the whole lot of them? Earl knew he was facing a revolt, and both thoughts probably crossed his mind.

After three full days Earl emerged from his den and gave quiet nods of approval to his surprised and relieved young cubs. He assented to their designs, and let them have their day. But that didn't exactly resolve GM's problem. It would take two years for GM's new styling to hit the market, and meanwhile Cadillac was stuck with cars that looked anything but streamlined. The fins on the 1957 models appeared slapped on as afterthoughts. The heavy front bumper and grille were layered with chrome. The torpedo-shaped chrome Dagmars got new black rubber tips quickly nicknamed "nipples." The contrast with the '57 Chryslers was stark.

The front ends of the 1957 Chryslers were three to five inches lower than the previous year, a change that had required Exner's designers to do some serious arm wrestling with the engineers.

The cars gradually gained height as they extended and expanded, wedgelike, toward the tall fins in the rear. "Reserved use of chrome . . . gives the Chrysler a dignified charm that will appeal to those who instinctively avoid the garish," the company said.[21]

The statement was a veiled slap at the Dagmars, nipples, and other ornamentation on Harley Earl's chrome-slathered Cadillacs. Only alongside those cars could tail fins be described as paradigms of understated elegance. But Chrysler's hype machine was just getting revved up. "Silver anodized aluminum Sportone trim . . . sweeps majestically back to the graceful Directional Stabilizers," boasted one company sales brochure.[22]

Who would have thought it? Having grown from infancy to adolescence and now into lofty adulthood, tail fins had become "directional stabilizers." It was like calling a jail an exclusive gated community, but Chrysler did it with a straight corporate face. "With the high upswept fins of our 1957 models," Exner said in a speech after he returned to work, "road-holding stability was improved by as much as twenty percent in strong crosswinds at normal driving speeds."[23] The corridors at Chrysler must have been windy indeed.

More honors kept coming Exner's way. In April 1957 Notre Dame's Detroit Alumni Club named him "Man of the Year," even though he had never graduated. In June the Industrial Designers Institute honored him. He accepted the award by declaring: "Tail fins are a natural and contemporary symbol of motion, appearing on nature's creatures and on aircraft . . . guided missiles and rockets."[24] A month later he was elected a vice president of Chrysler. And in December the Harvard Business School invited him to give an endowed lecture about his philosophy of design.

Exner chose the occasion to talk about not just styling and safety, but also about the advance of American civilization, which happily, as he saw it, was evident in its cars. "Our whole way of life in America is geared to the automobile," Exner said. "It has come to mean much more than transportation. It has become a status symbol. Since this particular status symbol tends to be brightest when it is new and easily distinguishable from other and older [status] symbols, automobile styling assumes a key role in creating sales appeal." [25]

The popularity of tail fins reflected "the growing artistic taste of the American consumer," he added. "I believe all American products will more and more reflect the spirit and character of our civilization." [26]

To some critics, Exner's bragging described the whole problem with America's wacko civilization, if "civilization" was the right word. In 1957 journalist Vance Packard published *The Hidden Persuaders*, casting the likes of Exner and his co-conspirators on Madison Avenue as manipulators of America's gullible-idiot masses, including professors at the Harvard Business School. Packard cast darts everywhere. He claimed that *Howdy Doody*, a popular children's television show, twisted young minds by presenting its adult characters, such as Chief Thunderthud and Mr. Bluster, as oversized imbeciles. (As if kids needed much convincing that most adults were dolts.)

As for Detroit, it had conducted secret psychological studies showing that "men saw the convertible as a possible symbolic mistress," Packard wrote. "The man knows he is not going to gratify his wish for a mistress, but it is pleasant to daydream." [27] (This was a secret?) Marketers of every product, explained Packard, used mind-control techniques to tap into the repressed emotional needs of American consumers.

The book was a best seller, as was *The Insolent Chariots*,

published a year later. Author John Keats picked up where Packard left off, concluding that Detroit created cars for "daydreaming nitwits." Keats added: "It is not sheer accident that most manufacturers put penial geegaws on the hoods of their cars, or that Cadillac's stylists speak of the 'bosoms' on their bumpers, or that Buick came up with its famous ring pierced by a flying phallus, or that Madison Avenue was quick to applaud the Edsel for its 'vaginal look' . . ."[28]

The "vaginal look" referred to the Edsel's enormous oval front grille, which drew comparisons to a horse collar, a toilet bowl, and, inevitably, the vagina. One joke held that if an Edsel should strike a tail-finned Chrysler or Cadillac from the rear, the result would be automotive sexual intercourse, right there on the highway. The Edsel abstained from tail-fin one-upsmanship, but the car would make its mark as a synonym for failure.

That wasn't really fair to Edsel Ford, for whom the car was named. The only son of the company's founder and the father of CEO Henry Ford II, he was long dead by the time the Edsel was introduced on September 4, 1957. Besides, as dull as it was, the name Edsel was a whole lot better than the alternatives suggested by Marianne Moore, a Pulitzer Prize–winning poet whom Ford hired to help name the car. Some of her suggestions, such as Ford Silver Sword and Varsity Stroke, seemed just as Freudian as Dagmars. Others, like Mongoose Civique and Utotopian Turtletop, were just weird.

The Edsel was a brand with four different models: the full-sized Citation and Corsair, and the smaller Pacer and Ranger. The idea was to give Ford owners an intermediate brand to step up to that wasn't as pricey as Lincoln. As Chevy owners gained wealth and stature, some 85 percent of them "graduated" to a premium GM brand, usually an Oldsmobile or Buick, and perhaps eventually to a Cadillac. But Ford's studies showed that just one-fourth of its owners moved up to Lincolns.

Ford invested some $250 million to create its own not-quite-luxury brand. Beyond the front grille, its notable feature was the push-button automatic-transmission controls in the hub of the steering wheel, where the horn usually was. Prices started around $2,500, about $300 more than the highest-priced Ford sedan. The timing of its introduction wasn't propitious. One month later the Soviet Union launched the first space satellite, Sputnik, which Nikita Khrushchev hailed as proof that Communism was outpacing the decadent West. America's buoyant postwar optimism got a needed jolt.

Nonetheless, on Sunday, October 13, Ford preempted the *Ed Sullivan Show* with a Sunday-night Edsel special featuring Frank Sinatra and Bing Crosby. Three weeks later, *Time* put George William Walker, the Edsel's designer, on its cover. "So far, Edsel sales have not lived up to Walker's hopes," the magazine observed, "but it will be months before he knows whether he has a lemon or lemonade." [29]

It was, for sure, a lemon. The Edsels suffered not only from their buffoonish front grille but also from shoddy quality. On top of that, in 1958 America dipped into recession, depressing sales of Cadillacs, Chryslers, and other premium-priced cars. Everything about the Edsels—their workmanship, styling, price, and timing—came out wrong. Ford had hoped to sell 200,000 Edsels in the first year, but wound up selling just one-third that many. In November 1959, after two years of futility and $400 million in losses, Ford killed the Edsel before the car could kill the company.

During the Edsel's death throes Cadillac sales also suffered. They dropped 17 percent in 1958, partly because of the recession, but also because Earl's styling was so obviously out of date. That August *Motor Life* fretted that "General Motors is in the position of not having any concrete styling philosophy for their rear ends," an awkward way of saying that GM was, um, lagging behind. "Cadillac will no doubt continue their now ten-year-old custom of using small fins . . ."[30] Nothing could have been further from the truth.

When the 1959 Cadillacs debuted, no second looks were necessary to identify them, but people took a lot of second looks anyway. The longest model, the Series 75, stretched more than twenty-one feet, more than three feet longer than GM's massive Hummer H2 nearly fifty years later. The cars were also several inches lower, requiring GM to drop the height of the front seats so people wouldn't scrape their heads against the ceiling. Every Cadillac had a 300-plus horsepower V8. The Eldorado Brougham cost some $14,000, equivalent to $100,000 today.

The rubber-tipped Dagmars disappeared, but copious chrome remained on the double-decked headlights and the mawlike front grille, which looked like it could inhale everything in its path. The long, concave sides, shaped like an airplane fuselage, swept back toward the round brake lights that looked like a jet engine exhaust. But the most prominent features were tail fins so enormous that they seemed lifted right off a rocket ship. Each fin was "accented by twin, projectile-shaped taillights," as a Chrysler "competitive assessment" report put it.[31] The twin taillights, soon nicknamed "gonads," briefly supplanted hubcaps as the theft item of choice for juvenile delinquents.

As big as they were, the tail fins on the 1959 Cadillacs might have been even bigger. At one point during the design process, Chuck Jordan and his colleagues sculpted a clay model and stepped back, only to discover that the fins stood taller than the roof of the car.[32] They went back to the drawing board and scaled the fins down, but only a bit. The fins displayed "Cadillac's realm of motoring majesty for 1959," one GM executive intoned.[33] An exultant Bill Mitchell later would put it more colorfully. "I say if you take the fins off the Cadillac it's like taking the antlers off a deer," he said. "You got a big rabbit."[34]

General Motors had made the point, physically and figuratively, that it would never again be out-finned by Virgil Exner and the upstarts over at Chrysler. Cadillac's sales surged in 1959, making up the prior year's decline. Other General Motors marques introduced their own distinctly shaped fins. Chevrolet had horizontal "bat-wing" fins, and Buicks sprouted angled "delta-wing" fins. GM's Firebird III "dream car" displayed at the Motorama in New York sprouted no fewer than nine fins—on the tail, the sides, and the trunk lid. For a brief, chrome-shiny moment, findom reigned supreme.

Harley Earl retired in late 1958, just as GM's 1959 cars debuted. Three years later Virgil Exner also retired, a victim of declining health and infighting within Chrysler's design department. By that time tail fins were on the way out. After peaking, as it were, in 1959, fins scaled back, becoming mere vestiges in 1964 and disappearing entirely in 1965.

Long after they vanished, however, tail fins continued to symbolize the sky's-the-limit ethos of America in the late 1950s. In 1974, a Texas millionaire named Stanley Marsh 3 took ten tail-finned Cadillacs and planted them in the ground—nose first, fins pointing to the sky—outside his hometown of Amarillo, near the confluence of Interstate 40 and the old Route 66. The "Cadillac Ranch," as the display was called, testified to the imprint tail fins had made on American minds. They made an imprint on non-

American minds, too, including that of Leif Kongso, a member of the Cadillac Club of Denmark.

Fifty years after the tail fin era, Kongso owned a Cadillac and several other American cars from the 1950s and 1960s. "I do not have the faintest interest in what type of car people use for grocery getting, nor do I care if BMW or VW has come up with a new model made from cheaper plastic somewhere in the Far East. To me, they are merely objects to get you from A to B. The dinosaurs I own are relics from past times, and such will never come again," he reflected at a meeting of Denmark's Cadillac Club. "We Danes have always been told, since we were children, that the U.S. is the country where dreams come true and where all is possible. Everybody has a dream of driving with the top down through the wide open, with some great music playing, toward the setting sun. So I choose a real car, not a mass-produced, modern, boring box. I am not a loud or bragging person, but I appreciate all the smiles and thumbs-ups that we get."

Perhaps the strangest thing about Kongso's affection for tail-finned Cadillacs is that he's confident that he wouldn't actually like *living* in America. "I doubt I could get used to such an enormous country," he said. "An American friend once told me he liked to go to bars in Denmark, because he always ran into somebody he had met before. That never happened to him back home. Simply too many new faces every time."[35]

So there it was. Cruising around in tail-finned Cadillacs allowed Danes to play American, as it were, like kids playing cowboys and Indians. They could experience the exciting America that attracts them while avoiding the impersonal America that repels them. It's a love-hate attitude that, in truth, many Americans share.

The most ironic legacy of tail fins is that they sparked a backlash. It was depicted comically in *Tin Men*, a 1987 movie in which

Danny DeVito and Richard Dreyfuss played competing aluminum siding salesmen whose rivalry included their respective tail-finned Cadillacs. But as the movie unfolds, their finned behemoths are shown parked next to small, simple cars shaped like little beetles. The sensibilities of some Americans had changed.

VOLKSWAGEN'S BEETLE AND MICROBUS: THE LONG AND WINDING ROAD FROM HITLER TO THE HIPPIES

I mean, I see you guys making it down the freeway in your Buicks . . . and then telling your old lady and your kids in their little Boy Scout army suits to "look at the freaks in the VW bus."

—*Car and Driver* satire, June 1970[1]

The funny thing about the Volkswagen Beetle is that while Americans thought the name was cute, the Germans hated it. Volkswagen even refused to use the name "Beetle" until the early 1970s. As if to prove Germans to be a humorless lot, the company called the car the "Volkswagen Sedan." That was better, at least, than its long-forgotten original name, the Kraft durch Freude Wagen.

The name, which meant "Strength through Joy Car," was taken from the name of Nazi Germany's national labor movement. It was announced on May 26, 1938, in Fallersleben at a cornerstone-laying ceremony to dedicate the factory to build the new car. Lots of people in the audience winced at the weird name, but no one audibly complained because Adolf Hitler made the announcement himself. It was "a rather unwieldy title," sniffed one British car magazine, "which has been abbreviated to KdF."[2] At least that was manageable.

Hitler wanted a practical car for average people to put his country on wheels, just as Henry Ford had done in America with the Model T thirty years earlier. The Führer's government dispatched representatives to Detroit to gain the aging American industrialist's advice on the project, and Ford was most helpful. A handful of first-generation German-Americans returned to Germany from Detroit to help build the new car. It wasn't the best career move, as things turned out.

Seven years later, in March 1945, with Russian armies advancing, the Führer traveled to the eastern front, about sixty miles from Berlin, to inspect his troops. To be less conspicuous he left his Mercedes-Benz limousine behind and made the trip in a KdF, returning before dark and promptly descending into his underground bunker. So it was that Adolf Hitler took his last tour of his crumbling empire in a Volkswagen Beetle.

The KdF fared better than the man who named it. It became enormously popular, especially among fans of an American musician who died in his sleep at age fifty-three in August 1995, a half century after Hitler's last ride, at the Serenity Knolls rehab center north of San Francisco. Jerry Garcia, leader of a band called the Grateful Dead, was a prophet of the long-haired, drug-addled American kids of the 1960s. America's hippies had nothing in common with Hitler. Their devotion to peace and personal gratification was the polar opposite of Nazi militarism and martial discipline.

But amazingly, Hitler's car became their car. The hippies were especially fond of the Volkswagen Microbus, a derivative of the Beetle developed shortly after the war and perfectly suited to their footloose, free-ranging lifestyle. A few weeks after Garcia's death, Volkswagen ran a full-page ad in *Rolling Stone* and other magazines, showing a Microbus, sparsely sketched in pencil, shedding a tear from one of its headlights. The caption simply read: "Jerry Garcia 1942–1995."[3]

The KdF's journey from its genesis as Hitler's industrial showpiece to its flowering as an American hippie icon is one of the strangest automotive journeys ever taken. The Beetle and the Microbus first came to prominence in America in the mid-1950s. They presented Americans of a certain mind-set with a distinct alternative to Detroit's ostentatious tail fins. The 1959 Cadillacs were more than five feet longer than the Beetle.

The counterculture appeal of the Beetle and the Microbus was cemented in the 1960s and 1970s. Volkswagen defied Detroit's glitter and glitz with unconventional advertising that was both witty and self-deprecating, which was beyond the comprehension of Detroit. Ironically, the people who worked at Volkswagen of America weren't much like their free-spirited customers. They were mostly like young Tom Rath, the lead character in *The Man in the Gray Flannel Suit*, Sloan Wilson's 1955 novel about a man's search for personal identity amid the straightjacket of corporate conformity.

Volkswagen's button-down corporate culture was obscured by an image of hip irreverence, created by advertising that struck a chord with freethinkers and dissidents—people Hitler would have detested. But beyond America's intellectuals and hippies, the bug-shaped little car and its boxy companion had broad appeal.

In 1962 an American Maryknoll missionary in Bolivia drove his Beetle to the remote village on Lake Titicaca where he had been assigned to serve. The Indians opened the hood to check out the car's engine but were shocked to find none. They didn't know that the Beetle's engine was in the rear. Word spread around the village that a powerful new witch doctor had arrived.[4]

Later, the Beetle and the Microbus gained a following among the brainy nerds who lived in and around San Jose, California. These young people were technology aficionados who loved the unpretentious elegance of Volkswagen's engineering and design. They were dismissive of mainstream corporate America and obsessed with the possibilities of computers. A bright young dreamer named Steve Jobs sold his Microbus to raise money to start a business in his garage, where he and a partner developed a personal computer that, like VW's Beetle and bus, became iconic in its own right. The Bible was almost right. The geeks did inherit the earth.

The Beetle's story began not with Adolf Hitler but with another man who, like him, grew up in a little village in Austria. Ferdinand Porsche, born on September 3, 1875, in Maffersdorf, displayed his engineering genius as a teenager when he wired his parents' home for electricity. His talent landed him jobs with several car companies, among them Austro-Daimler and Steyr in Austria and Daimler-Benz in Germany. He developed a prototype gas-electric hybrid car along the way, a century ahead of its time, but it was too complicated and expensive to produce.

At Daimler-Benz he became chief engineer and, at a racing event in 1925, met Hitler, then a budding politician. Hitler knew of Porsche's reputation. But Porsche's genius came with a stubborn, difficult temperament that caused him to part ways with a succession of employers. In 1931, at age fifty-six, he started his own business, Porsche-Konstruktionsbüro, a tiny engineering shop in Stuttgart, the hometown of Mercedes-Benz, which specialized in designing racing cars.

That September Porsche called his employees together to announce something new, called Project 12. There weren't eleven other projects at the fledgling company, but Porsche figured the name Project 12 would create the illusion for potential customers that his company had a backlog of orders. Project 12 would depart from the company's racing roots and design something different: a small car affordable to average Germans.

Porsche wasn't alone in pursuing that idea. In 1931 Josef Ganz, editor of a German car magazine, produced plans for a simple little car called the *Maikaefer*, or "May Bug." It had mechanical similarities to the eventual Beetle design, though also important differences. The engine was mounted under and behind the driver's seat, instead of in the rear. Also, with an open-air roadster style and seating for just two, the *Maikaefer* wasn't practical family transportation. But the car's real problem was that Ganz, its inventor, was Jewish, and thus hounded out of Germany in the mid-1930s.[5]

At the time, Germany was reeling from the global economic Depression, and Porsche, who had to borrow money to meet his payroll, needed more business. He didn't have a contract to design a small car, but intended to do the preliminary engineering work and sell it to an established company. He didn't get very far. By January 1933 Porsche was desperate.

That was the same month that Adolf Hitler became Germany's chancellor. On February 11, 1933, less than two weeks after taking office, Hitler delivered the chancellor's customary opening speech at the annual Berlin Automobile Show. He arrived wearing a reassuring morning coat instead of his usual Nazi brown-shirt uniform, and pledged prompt, effective action to revive the German economy. He specifically promised to put a car within reach of every German family. Later that month, after the Reichstag burned in a suspicious fire, Germany's parliament gave Hitler dictatorial powers.

It isn't clear what Porsche thought of Hitler's quick consolidation of power, but it meant opportunity for him. On January 17, 1934, Porsche submitted a lengthy proposal for a people's car to the Ministry of Transportation. He sent a copy to the Führer himself for good measure. "By the people's car I do not [mean] merely a small edition of what exists already," Porsche wrote. "If the traditional car is to be made into a people's car, then fundamentally new solutions will be necessary."[6]

His presentation called for a rounded shape to reduce aerodynamic drag and an unconventional rear-mounted, air-cooled engine—just the opposite of the front-mounted, water-cooled engines of most cars. The design would save weight and space by eliminating what conventional cars needed: liquid engine coolant and a heavy drive shaft to connect the engine in front with the wheels in back.

By putting the engine in back, weight would sit over the drive wheels and enhance traction in rain and snow. Porsche's concept and his drawings, in short, resembled the Beetle's eventual design. In early March Hitler opened the 1934 Berlin Automobile Show, this time wearing a Nazi uniform. He called for "a German car mass-produced so it can be bought by anyone who can afford a motorcycle. . . . We must have a real car for the German people, a *volkswagen.*"[7] "People's car" was a project description, not a brand name, but it was the first time he used the word *volkswagen* in public.

On June 22, the German Automobile Manufacturers Association awarded the people's car contract to Ferdinand Porsche. Germany's automotive establishment didn't like the cantankerous "professor," but the bosses knew Hitler wanted the car, and wanted Porsche to develop it. With the people's-car contract, Porsche-Konstruktionsbüro, like Germany itself, marched from poverty to prosperity in just eighteen months' time.

The following years were momentous for the Third Reich. On June 30, 1934, just eight days after Porsche got his small-car contract, Hitler eliminated his opponents during the Night of the Long Knives. During the next three years his troops marched into the Saar, reoccupied Germany's Rhineland, and seized the German-speaking regions of Czechoslovakia. Amid all this, his regime took progressively harsher measures against Germany's Jews, some of whom sensed the ugliness to come and fled. One was Adolf Rosenberger, a partner of Porsche's with 15 percent of the company, who sold his shares and departed for Paris and then the United States.

The Nazis' ruthlessly efficient power grabs stood in contrast to the delay and disarray in Porsche's work on the people's car. Porsche kept tinkering with the design, "like a woman building a house," one associate sniffed.[8] In road tests, the car's crankshafts kept breaking and the brakes kept failing. The engine proved a particular problem. Hitler insisted that the car be priced no higher than 1,000 reichsmarks (about $400), 30 percent below any other German car of its day. The engine had to be inexpensive to build, as well as modestly powerful and reliable. These difficulties and others meant that the first prototype cars weren't delivered until October 1936, eighteen months behind schedule.

Hitler seemed unfazed, preoccupied with more sinister matters, but the Volkswagen project surfaced in his public pronouncements on occasion. At the Berlin Automobile Show of February 1937, he praised Porsche's work. Thirteen months later he annexed Austria and touted his new car during his triumphant visit to Vienna. "The German volkswagen will make a wish come true," he told the cheering Viennese throng, "harbored by so many hard-working and underpaid Austrian people."[9]

Meanwhile, because buying cars on credit was an unknown practice in Germany, the government started a layaway-type plan for workers to buy cars. After paying five reichsmarks (about $2) per week for four years, workers would travel to the factory in Fallersleben, which in 1938 was still being built, and drive their car home. Eventually 336,668 Germans would enroll in the plan.[10]

Throughout these years, well before production began, the KdF Wagen was a public-relations coup for Hitler, both in Germany and abroad. "German Car for Masses: First of $400 'Strength Through Joy' Autos Is Expected in 1940," declared a *New York Times* headline on Sunday, July 3, 1938, just weeks after the cornerstone ceremony in Fallersleben. The article explained:

> Der Führer is going to plaster his great sweeps of smooth motor highways with thousands and thousands of shiny little beetles purring along from the Baltic to Switzerland and from Poland to France, with father, mother and up to three kids packed inside and seeing their Fatherland for the first time through their own windshield. The new automobile is already nicknamed the 'Baby Hitler.' . . . Judging by the pace at which things get done in the Third Reich today, citizens will not be long in getting theirs.[11]

The Reich's citizens would get theirs, all right, but it wouldn't be a car. The *Times* editors later would wince, no doubt, at the fawning article, which had but one redeeming feature: it was the first to use the word "beetle" to describe the people's car in print.

The *Times* prediction that thousands of beetles soon would pack Hitler's highways was undone by events. In early 1939, Hitler seized Czechoslovakia. On September 1 he invaded Poland. Two days later, Britain and France declared war. Production of the KdF halted.

In 1944 an American engineering journal carried a lengthy technical review of a captured vehicle it described as "the military adaptation of the original volkswagen."[12] Germans called it the Kübelwagen. American soldiers called it the German jeep. It didn't have four-wheel drive, which allowed the American jeep to go almost anywhere. Instead the Kübelwagen was a two-wheel-drive vehicle. But because the engine was mounted in the rear, right over the drive wheels, it nearly matched the American jeep in mobility.

For trickier terrain the German army used the Schwimmwagen, or "swimming car," an amphibious adaptation of the KdF. Both the Kübelwagen and Schwimmwagen were built in the factory at Fallersleben, along with army field stoves, tank components, and eventually components for the V-1 rocket. The KdF factory had been converted to wartime production, just like the factories in Detroit. The Germans relied on slave laborers, however, instead of Rosie the Riveter.

In the spring and summer of 1944, Allied bombing raids did extensive damage to the Fallersleben factory and killed several dozen workers. One bomber scored a bull's-eye on the turbines in a nearby power plant that provided electricity to the KdF factory. But the bomb failed to explode, leaving the power supply intact. It would prove to be one of history's most fateful duds.

On April 10, 1945, not long after Hitler's final ride in a KdF, the American 102nd Infantry Division occupied Fallersleben. Most townspeople huddled in their homes, frightened and hungry, hoping that the Russians wouldn't come. They didn't. When the Americans pulled out a few weeks later, British troops relieved them. Among them was Ivan Hirst, a twenty-nine-year-old major who had been evacuated from the beach at Dunkirk in the bleak spring of 1940. Five years later he was running a tank repair unit of the Royal Electrical and Mechanical Engineers.

The machinery at the bomb-scarred KdF factory was to be dismantled and shipped to the Soviet Union, as war reparations. But despite the gaping holes in the roof the factory had electric power, thanks to the dud bomb, and most of the machinery was intact. To Hirst, the machinery was worth keeping to repair British tanks and trucks, after patching the factory's roof holes with branches and canvas.

Hirst found one of the few remaining KdFs in the plant, sitting abandoned. He had it repainted in British army colors and started driving it. He liked the little buglike cars, and in September convinced the British army to order 20,000 of them for occupation troops and officials. Ordering the cars was one thing. Building them was quite another. The biggest problem was finding parts. The factory's workers, German civilians, started scrounging. By the end of the year, their crisis-prone factory had turned out 1,785 cars.

It was a remarkable feat under the circumstances, but it wasn't enough for the British army brass. As 1946 began they threatened again to ship the machinery to the Russians unless the factory increased production to 1,000 cars a month. That March the factory built 1,003 Volkswagens, just enough.

But the factory's future remained fragile even though Hirst was gaining faith in the little car, which he found to be durable, reliable, and economical. He tried to convince Britain's leading auto executive, Sir William Rootes, to take over the plant, but Rootes replied: "If you think you're going to build cars in this place, young man, you're a bloody fool."[13] Another setback came from an engineering evaluation by the British military, which concluded "the vehicle does not meet the fundamental technical requirements of a motorcar."

Hirst, however, remained undeterred. He decided that if the British didn't want to build the little car, he'd find a German to lead the effort. The obvious choice was Ferdinand Porsche, who was imprisoned by the French, but he was ruled out because of his involvement with Hitler. Another name that surfaced was Heinrich "Heinz" Nordhoff, a former executive of Opel, a German automaker acquired by General Motors a few years before the war.

Nordhoff had never joined the Nazi party, but during the war he had managed an Opel factory near Berlin that made trucks for the Wehrmacht. He wanted to return to Opel after the war, but the American army banned him from holding any managerial job in its zone, where Opel's headquarters was located.

Hirst, meanwhile, was running out of options. The KdF factory was in the middle of nowhere and faced an uncertain future. The roof remained full of holes. The engine assembly area in the basement flooded during heavy rains. And workers started fires in open steel drums to keep hydraulic fluids from freezing in cold weather. Even amid the desperation of postwar Germany, qualified candidates were scarce.

Heinz Nordhoff wasn't thrilled with the thought of dragging his wife and two daughters from Berlin to remote, rural Fallersleben, even if it had been renamed Wolfsburg, after the local castle, to burnish its image. But Nordhoff had lost everything during the war. He was hungry, jobless, and desperate.

After some cajoling by Hirst and other British officers, he took the plant manager's job on January 1, 1948, the day before his fiftieth birthday. He left his family in Berlin until he could find suitable local housing, and set up a sleeping cot in an empty room next to his office. It would be his home for six months. He often was kept awake at night by the sounds of rats scurrying for food. "To be poor in the fullest sense," Nordhoff would recall years later, "has certain compensations. It strips the soul clean."[14]

Nordhoff took a no-nonsense approach. He informed his ragtag workforce that quick improvements were critical, because the factory was inefficient and its cars were shoddy. Some of the first cars built were bartered for parts to make more cars, as materials were scarce. Nordhoff ordered engineers to tinker with Porsche's design to make the engine quieter, the ride smoother, and the brakes more efficient. But he also gave his employees an extra meal each day and, as soon as he could, built worker housing. The moves resonated with men who were scrambling to find food and shelter.

Nordhoff's tenure as plant manager almost came to an early end anyway. In February and March he met with Henry Ford II, then the president of Ford Motor, at the behest of the British authorities. They wanted to give his troubled little factory to Ford, free of charge. During the second meeting the young scion turned to Ernest Breech, a Ford vice president and seasoned executive, who said: "Mr. Ford, I don't think what we are being offered here is worth a damn." [15] Now the Americans, like the British, let opportunity slip away.

Not long afterward, fortune smiled on Nordhoff and his ramshackle factory. On June 20, 1948, the German government instituted a radical financial reform. Germans had to swap their inflated wartime currency, the reichsmark, for a new currency called the deutsche mark, at an exchange rate of ten to one. It was an austere conversion rate, but it quickly established sound money as the basis for the new German economy and ended the need for rationing.

Demand for cars jumped. Many Germans feared the return of inflation and rushed out to buy durable goods. The new government remained steadfast, and inflation didn't return. Nordhoff sold all the cars stored in his factory's basement and boosted his production plans.

In 1949 Ivan Hirst departed Wolfsburg, refusing to take the Volkswagen that Nordhoff and the workers offered him as a gesture of thanks. (He accepted a scale model instead.) [16] The British transferred control of the factory, now named Volkswagenwerk, to the German government, which nonetheless didn't claim actual ownership because it wasn't clear who had legal title to the place.

Nordhoff was running an enterprise with no owners, stockholders, or board of directors. West Germany's courts would take decades to sort it all out.

Nonetheless, in a triumph for Hirst's vision and for Nordhoff's acumen, Volkswagenwerk produced 46,154 cars that year, five times the number it had made two years earlier. Two of those were registered for ownership in the United States, giving the Beetle a tiny American beachhead. The next year the factory shipped 157 more Volkswagens there.

Among them were a couple of breadbox-shaped "Microbus" models. The Microbus concept was sketched on notebook paper after the war, in 1947, by the company's first distributor outside Germany, a Dutchman named Ben Pon. Nordhoff liked it. The vehicle used the same mechanical underpinnings as the Beetle, which made it efficient to build. He started making the Microbus in 1950.

That same year Ferdinand Porsche, whose son had started a separate sports car company under the family name, made his only visit to the Wolfsburg factory. On November 19, the day after his visit, the elder Porsche suffered a debilitating stroke. He died six weeks later, on January 30, 1951.

Even if the Beetle's story had ended here, it would have been remarkable. Ferdinand Porsche had invented the car. Adolf Hitler had adopted it. Ivan Hirst had saved it, "by one of the ironic jokes history is sometimes tempted to produce," as Nordhoff would observe.[17] Now Heinz Nordhoff would make it an icon.

Nordhoff had many needs during the early postwar years, but the most pressing was modern production machinery. At that time, America was the best place to buy it. Nordhoff needed dollars, and the surest way to get them would be to sell cars in the United States. Thus the Volkswagen sedan, as the company called it, officially went on sale on July 16, 1950, at New York's Hoffman Motor Car show-room on 59th Street and Park Avenue. Maximilian E. Hoffman, a German-American dealer, invited favored customers to a 5 p.m. cocktail party to celebrate "the occasion of the American premiere of the famous 'Volkswagen' popular-priced West German car." [18]

Soon afterward, *Popular Science* praised the car's "homely virtues," in an article headlined "Hitler's Flivver Now Sold in the U.S." [19] It wasn't exactly the headline Volkswagen wanted. The Volkswagen had a 25-horsepower engine, about the same as today's John Deere lawn tractor, and took more than 37 seconds to go from zero to 60 miles an hour, slower than the average cheetah.

The people's car was too plebeian for Maxie Hoffman. He specialized in selling high-end imports—Jaguars, Lagondas, Dela-hayes, and Citroëns—to wealthy customers. Volkswagens didn't fit his business or his clientele. The mismatch prompted Volkswagen to separate from Hoffman in 1954. A year later the company set up a U.S. subsidiary, headquartered in New York, and started enlisting dealers committed to selling the quirky little car.

Meanwhile, back in Germany Volkswagen sales were booming along with the postwar German economy. On February 15, 1954, just eight years after his desperate and hungry early days at Volks-wagen, Heinz Nordhoff appeared on the cover of *Time* with the headline: "Germany: The Fabulous Recovery." Volkswagen was the centerpiece of the article, and of the German economic miracle. Nordhoff told *Time* he wanted to sell 4,000 Volkswagens in the United States that year, four times the number of the prior year. VW would sell 6,343, topping his goal by more than 50 percent.

But the company still faced significant handicaps in America. For all its virtues, homely or otherwise, the car looked weird to most people. Besides the buglike shape and tiny engine (which VW increased . . . to all of 30 horsepower), nothing was where it was supposed to be. The engine was in the rear. The trunk was in the front. There wasn't any coolant because the engine was air-cooled. One joke described an American driver looking under the hood of his stalled Volkswagen and saying: "No wonder it won't run. I must have lost my engine." To which another driver, approaching the scene in his own VW, replied: "Don't worry, you're in luck. I just looked in the trunk, and they've given me a spare."

There were other jokes like that. One described a donkey saying to a Volkswagen, "Hey, what are you?"

"I'm a car," the VW replied. "What are you?"

"In that case," said the donkey, "I'm a horse."

"Beetle stuffing" became part of the fad-prone Fifties, along with Hula-Hoops and Davy Crockett coonskin caps. In 1958 students at an Oregon high school crammed thirty-three students into or on top of a Beetle. As Beetles became more popular, young children took to playing "Slug Bug," in which the first person to spot an approaching Volkswagen would punch the other players, playfully or otherwise. Remarkably, the car championed by Adolf Hitler was becoming "cute" in the eyes of Americans.

The perception drove the Germans nuts. The terms "Beetle" and "Bug" remained verboten in official Volkswagen terminology. The Beetle would remain the "Type 1 sedan" in company pronouncements, and the Microbus the "Type 2 station wagon," until 1971. Though it might have been demeaning, as time went on the "cute" image became endearing. Both the Beetle and the Microbus got helpful word-of-mouth publicity, especially from former GIs who had been in Europe. They had gotten to know the Kübelwagen and later the Beetle during their years in Germany, and many had come away impressed.

One such man was Holman Jenkins Sr. He landed in France in the fall of 1944 and marched into Germany the following spring as part of the U.S. Third Army. A decade later he was getting his doctorate from the University of Pennsylvania, about to launch a career as a professor of political science and start a family. "There was antagonism for things German at that time, but I just liked the car," Jenkins recalled decades later.[20]

The Beetle fit the ethos of a professor put off by the ostentation of tail fins, as well as the budget of a family on an academic salary. A new 1958 Beetle cost just $1,545, compared to $2,428 for a Ford Fairlane. From the mid-Fifties until the mid-Sixties, the Jenkins family made do with one car, owning three Beetles in succession until the three children outgrew the backseat.

For the first half of the 1950s, the top-selling imported car in America was Renault. But Volkswagen rolled over Renault like the Wehrmacht routing the French Army, and the secret was the same: logistics. Renaults in need of repair waited weeks for replacement parts from France. Nordhoff, though, insisted that Volkswagen's American dealers stock parts. That way owners could get repairs promptly. As a result, Volkswagen's U.S. sales surged to nearly 31,000 cars in 1955, 80,000 in 1957, and more than 150,000 in 1959. In fact, American sales accounted for nearly one-fourth of Volkswagen's total production that year. In late 1959 Volkswagen of America made another bold move. It took a step that, over the next decade, would transform the Beetle and the Microbus from successful products into cultural phenomena. The grimly efficient car company from Germany embraced that most decadent of American business practices: advertising.

The decision to advertise didn't come naturally to Carl Hahn, the young German running VW's American operations (he would later become Volkswagen's CEO). The company's U.S. sales were setting new records, and demand for Volkswagens exceeded available supply. But by 1959, the landscape was starting to change. GM, Ford, and Chrysler were all impressed by the success of the Beetle and of the little compact Ramblers being sold by American Motors. Each prepared to launch its first compact car: the Ford Falcon, the Plymouth Valiant, and the Chevrolet Corvair. The Corvair would have a rear-mounted, air-cooled engine, just like the Beetle. Hahn decided advertising might help VW defend the Beetle against Detroit's foray.

Hahn and his colleagues sat through droning presentations from more than a dozen agencies before they settled on a small, ten-year-old New York outfit called Doyle, Dane and Bernbach. After DDB landed the account, one of its executives moaned: "We have to sell a Nazi car in a Jewish town." [21] That wasn't DDB's only challenge. The agency also had to find selling points for a quirky little car with the engine in back. That meant taking a trip to the factory in Wolfsburg, which wasn't exactly an ideal European junket. It was a dull little place on the North German Plain, far from any big-city lights.

William Bernbach, one of the agency's founders, led the group. It spent time with engineers, manufacturing managers, and even assembly-line workers, getting a crash course in the intricacies of the Beetle. The workers' Germanic attention to detail, and the engineers' sober emphasis on practicality instead of glitz, impressed the visitors. Being humorless had its advantages. "Yes, this was an honest car," Bernbach later wrote. "We had found our selling proposition." [22]

"Honest car." Well, that was something novel, or so it seemed compared to Detroit's advertising scenes of gorgeous women, muscular boy-racers, and exotic winding roads. Bernbach concluded that Volkswagen's advertising shouldn't try to out-Detroit Detroit. The Volkswagen ads instead would be witty and self-deprecating, befitting a car that looked, well, funny. One ad showed a Beetle, head-on, under the heading: "Ugly is only skin-deep." He also decided to use logic instead of emotion, flouting the conventional wisdom that effective advertising sold to the heart instead of the head. In mid-1959 an early DDB ad in *Life* magazine asked, "Why the engine in the back?" The answer: "Greater efficiency, better traction and an honest 32 miles to the gallon—regular gas."[23]

The agency soon followed with a far bolder advertisement. "What year car do the Joneses drive?" asked an ad showing a Beetle sitting in front of a comfortable, prosperous suburban house. The answer, of course, was that nobody could tell. "The Jones drive a Volkswagen and Volkswagens look alike from year to year," the ad explained. "A Volkswagen is never outmoded."[24]

DDB took dead aim at the American urge to "keep up with the Joneses," a term popularized by a newspaper comic strip of that name that appeared between 1913 and 1939. The impetus to "keep up" propelled Detroit's annual styling change, indeed its entire business model, not to mention the conspicuous consumption driving the whole American economy. By implication Volkswagen buyers were different—heck, not just different, but better—than their conventional, status-seeking neighbors. It was subversive. "The ad coincided with the beginning of a bigger social change: dissent," a British advertising historian, Clive Challis, would later write. "A growing number of consumers weren't going to take what corporate America thought was good for them anymore."[25] By the late 1950s these people had a voice, and not just thanks to Vance Packard and John Keats. In 1957 Beat writer Jack Kerouac published *On the Road*, a novel describing his journey of personal discovery far removed from the middle-class Jell-O mold of American conformity. The novel was an instant hit.

That a straitlaced German corporation was finding success with an antimaterialist, nonconformist sales pitch was ironic. So was the background of the DDB art director who helped shape the Volkswagen campaign. Helmut Krone was a first-generation German-American. During the agency's initial trip to Wolfsburg, he took a side trip to see relatives he had never met. Krone was raised in a working-class New York family and grew up aware of his heritage. "A German son is always wrong until he's proved himself right," he would recall of his childhood. "It gives you a certain insecurity which is the opposite of chutzpah." [26]

Krone's ads used arresting photographs that filled the page, along with intriguing headlines and intelligent text. In August of 1959, after his first Beetle ads appeared, Krone left for vacation beset with self-doubt about the adequacy of his work. When he returned two weeks later, he was stunned that his work had drawn acclaim, and he had become a star.

Emboldened, he produced another ad showing a Beetle standing above a one-word headline: "Lemon." The headline might as well have said "Shit." To car company executives, the word "lemon" was *never* to be used in polite speech, much less in advertising. But Krone's ad described the work of Volkswagen's 3,389 quality inspectors, whose job was to find lemons and reject them before they got to customers. "We pluck the lemons," the ad explained, "you get the plums." Andy Warhol did a painting of the ad.

Another Krone ad carried another impudent headline: "Think small." Americans were supposed to think *big*, of course, about cars, houses, bank accounts, and everything else. Thinking small was heretical, but Krone portrayed it as good common sense. Small was a virtue, the ad explained, "when you squeeze into a small parking spot. Or renew your small insurance. Or pay a small repair bill . . . Think it over." [27]

Pointed, wry humor also appeared in DDB's television

commercials, including one in 1963 that touted the Beetle's remarkable traction. It showed a Beetle driving through a heavy snowstorm, then stopping in front of a shed with a big machine inside. The announcer then intoned: "How do you think the snowplow man gets to the snowplow?" (In 1999 it would be voted the "Best Television Commercial of the Century" at an advertising convention in Cannes.)

A Volkswagen advertisement from 1966 broke a couple of taboos: making fun of your own product and featuring a black man. In the ad Wilt Chamberlain, a seven-foot-one-inch basketball star, tried to climb into a Beetle with the headline: "They said it couldn't be done. It couldn't." The ad then explained that anyone six-foot-seven or shorter would easily fit into the car. A year later Krone used an arresting photo of a Beetle floating in water under the headline: "Volkswagen's unique construction keeps dampness out." That ad would have unintended consequences several years later.

As time went on, some Volkswagen owners aped the company's irreverent wit when they advertised their used Beetles for resale. "Family growing. Car isn't," read one classified ad. "Must sell 2 boys or Volkswagen. Your choice, $926.26." Another classified ad said, "For sale: VW '62. Cheerful, reverent, brave, kind, careful and obedient. Dependable, too."[28]

Some Beetle owners wrote to DDB, suggesting ideas for future ads. One writer described the Hinsleys, a couple who lived in a log cabin in the Ozarks. The Hinsleys had bought a used Beetle after their mule had died. Bob Kuperman, one of Krone's successors as DDB's art director (and later the agency's chairman) embarked to the Ozarks and asked the couple to pose for an ad. Initially they refused, figuring Kuperman was a city slicker out to get their land.[29] But Kuperman persuaded them, and created an ad showing the couple in front of their Beetle and their cabin. Mr. Hinsley wears overalls and holds a pitchfork, *American Gothic*–style, with his wife seated beside him. The headline: "It was the only thing to do after the mule died."[30]

VW's press releases, like its advertising, cut against the grain of Detroit's hyperbole. "The familiar Volkswagen goes on as before," declared a press release for the 1962 model.[31] The big news that year was the addition of a gas gauge. Previously drivers had relied on guesswork, backed up by a lever that released a small reserve supply if the engine started sputtering.

That same year Volkswagen of America launched a quarterly owners' magazine called *Small World*, with adventure stories from owners. One was about a young Miami woman who left her keys in her Beetle, returned from shopping, hopped in, and drove away. Minutes later she realized she had driven off in a car that belonged to another Beetle owner who, meanwhile, had mistakenly taken her car and driven a mile down the road.[32]

Another marketing gambit was Bonds for Babies Born in Beetles, launched in 1964. After hearing stories about babies born in Beetles en route to the hospital, Volkswagen began awarding $25 U.S. savings bonds to newborns, whose method of entering the world proved—as the company put it—that there was always room for one more in a Beetle. By 1969, 125 "Beetle babies" had been awarded bonds.[33]

As a place to work, Volkswagen of America wasn't the hip, iconoclastic corner of corporate America its advertising implied. To some who worked there, there were two classes of employees: the Germans and everybody else. One American passed over for promotion in favor of a German employee told his American colleagues: "We both raised our hands for the job, but I forgot to say 'Sieg Heil.'"[34] Such tensions, though, remained safely out of public view.

Initially, Hahn had given DDB only the Beetle account,

choosing another agency for the Microbus ads. But the response to the Beetle ads prompted him to give that account, too, to DDB, with much the same result. When research showed that Microbuses didn't appeal to women, DDB responded with a 1963 ad headlined: "Do you have the right kind of wife for it?" The headline surely would have raised feminist ire just a few years later, but the ad's text was surprisingly profeminist for its day. "Can she get a kid's leg stitched and not phone you at the office until it's all over? Does she worry about the bomb? Invite 13 people to dinner even though she only has service for 12? Let you give up your job with a smile? And mean it?"[35] Strong, independent-minded women, the message implied, would like the Bus.

Another ad touted the Microbus's roominess by showing one with its side doors open and an enormous, T-shaped package, wrapped in brown paper, protruding through the sunroof. "What's in the package?" the text asked. "8 pairs of skis, the complete works of Dickens, 98 pounds of frozen spinach, a hutch used by Grover Cleveland, 80 Hollywood High gym sweaters, a suit of armor and a full-sized reproduction of the Winged Victory of Samothrace."

Many Americans probably thought Winged Victory of Samothrace was the latest ride at Disneyland. But that was the point. Volkswagen and DDB aimed at buyers who knew better, and were *proud* of it. What they didn't know, in this case, was that the package in the ad contained just empty boxes. The advertising applauded VW owners for being smarter than most Americans. The owners were wealthier and better-educated, too.

Actor Paul Newman bought his first Beetle in 1953, and eventually owned five.[36] Britain's Princess Margaret owned one, as did Beatle (with an "a") John Lennon, who put his white Beetle—as opposed to his Rolls-Royce or his Ferrari—on the cover of the *Abbey Road* album. It's unlikely any of them would have bought a

Beetle had Volkswagen marketed it as the perfect car for poor people, which in fact it was. But Beetles were anathema to poor people, who viewed them as a public display of poverty and wanted the biggest piece of Detroit iron they could buy. Poor people didn't want to look poor, and rich people didn't want to look rich.

The pretensions of well-heeled Beetle owners were described in *Playboy's* July 1964 "Snobs' Guide to Status Cars." To be a Beetle owner, it said, you should "have at least three children and name them after characters from *Winnie-the-Pooh* . . . Use a lot of Freudian terminology in your speech . . . Read the *New Yorker* and check off all the movies in the front of the magazine after you have seen them. Read *Time* but hate it . . . Tell people you voted for Stevenson the first time he ran but not the second . . ."[37]

Microbus owners, the Snobs' Guide continued, "call food 'grub,' sleep 'raw,' wear blue denim shirts to the opera and have sex in a sleeping bag. Grow a bushy mustache. Get haircuts that don't look like you went to a barbershop, even if you did. Enjoy all natural body smells, especially your own . . . It is all right to take a microbus to a surplus store or a peace march. It is not all right to take a microbus to Bloomingdale's."[38]

In truth, many Microbus drivers weren't hippies but early versions of suburban soccer moms. But they weren't the ones who got the attention. How could they, after hippie guru Ken Kesey and his Merry Pranksters followers rolled into New York in the summer of '64 after an LSD-laced cross-country trip (pun intended) in their psychedelic bus? It was a customized 1939 International Harvester school bus named "Further."

It didn't take Kesey's fellow travelers long to learn that the Volkswagen Microbus provided a cheaper, more practical alternative for their lifestyle than converted school buses. The VW bus became the unofficial hippie-mobile. *Car and Driver* affirmed its status with a satirical "review" of the vehicle by a make-believe hippie named Tom Finn, identified as cochairman of the Leon Trotsky Socialist Purge Committee.

"I will admit to being pretty zonked—but nothing like the guys and the chick in the back," wrote the fictional Finn. "Shiek and Mona are balling by the engine, while Murph is reading an astrology table . . ." When it came to a new paint job, Finn didn't want "any of that fake psychedelic freakout shit on it because every housewife in Sausalito is driving around in a bus painted so it looks like the Merry Pranksters went after it with stencils."[39]

As the Sixties unfolded, Volkswagen cartoons became a staple in magazines. One in the *New Yorker* showed a tuxedo-clad husband and his evening-gowned wife arranging the seating for a dinner party. "It should be a scintillating conversation," she said. "I'm placing a VW between a Cadillac and a Lincoln Continental."[40]

Volkswagen's advertising and marketing reached their zenith at the end of the Sixties. The highest-grossing movie of 1969 was *The Love Bug*, a Disney flick about an almost-human Beetle named Herbie that helped his owner, a washed-up race car driver, win again. Volkswagen executives were reluctant to cooperate with Disney, but were shocked and delighted by the movie's success. Three Herbie sequel movies and a short television series followed.

In July 1969, after the historic landing of three American astronauts on the moon, DDB produced an ad for VW with no car and no text. It just showed the ungainly lunar landing module, the VW corporate logo, and a headline: "It's ugly, but it gets you there."[41] DDB had broken the rules again, and produced not just advertising but art.

Three months later, *Newsweek* described a recent Beetle press preview in Vermont that parodied the horsepower-and-hype events in Detroit. The Volkswagen preview began with a tongue-in-cheek "economy run" in which journalists were allowed only enough gas to fill a whiskey jigger. Arthur Railton, Volkswagen's American PR chief, then bragged about the, um, big styling change for 1970: "A few ventilation slots have been added to the rear engine-compartment lid." He whipped out an electric drill and, with a straight face, drilled slots into the prior year's trunk lid, explaining that owners of "old VWs can make them look like 1970 models." Noting that Volkswagen had boosted the Beetle's engine to 57 horsepower from 53 (still some 300 horsepower less than Detroit's muscle cars), Railton sighed. "I just don't know how long this horsepower race will go on."[42] The journalists chortled.

The Sixties were a decade of nonstop success for Volkswagen in America. In 1962 Heinz Nordhoff toured the United States, dedicating a new American headquarters in New Jersey and meeting with dealers. He then traveled to Detroit, where he addressed the city's Economic Club.

His long career spent against the backdrop of history had made Nordhoff a philosopher as well as a businessman. In Detroit he spoke about freedom. "As a man who lives only five miles from that tragic dividing line, the iron curtain, I think about this," he said. "At my house, when the wind blows from the east, I can smell the heavy air of slavery first hand."[43] Later, Nordhoff was feted at a dinner of American auto executives. The next day he had lunch with Henry Ford II, the same man who had rebuffed a chance to buy the struggling VW factory fourteen years earlier.

All the while the Beetle and the Microbus were becoming cultural totems. They were the perfect vehicles, functionally and symbolically, for Americans who wanted to be different, even though being "different" became a new form of conformity as the Sixties unfolded. Volkswagen's counterculture-car image came not only from the company's own advertising but also from serendipitous sources.

One of them was, of all things, a repair and maintenance manual, *How to Keep Your Volkswagen Alive: A Manual of Step by Step Procedures for the Compleat Idiot.* It began by exhorting, "Now that you've spent your bread for this book—read it!" Written by a Volkswagen-loving mechanic from Santa Fe, the book combined detailed technical instructions for laymen with Age-of-Aquarius philosophy. "Feel with your car," the book implored, "use all of your receptive senses and when you find out what it needs, seek the operation out and perform it with love."[44] This wasn't exactly how the manuals read for the Mr. Goodwrench mechanics at the GM dealerships. Within ten years of being published the *Compleat Idiot* got twenty-two printings.

Toward the end of its decade of destiny, however, Volkswagen suffered one overarching loss. Nordhoff, who had led the company to growth and greatness, died on April 12, 1968, at age sixty-nine. He had steered Volkswagen through the terrible years after the war, when nothing was certain for the company or the German people or even the world. He had transformed VW into the world's third-largest car company, after GM and Ford, and into the symbol of Germany's national rehabilitation. Governments and universities around the world had honored him, with good reason.

Nordhoff had said that his greatest achievement was resisting the pressure to change the Beetle's basic design from the one conceived by Ferdinand Porsche thirty-five years earlier. Nordhoff's successors wouldn't—and, in truth, couldn't—be so steadfast. The Seventies wouldn't be so kind to Volkswagen, or to the company's car.

Volkswagen had a reputation for witty, irreverent ads, but one that appeared in the fall of 1973 was over the top. It showed a Beetle, floating in water, with the headline: "If Ted Kennedy drove a Volkswagen, he'd be president today." The ad mentioned, by name, Mary Jo Kopechne, the young woman who drowned at Chappaquiddick in 1969 in a car driven by Senator Kennedy.

Angry calls and letters, many from loyal VW customers who vowed never to buy a Volkswagen again, besieged Volkswagen. The real problem, though, was that the advertisement didn't come from Volkswagen or DDB, and it wasn't even an ad. It was a spoof run by the *National Lampoon* magazine. What made the spoof credible was VW's memorable 1967 ad that had showed the Beetle floating in water. Volkswagen filed a $30 million lawsuit prompting *National Lampoon* to publish a retraction. But the brouhaha was all too symbolic of the troubles Volkswagen would face in the 1970s.

By the time the *Lampoon* "ad" appeared, Volkswagen was moving away from the Beetle and from rear-mounted, air-cooled engines. The company had introduced the Super Beetle in 1971, with better suspension and more storage space than the regular Beetle. But Beetle sales dropped for the next three years.

Events were outrunning the doughty little car. New clean-air laws required major reductions in tailpipe emissions, which were difficult and costly with an air-cooled engine. And during the 1973 Arab oil embargo, more Americans started trying small cars imported not from Germany but from their other World War II foe—Japan. In late 1972 a Japanese motorcycle maker, Honda, introduced its first car, the subcompact Civic. It used a water-cooled engine mounted transversely, or crosswise, under the front hood.

But it had front-wheel drive, providing Beetle-like traction in snow and making the car surprisingly roomy. The Civic was more than 20 inches shorter and 100 pounds lighter than the Beetle, providing better maneuverability, fuel economy, handling, and acceleration. The Civic and other little cars from Japan provided all the advantages of the Beetle without the flaws: a high noise level, sluggish acceleration, and a heater so weak that it was, as one joke put it, "like a mouse panting on your ankles." After forty years, Ferdinand Porsche's timeless design was dated.

In early 1974 Volkswagen launched the Dasher, its first car with a front-mounted, water-cooled engine. But the Dasher was priced just under $4,000, some 35 percent higher than the Civic. Volkswagen said the Dasher was supposed to supplement the Beetle, not replace it, but nobody believed that.

The Smithsonian's National Museum of American History in Washington enshrined a Beetle in 1976. That was also the last year the Beetle hardtop was sold in America. Volkswagen sold the Beetle convertible in the United States until 1979. The Microbus soldiered on for another decade in America before the VW EuroVan replaced it. Volkswagen gradually moved away from its heritage of providing economical but reliable cars. It moved "upmarket," but downward on the sales charts.

The Beetle and the Bus continued to be produced and sold outside the United States, however. By the time the last air-cooled, rear-engine Beetle was built in Puebla, Mexico, in 2003, more than 21.5 million had been sold, easily surpassing the 15 million Model T Fords, and making the Beetle the best-selling car ever.

The British officer who had saved the Beetle, Ivan Hirst, had died three years earlier at age eighty-four, having spent his career in public service. Before he died he drove one of the first New Beetles, which Volkswagen introduced in 1998. Mechanically it had a conventional front-mounted, liquid-cooled engine and a front-wheel-drive platform, but it replicated the original Beetle's buglike shape.

The New Beetle tapped the nostalgia of aging baby boomers,

whose minds were getting wistful as their hair was getting grayer. Many of them couldn't remember the license plate number on their fancy new BMW, but had no trouble recalling the number on the bare-bones Beetle they had driven in high school or college. Bill Campbell drove a green one during the mid-Sixties as a student at Columbia, fighting high crosswinds on highways en route to football games. Later he would become chairman of Intuit.

At Sun Microsystems, a young vice president named Eric Schmidt walked into his office on April Fool's Day 1986, only to find it empty except for a Beetle. His employees had whisked away the furniture during the night, brought in the Beetle piece by piece, and reassembled it on the spot. Their film that captured Schmidt's wide-eyed surprise later found its way to YouTube, which preserved the prank for posterity. Schmidt later became CEO of a company called Google, which bought YouTube.

One of the Beetle's many legacies was its impact on Detroit. The Beetle's success inspired Chevrolet to launch a competing car in late 1959 with a rear-mounted, air-cooled engine. "Volkswagen will be out of business in this country in two years," predicted Chevrolet's general manager, Ed Cole.[45] The prediction proved utterly wrong, like so many other things about the Chevrolet Corvair.

5

THE CHEVY CORVAIR MAKES RALPH NADER FAMOUS, LAWYERS UBIQUITOUS, AND (EVENTUALLY) GEORGE W. BUSH PRESIDENT OF THE UNITED STATES

Kick the hell out of the status quo!

—Ed Cole's favorite exhortation to his troops at GM[1]

The October 5, 1959, cover of *Time* showed a man flashing the broad grin and bright eyes of personal triumph. He was Ed Cole, whose Horatio Alger life had brought him to the top of Chevrolet, the biggest automotive nameplate in the world. "Not since [the] Model T has such a great and sweeping change hit the auto industry," the article began. "Out from Detroit and into 7,200 Chevrolet showrooms this week rolled the radically designed Corvair . . . like no other model ever mass-produced in the U.S.; its engine is made of aluminum and cooled by air, and it is mounted in the rear. To Chevrolet's folksy, brilliant General Manager Edward N. Cole, 50, the new car marks the fulfillment of a 15-year dream. Says Ed Cole jubilantly: 'If I felt any better about our Chevy Corvair, I think I'd blow up.'"[2]

It would be hard to top those words for irony. Within just a few years the Corvair would blow up right in the face of General Motors, the parent of Chevrolet and then the world's biggest and most profitable corporation. The reverberations would reshape the U.S. legal system and the American government's relationship with business. The car's ultimate impact would bear little resemblance to what the überconfident Ed Cole had intended.

The Corvair was GM's answer to the Beetle, mechanically similar but bigger. For years, Detroit couldn't comprehend the appeal of the strange little car from Germany. GM executives regarded Beetle buyers as weirdos, pinkos, or cheapos, or maybe all three. Real Americans bought chrome-bedecked six-passenger sedans or full-size station wagons. And they bought them as often as possible to keep up with Detroit's annual styling changes, lest they fall behind the Joneses or the Smiths or the Whatevers. But by the late 1950s the Beetle's growing popularity was getting harder to ignore. And Detroit's then infant (and often infantile) market research disclosed that many Beetle buyers were well-heeled, well-educated people who could afford bigger cars, but actually *chose* not to buy them.

One reason was the "Eisenhower recession" of 1958 that had helped doom the Edsel but prompted a surge in the sale of "compact cars," a term coined by George Romney, the CEO of tiny American Motors. AMC's Rambler jumped to the industry's third-best-selling nameplate that year, behind only Chevrolet and Ford and ahead of such longtime stalwarts as Pontiac, Oldsmobile, Buick, Plymouth, and Dodge. Big Three bosses were shocked. Except Ed Cole.

He had known this day would come, even without a recession. Two-car families were a minority in America, but they were increasingly common. Such families didn't need two full-sized vehicles. One big car could hold the whole family. Dad could drive a smaller, sportier, more economical car to work. That would be a better combination.

In 1957, even before the recession hit, Cole used that reasoning to sell his pet project to Harlow "Red" Curtice, the president of General Motors. Curtice wasn't an easy sell, but Cole wasn't deterred. For years he had bootlegged the Corvair project, developing the car in secret and covering the cost with other budgets without corporate approval. He believed in the car and figured that the risk-averse top brass wouldn't buy it unless presented with a fait accompli.

The Corvair's air-cooled engine eliminated the need for a water pump, radiator, and antifreeze, all of which added cost and weight to conventional cars. With the engine mounted in the rear, and joined to the transmission right over the drive wheels, there was no need for a drive shaft to connect the engine and transmission in front to the drive wheels in back. That saved still more cost and weight, and created extra interior space by eliminating the floor hump for the drive shaft. Cole's new car got nearly 30 miles a gallon at a time when most other cars got barely half that. Thanks to the car's modest weight, its 80-horsepower engine could hit and hold a top speed of 88 miles an hour, higher than the legal speed limit on any American road. Unlike the Beetle, the Corvair had room for six.

In essence, it was a Beetle for Americans, big enough to haul a family and stylish enough to appeal to mainstream buyers. The car was nicknamed "America's folks wagon," or, alternatively, "the Waterless Wonder from Willow Run," the name of the GM factory in Ypsilanti, Michigan, where it was built.

The press was awestruck. Nobody expected such daring from Detroit. The Corvair is "the most profoundly revolutionary car . . . ever offered by a major manufacturer," wrote *Sports*

Car Illustrated.[3] Another publication, *True: The Man's Magazine*, added: "After nearly three decades of selling the conventional, Detroit's biggest company is making a daring gamble."[4] Indeed, the Corvair was a rousing response to those who dismissed American cars as long on styling but short on engineering, devoid of innovation. In many ways it would be, at least in concept, just the sort of revolutionary car America would need a half century later in the era of jammed freeways and $100-a-barrel oil. Which made the impact of the car's ultimate failure all the more tragic.

Like the Ford Model T, the Chevrolet Corvair sprang from the vision and determination of a Michigan farmboy. Edward Nicholas Cole was born on September 17, 1909, in Marne, Michigan, a tiny town west of Grand Rapids where the ethos was dominated, like that of the surrounding region, by the Calvinist sensibilities of the Dutch Reformed Church.

His parents were dairy farmers. As a boy Cole milked the cows and delivered fresh milk every morning. The experience, he later said, trained him to walk quickly, as his load would get lighter after every delivery. At age five he climbed behind the wheel of the family's 1908 Buick, started the car, and smashed it into a tree. At sixteen Cole rebuilt two damaged cars, making him one of the rare two-car owners of his day. One was a four-cylinder Saxon, whose engine Cole tuned so aggressively that it could outrun any other car in Ottawa County.

During high school Cole wanted to become a lawyer, not knowing that he later would spend years fighting one particular lawyer. But after attending Grand Rapids Community College he moved on to the General Motors Institute in Flint. GMI was a company college, focused on engineering and organized so students could alternate periods of work and study to support themselves through school. Young Cole drew duty at Cadillac, where his obvious talents landed him a full-time engineering job in 1933.

Cole confirmed Cadillac's faith in his mettle, though sometimes in ways that gave his bosses grief. He once outran a new Cadillac V12 convertible in a public drag race, using an old Chevrolet whose engine he had "hopped up" with three carburetors.[5] But during the 1930s Cole succeeded in making Cadillac's engines much quieter, and its engine-cooling systems more efficient.

The latter achievement drew attention from the U.S. Army. Its tanks had a habit of breaking down when the engines overheated, due to driving in conditions that made potholes look tame. Shortly before World War II Cole's GM bosses assigned him to make the engine on the army's M5 tank more reliable. He beat his ninety-day deadline, winning Cadillac a big tank-production contract during the war.

When peace returned, Cole developed an experimental rear-engine Cadillac. It was a defensive move by GM to head off maverick entrepreneur Preston Tucker's effort to build a full-sized rear-engine car. Cole's "Cadibacks," as they were termed, looked odd. They used oversized truck tires in the rear to handle the weight load of the engine. But having the engine's weight atop the drive wheels produced traction that enabled Cole to glide through the ice and snow of Michigan winters, even as his neighbors skidded into ditches. Rear-engine cars began to fascinate Cole.

Tucker's venture collapsed because his cars weren't reliable, and GM abandoned Cole's project. The company was selling every Cadillac it could make thanks to the postwar buying boom, and it didn't need to spend money on experimental projects. Cadillac was also making enormous strides in improving its regular engines, with Cole overseeing the effort.

Cole led the development of a new "high-compression" V8

that was 25 percent lighter than its predecessor, but boosted horse-power and increased fuel economy. Cadillac executives forgot rear-engine cars, except for Ed Cole. While managing Cadillac's sprawling factory in Cleveland he tinkered with plans for a rear-engine small car.

Then in April 1952, Cole was recalled to Detroit. Chevrolet, GM's biggest brand, was saddled with aging cars and outmoded engines, and reeling from a sales plunge of 40 percent in just three years. Company president "Engine Charlie" Wilson, who later would famously tell Congress that what was "good for our country is good for General Motors, and vice versa," offered Chevy chief Thomas Keating anything he needed to fix the division's woes. "I want Ed Cole," Keating replied. Cole was told to drop the keys to his Cleveland office on his boss's desk and report immediately to Detroit. At age forty-one the man from Marne was Chevrolet's chief engineer.[6]

Cole tripled the size of Chevrolet's 851-man engineering staff, and ordered the escalators in the engineering building speeded up by 30 percent. The engineers called them "Cole's turnpikes." Under Cole's hands-on supervision, the bigger, faster-moving engineering staff redesigned Chevy's V8 to be lighter and more powerful. Within a year, Chevy's sales started to recover.

In 1955 Chevrolet boosted its fortunes further with a revamped Bel Air. Cole said the car had "intrigue," the euphemism he used for "sex appeal." It featured sleek lines, a toothy front grille, and Cole's new "small block" V8, which was smaller than the massive V8s used in Cadillacs, but still delivered 162 horsepower, potent for its day. "Try this for sighs," Chevy's advertising proclaimed. It was the same year that Chevy put the first V8 in the Corvette.

Unbeknownst even to Keating, Cole kept a small team of engineers secretly experimenting with ideas for a new small car. They tried front-engine cars with front-wheel drive and rear-engine cars with front-wheel drive. They even fiddled with conventional front-engine, rear-drive cars. But Cole kept plugging his rear-engine, rear-wheel-drive dream, just like his experimental "Cadiback." In the spring of 1956 he tested a prototype, disguised with a Porsche body, by tearing around Chevy's engineering center at 80 miles an hour, flouting the posted speed limit of 25. His underlings could barely watch, for fear he would crash. But Cole screeched safely to a stop in front of them and blurted out: "This is it."[7]

That July Cole succeeded Keating as general manager of Chevrolet, becoming at age forty-six the youngest man ever to hold the job. The boss who sped up the escalators kept a blistering pace himself. "Cole tears down GM's corridors like Patton went through southern Germany," one journalist observed. "Stenotypists who must keep track of his off-the-cuff remarks at press conferences often run paragraphs behind . . . He claims to fish for relaxation, but associates say that very few fish have been quick enough to nab his hook."[8]

Nor were GM's bosses nimble enough to track Cole's activities. Work on the as yet unnamed Corvair remained secret, even as Cole accelerated the project. In the spring of 1957 Cole produced detailed engineering drawings and built a working prototype without informing either CEO Red Curtice or the corporation's powerful Engineering Policy Committee. The decision was bold if not reckless, but Cole believed his car had to be seen and driven to be believed. Late that summer he steeled himself to tell Curtice what he had done.

The boss didn't explode, but he gave Cole a two-hour grilling. Was he sure there was a market for this car? Could GM procure enough aluminum at a reasonable price to manufacture the lightweight engine? Yes to both, Cole answered. He further explained that the engine would easily fit into the rear space normally reserved for the trunk because of its innovative design. The pistons and cylinders wouldn't stand upright, as on normal engines, but would lay flat, three on each side of the engine, in "pancake" style.

Another question was whether the new car would cannibalize GM's highly profitable bigger cars. "If we don't build this car, Red, someone else will."[9] After an impromptu test drive Curtice was impressed. He even floated the idea of creating a sixth GM division—in addition to Chevrolet, Pontiac, Oldsmobile, Buick, and Cadillac—to manufacture and market the car. Cole, however, wanted it for Chevy alone.

By the time Cole ran his project through the gauntlet of the Engineering Policy Committee, its budget was *increased* to $150 million from the mere $85 million that Cole, with uncharacteristic frugality, had proposed. In December 1957, GM's board approved Cole's car. A month later the board members watched Cole climb into a prototype car and tear around a GM test track. As he "tooled down off the high speed bankings and braked to a stop in front of the corporation brass," reported *True*, "two things were solidly apparent: Chevrolet had one helluva new car in the Corvair, and one helluva leader in Ed Cole."[10]

General Motors had prior experience with an air-cooled engine, but it wasn't good. In 1922 GM's Charles "Boss" Kettering, the inventor of the electric starter, had developed a similar engine, calling it "copper-cooled." It relied on copper fins to draw away heat without a radiator. Kettering had pushed the engine as a way to leapfrog Henry Ford, but CEO Alfred Sloan questioned its reliability and killed it. Sloan retired as GM's chairman in 1956, only a year before the company embraced the Corvair.

As the Corvair's development continued, now with the corporate imprimatur, two competing concerns surfaced: the need for special tires and GM's desire for absolute secrecy. It would be difficult to develop the former without compromising the latter. The Corvair's unusual design, with only 40 percent of the car's weight in the front and 60 percent in the rear, meant that ordinary tires wouldn't provide enough stability. This wasn't an issue with the Beetle because its body was so small. The Corvair was longer, lower, and heavier.

Chevrolet asked U.S. Rubber Company (later Uniroyal) to develop special tires to handle extra heavy rear loads, but Chevy engineers took pains to hide the real reason for the request. They told U.S. Rubber that the unusual tires were for a new car at Holden, the company's subsidiary in Australia, a ruse that grew more elaborate over time. Chevrolet's purchasing managers printed procurement forms designating certain components as "boxed for export to Holden," though they never were. One Chevy executive, with a reporter sitting in his office, placed a fake phone call to Australia to discuss a rear-engine Holden car, to convince him that rumors about a domestic rear-engine car were untrue.

The U.S. Rubber team, meanwhile, designed tires that used a new rubber compound and rode on wide rims, like those on much bigger cars, to better support extra weight in the rear. The company specified that the rear tires be inflated with air pressure of 26 pounds per square inch, compared to only 15 psi for the new car's front tires. The higher air pressure in the rear tires was supposed to increase the Corvair's rear-end stability and keep it from spinning out around corners. But critically, the specification also required drivers to monitor their tire pressures constantly. Few people would take the time to do that when dashing off to work or rushing to a Little League game. But that would be their problem, wouldn't it?

In the summer of 1959, with the Corvair's debut just months away, Chevy's PR staff summoned some thirty automotive writers to the Detroit Athletic Club for a presentation by Maurice Olley, the straitlaced engineer who had bedeviled Zora Arkus-Duntov. Olley had retired recently, but reporters respected him.

The circumstances surrounding his presentation, however, were clumsy. Chevrolet had acknowledged it was developing a new model called the Corvair, but still refused to confirm rumors about its rear-mounted engine. Olley's task was to extoll the advantages of rear-engine design without saying GM would build one. "All the machinery with its attendant heat and noise is at the rear," Olley explained. He added that a rear-engine layout would improve traction and handling by putting more weight over the drive wheels. After his remarks he dutifully ducked questions by adding: "I really don't know what the Chevrolet people plan to do." [11]

A couple days later, the journalists who had attended the lunch received mysterious, unmarked envelopes in the mail. Inside was a technical paper that had been presented to the Society of Automotive Engineers in 1953, questioning the safety of rear-engine cars. "It has always appeared to me that rear-engine cars are a poor bargain," the paper stated. "It appears impossible to make a car with heavily loaded rear tires handle safely in a wind even at moderate speeds." [12] The presentation, it turned out, had been from Maurice Olley himself.

This maneuver was classic corporate spin control, before the term had even been invented. Amid their guffaws, the reporters suspected the anonymous envelopes had come from the flacks over at Ford, or maybe Chrysler. Both companies were preparing to introduce their own compact cars, the Ford Falcon and the Plymouth Valiant, with conventional designs: engines in front and drive wheels in the back. It would be beneficial, and a whole lot of fun, for both companies to undermine the competing car from Chevrolet. Chevy was still obsessed with secrecy about the Corvair and thus couldn't respond, but its silence didn't last long. Just a few months later, when Detroit unveiled its cars for the 1960 model year, the Chevrolet marketing machine revved into high gear.

"Our cities have been straining at their seams. Traffic is jam-packed. Parking space is at a premium. And our suburbs have spread like wildfire. People are living farther from their work, driving more miles on crowded streets."[13] The words could be from a Toyota Prius advertisement in 2012. Instead, more than fifty years earlier, they appeared in a two-page advertisement that Chevrolet ran in newspapers across America on Sunday, September 27, 1959. The ad touted a vehicle "unlike any car we or anybody else ever built—the revolutionary Corvair, with the engine in the *rear*, where it belongs in a compact car."[14]

To highlight the rear-engine design, the Corvair didn't have a fake air-intake on the front grille. Instead, the famous Chevrolet bow tie badge was mounted on a flat metal sheet running between the headlights. The lines were clean and simple, just the opposite of Detroit's usual styling, with a high "belt line" that accentuated the Corvair's roominess. Instead of selling just styling sizzle, the Chevrolet was selling the steak of engineering innovation.

"This is a *six-passenger* compact car"—in contrast to the Beetle, the ad didn't have to add—"with a really remarkable performance . . . Among the basic advantages resulting from this engine location are better traction . . . and a practically flat floor. But to be placed in the rear, the engine had to be ultra light and ultra short. So Corvair's engine is totally new—mostly aluminum and *air cooled*; it weighs about 40 per cent less than conventional engines . . . The ride is fantastic."[15] Because the engineless front end was so light, Chevrolet boasted, the Corvair was easy to drive without power steering.

Chevrolet followed the newspaper ads with a blitzkrieg of press releases, leaflets, and sales brochures. One brochure described how, at an off-road test track in Lime Rock, Connecticut, drivers "took the new Corvair up a steep bluff, down into gullies, slogging through mud up to its hubcaps, wading across a stream . . . terrain no front engine compact car could hope to navigate." It was a direct slap at the Falcon and Valiant, and an attempt to even the score for hapless Olley. The brochure boasted: "The Corvair's rear-engine design can take another bow for the car's exceptional road-hugging stability."[16] A Chevy sales leaflet called the Corvair "more sure-footed than a polo pony."[17] Both claims would prove portentous in years to come.

The publicity barrage generated extensive press coverage. Besides gracing the cover of *Time*, along with Ed Cole, the Corvair landed the Car of the Year award from *Motor Trend*. *Mechanix Illustrated* declared the Corvair a "delight." Most articles praised the car, albeit with some reservations. *Time* criticized the Corvair's lack of luggage space, just 15.6 cubic feet compared to nearly 25 cubic feet for the Falcon and the Valiant. Other publications noted that the Corvair's base price of $1,860 was just $196 less than the full-sized Chevy Biscayne, and about $250 more than import compacts. Those were big differences back then.

From the beginning, however, there was more ominous concern about the Corvair's stability, even though Chevy touted it as one of the car's major strengths. "Front-end advocates assert that an engine in the back of the car produces wrong weight distribution," observed the *Saturday Evening Post*, "and exerts a spin-out force similar to that on the end ice skater in a crack-the-whip line."[18] The vivid imagery didn't take an engineering degree to understand.

The car magazines used more technical language, saying the Corvair had a tendency to "oversteer." *Sports Car Illustrated* suggested that "one of the first things that should be done" is to put a weight-balancing "anti-roll" bar under the car's front end.[19] It wouldn't have to be fancy; just a heavy metal bar bolted under the front of the car to even out the weight. As it happened, Volkswagen had just installed one on the Beetle. GM had used an anti-roll bar on prototype Corvairs, but deemed it unnecessary when the car went into production.[20]

An internal Chrysler engineering report, meanwhile, cited the difference in the recommended tire-inflation pressures, front versus rear, as "further evidence of Corvair's concern over handling."[21] The *Sports Car Illustrated* piece dryly observed that "it's unlikely that most Corvair owners will ever maintain the pressures recommended."[22]

Ed Cole wasn't buying any of it. "Critics know nothing about this car," he snapped to one interviewer. "We know what it has done in millions of miles of testing."[23] When he saw Henry Ford II at an industry gathering he challenged him to a duel, Detroit-style. "Henry," said Cole, "I'll corner inside you against anything you've got."[24] "Hank the Deuce," as he was called in Detroit, didn't take the challenge.

Chevrolet's foot soldiers, meanwhile, mobilized to defend the Corvair. In April 1960, just six months after the car launched, Chevy sent a Corvair across deserts, mountains, and windswept farm country on the 2,061-mile Mobilgas Economy Run from Los Angeles to Minneapolis, averaging an impressive 27.03 miles per gallon.

The same car then headed back west to make an early-spring run up the icy slopes of Pikes Peak, reaching the 14,110-foot summit without snow tires or tire chains. Chevy's publicists touted both feats, highlighting the Corvair's combination of fuel economy and traction, in an elaborate sales brochure. (The only glitch was that Chevy's brochure reversed the locations of Wichita and Des Moines on the route map.)[25]

As if climbing Pikes Peak wasn't proof enough, Chevy's marketers then sent three Corvairs on a 6,000-mile trek from Chicago to Panama City, Panama, that they dubbed "Operation Americas." Yet another glossy sales brochure, this one with a geographically correct route map, described how the Corvairs "pounded out the miles over every conceivable kind of road. Rutted. Rocky. Dust-choked. Twisting. Rain-swept."[26] No adjective was left behind.

For good measure, the brochure reprinted letters from happy Corvair owners around the country: a housewife in Baton Rouge, a dentist in Minnesota, an insurance agent in Stockton, California, and more. Cole and his minions wanted to prove that the Corvair was reliable and safe.

Meanwhile, three Chevrolet engineers submitted a detailed technical paper to the Society of Automotive Engineers, the same group that had heard Maurice Olley's criticism of rear-engine cars seven years before. "The Corvair, with its low center of gravity and swing-axle rear suspension, has excellent roll stability," the paper stated. "After exhaustive testing it was found that a stabilizer was not required."[27]

The paper, however, contained a dry chart with some revealing numbers. It showed that while the target weight for the Corvair's rear suspension had been 148 pounds, the actual weight was 175 pounds. The Corvair's engine also contained some parts originally intended to be aluminum that were made from iron instead, boosting the engine's weight to 366 pounds, well above the target of 288.[28] The Corvair, like a person struggling with a diet, had packed on an extra 105 pounds of rear-end weight. And the car's weight distribution, originally intended to be 40 percent in the front and 60 percent in the rear, actually came in at 38–62.[29] The difference seemed small at the time, but it wasn't.

Chevrolet sold 250,000 Corvairs in 1960, a respectable number, but far fewer than the 457,000 Falcons sold by Ford. At first the Corvair came in two versions, a basic sedan and a more upscale version. But before long Chevy found a way to put new impetus behind the Corvair.

At the Chicago Auto Show in February 1960, Chevy displayed a prototype sporty version called the Monza coupe, with two doors instead of four, standard bucket seats, a four-speed, floor-mounted manual transmission, and a 95-horsepower engine (compared to the standard Corvair's 80). The public's reaction was enthusiastic, and Chevy put the Monza into production just three months later.

In October 1960, the start of Detroit's 1961 model year, Chevy added still more versions of the Corvair: a Monza sedan, the Lakewood station wagon, the Corvan commercial van, the Rampside pickup truck, and the Greenbrier Sports Wagon. The latter was available with a car-top sleeper tent and a "camper unit" that converted the vehicle into a mini–mobile home, just like VW's Microbus.

Thus the Corvair became a family of rear-engine vehicles, a brand within a brand. A Chevy sales brochure showed custom fly-fishing patterns named for each vehicle in the Corvair clan. It wasn't clear whether the fish liked the flies, but the public liked the lineup. Corvair sales surged 32 percent to 329,632 cars, trucks, and vans. In 1962 Chevy added the hot new Corvair Monza Spyder coupe and convertible, with a 150-horsepower turbocharged engine. The Corvair was gaining momentum.

• • •

On January 12 of that year, at around 1:30 a.m. in Los Angeles, well-known television comedian Ernie Kovacs was driving his new Corvair station wagon home after a party at the home of director Billy Wilder. Kovacs was following his wife, singer Edie Adams, who was driving home in the couple's Rolls-Royce. (He had arrived at the party in the Rolls and she in the Corvair, but the couple switched cars for the ride home.) When Kovacs turned left onto Santa Monica Boulevard, the car spun out on the rain-slickened street and slammed sideways into a steel utility pole. His broken ribs ruptured his aorta, killing Kovacs at age forty-two. He became the most famous victim in a list of Corvair accidents that was quietly but steadily growing.

Two years later *Car and Driver* named the Corvair the "Best Compact Sedan,"[30] but the real news was underneath the car. Chevrolet added an anti-roll stabilizer bar under the front end, just as *Sports Car Illustrated* had recommended four years earlier. Chevy also changed the suspension system to limit the tendency of the rear outside wheel to tuck underneath the car during hard turns. Both changes improved the Corvair's handling and stability.

Such improvements usually would have been grist for Chevrolet's publicity mill, but not this time. Chevy issued a terse press release noting "a new suspension design will enhance handling characteristics of all Corvair models."[31] Making a big deal of the new suspension and the stability bar would have risked conceding that the 1960–63 Corvair models were unstable.

There was no point in adding fuel to the brushfires of Corvair lawsuits that were popping up around the country. Besides, the 1965 Corvair would get further suspension improvements and bigger brakes, along with curvy styling to replace the original clean-but-boxy design. Corvair sales had been slumping, but Chevrolet was optimistic about the car's prospects. Events elsewhere, however, would make the optimism unwarranted.

Early in 1965, a thirty-one-year-old Washington lawyer took an unusual career detour. He left his paid position with the Department of Labor to become an unpaid advisor to the U.S. Senate Subcommittee on Executive Reorganization chaired by Senator Abraham Ribicoff, a Democrat from Connecticut. The young lawyer was from Connecticut, too. His name was Ralph Nader.

Nader was born in 1934 to Nathra and Rose Nader, immigrants from Lebanon, and raised in Winsted, in the northwest part of the state. Nathra owned a restaurant there, the Highland Arms, where he served meals along with generous helpings of his opinions on local issues and social justice. The Nader family dinner table also served double duty as a discussion forum, where Ralph, the youngest of four children, learned the techniques of rhetorical give-and-take.

He was precocious, given to reading the dry and windy *Congressional Record* while other kids read the Hardy Boys. He loved playing baseball and declared allegiance to the Yankees in a town divided between them and the Boston Red Sox. He showed no interest in dating or high school social life, but his academic record was outstanding.

Nader graduated from Princeton University with Phi Beta Kappa honors in 1955. He went to Harvard Law School, where he was, for the first time in his life, an indifferent student, bored by the minutiae of his coursework. In his third and final year, however, he wrote a paper titled "Negligent Automobile Design and the Law." The subject captivated him.

Following a short stint in the military, Nader practiced law for four years in Connecticut and Massachusetts. After moving to Washington in 1965 to work for the Labor Department he wrote occasional articles on automobile safety for the *New Republic* and other magazines. A letter from a disgruntled General Motors worker brought the Corvair to his attention.

By that time Nader was thinking about writing a book on the defects he saw in many cars, and he came to the attention of Senator Ribicoff's staffers. Auto safety fell under a slew of government agencies, so Ribicoff's Executive Reorganization Subcommittee took the lead on the issue. Nader was happy to help, even without pay. He had few living expenses: no car and rent of just $80 a month at a Washington rooming house. He didn't even have a television, the medium that would soon make him famous.

Nader didn't have to find a publisher for his book idea. A publisher found him. Richard L. Grossman, head of three-year-old Grossman Publishers of New York, wanted to publish a book on auto safety. He got Nader's name from an editor at the *New Republic* who had tapped Nader as a reliable source on the subject. In November 1965 Nader's book *Unsafe at Any Speed* became the fledgling publishing company's sixteenth book. "For over half a century the automobile has brought death, injury, and the most inestimable sorrow and deprivation to millions of people," the preface began.[32] This might have been news to the farmers liberated by the Model T. But Nader's book was a sweeping indictment of car companies and their utter neglect, as he saw it, for the safety of the motoring public.

Only the first chapter was about the Corvair, but that was enough. Nader described the plight of Mrs. Rose Pierini of Santa Barbara, California, whose left arm was severed in September 1961 when her Corvair, traveling at about 35 miles an hour, overturned. Nearly three years later, Nader recounted, "General Motors decided to pay Mrs. Pierini $70,000 rather than continue a trial which for three days threatened to expose on the public record one of the greatest acts of industrial irresponsibility in this century."[33]

He recounted how automotive-accessory makers did brisk business selling kits, including stabilizer bars, designed to reduce the Corvair's penchant for rear-end spinouts. He added that GM itself had started selling an optional "stability enhancement" kit in 1961, but hadn't advertised it. Nader wrote how car magazines had described the controversy over rear-engine design when the Corvair was first introduced, but had muted criticism in deference to GM's advertising dollars, in his view.

There was more. Making the driver responsible for the Corvair's stability, Nader wrote, "by requiring him to monitor closely and persistently tire pressure differentials cannot be described as sound or sane engineering practice." [34] He noted the improved suspension on the 1964 model, but added: "It took General Motors four years of the model and 1,124,076 Corvairs before they decided to do something . . ." [35]

His language could be impenetrable in places. He described how on the early Corvairs "the rear wheel is mounted on a control arm which hinges and pivots on an axis at the inboard end of the arm near the center of the vehicle." [36] But for the most part, Nader was clear and compelling. "The Corvair was a tragedy, not a blunder," he wrote. "The tragedy was overwhelmingly the fault of cutting corners to shave costs. This happens all the time in the automobile industry, but with the Corvair it happened in a big way." [37]

Even before the book was published, Nader had come to the attention of GM's lawyers. By the fall of 1965, GM faced 106 Corvair liability lawsuits around the country. Nader's name had surfaced in several as an expert witness. The newsletter of the American Trial Lawyers Association, whose members represented plaintiffs against GM, had identified Nader as a man worth consulting in Corvair cases. Nader himself, meanwhile, had written about the Corvair in the *Nation*. His article, titled "Profits vs.

Engineering—The Corvair Story," was basically a summation of his book's first chapter.[38]

It was little wonder, then, that Aloysius F. Power, GM's general counsel, asked his staff to find out more about Nader. It was, at the outset, a reasonable request. But few Corvairs would spin so wildly out of control as the investigation that followed.

For all its far-reaching consequences, GM's investigation of Nader started comically. The first investigator hired went to Winsted and "discovered" that Nader frequented a local restaurant called the Highland Arms, not knowing it was owned by Nader's father. That and other banal bits of noninformation didn't satisfy GM's attorneys. They figured Nader had to have a financial stake in the work he was doing for the trial lawyers. The company's lawyers, deciding they needed more than a third-rate gumshoe, turned to Richard Danner, an attorney with Alvord & Alvord, a Washington law firm. Danner, in turn, retained Vincent Gillen, a former FBI agent who ran a private detective agency in New York. And Gillen, in turn, farmed out some of the field work to his contacts in Boston and Washington.

This wasn't a short and crisp chain of command, but at least Gillen's instructions to the field team were clear. "Our job is to . . . determine 'what makes him tick,'" he wrote in a memo, "such as his real interest in safety, his supporters, if any, his politics, his marital status, his friends, his women, boys, etc., drinking, dope, jobs—in all facets of his life."[39] By late January 1966, Gillen's detectives were asking about Nader everywhere: in Washington, in New York publishing circles, in Winsted, and at the University of Hartford, where Nader had briefly taught. Their pretext, when queried, was that their client was considering the promising young man for a sensitive job.

Had that been the case, the prospective employer would have learned that Nader was smart, dutiful, churchgoing, sensitive, hardworking, and without apparent vices. The detectives, though, probed further, for potential anti-Semitism stemming from Nader's Lebanese ancestry or for any steady companions, male or female. They found nothing. Their surveillance of his day-to-day movements likewise turned up empty.

The detectives were getting desperate to find something, *anything*. In February, as Nader prepared to testify before Ribicoff's committee, suspicious incidents occurred. Nader started getting strange phone calls, sometimes well after midnight, with obviously phony inquiries about freight packages or airline tickets. On the evening of February 20, when he was browsing through magazines at a drugstore near his apartment, an attractive young woman asked him to come to her apartment to discuss "foreign affairs." Not long after, another woman approached him in a Safeway supermarket in Washington and asked if he would help carry heavy items to her apartment. Nader recounted these events to James Ridgeway, a young reporter for the *New Republic*. Ridgeway wrote an article headlined "The Dick" in the magazine's issue dated March 12, which hit newsstands March 4.

The article caught the eye of Walter Rugaber, a reporter at the *New York Times*, who did reporting of his own to verify and amplify Ridgeway's work. Rugaber's article appeared on March 6, on the front page of the *Times*' second section, under the headline: "Critic of Auto Industry's Safety Standards Says He Was Trailed and Harassed."

Most corporate types disdained the *Nation* and the *New Republic* as little liberal rags, if they even heard of the magazines. The *New York Times*, though, was the nation's leading national newspaper. It couldn't easily be ignored. Ford responded, issuing a statement denying involvement in any investigation of Nader. On March 8, two days after the *Times* article, GM president James M. Roche directed his PR department to prepare a similar statement. Later that day, however, Roche learned to his chagrin that General Motors indeed was involved. On the night of March 9, General Motors issued a statement that was altogether different from Ford's. GM said "the office of its general counsel initiated a routine investigation through a reputable law firm to determine whether Ralph Nader was acting on behalf of litigants or their attorneys in Corvair design cases. . . . The investigation was limited only to Mr. Nader's qualifications, background, expertise, and association with such attorneys. It did not include any of the alleged harassment or intimidation recently reported in the press."[40]

Hell broke loose. The next day, Roche received a telegram informing him that Senator Ribicoff's subcommittee "will hold hearings March 22 on Nader-GM matter. Respectfully request your attendance as a witness."[41] Roche couldn't refuse. The hearing combined drama, humor, and hubris in delicious detail. Power, GM's general counsel, sparred with committee members over the definition of harassment, though he conceded he himself might have felt harassed were he Nader. When Detective Gillen insisted that the questions his people asked about Nader's personal life were simply "in fairness to Ralph," Senator Robert Kennedy, a member of the committee, rebuked him angrily. "What the hell is fairness to Ralph?" Kennedy snapped. "You have to keep proving he's not queer and he's not anti-Semitic?"[42]

The hearing was humiliating for General Motors. Roche denied any attempt by GM to silence or discredit Nader, but told the subcommittee: "I want to apologize here and now." Nader wasn't there to hear the apology. The man who didn't own a car arrived at the hearing late because he had trouble hailing a cab. "I almost felt like going out and buying a Chevrolet," he cracked when he arrived.[43]

The Ribicoff hearing caused a sensation. *Time*, which had gushed over the Corvair six years earlier, headlined an article, "Why Cars Must—and Can—Be Made Safer."[44] *Unsafe at Any Speed*, which had been languishing on bookstore shelves, surged onto the best-seller list. Nader became a celebrity. Almost anything he said made news, including his attacks on the Volkswagen Beetle for its own rear-engine design, which VW successfully rebuffed. The thirty-two-year-old unemployed lawyer was invited to address the British parliament in London, and after that the Swedish parliament in Stockholm.

In November 1966, eight months after the hearing, Nader sued General Motors for invasion of privacy, seeking $26 million, according to GM, or $12 million, according to Nader. The company and the crusader couldn't even agree on the amount.

For its defense GM hired former JFK aide Theodore Sorensen, a New York lawyer, and one of Sorensen's law partners, a former federal judge. As the lawsuit wound through the courts, the Corvair controversy began to exert a profound impact on American public life. Congress passed the National Traffic and Motor Vehicle Safety Act of 1966, and President Johnson signed it into law. It required automakers for the first time to announce recalls of their vehicles publicly. But the new regulatory zeal wasn't limited to car companies.

In 1967 Congress passed a law imposing more-stringent health standards and regular government inspections on meat packers. The following year brought the Natural Gas Pipeline Safety Act, the Radiation Control for Health and Safety Act that regulated medical X-rays, and the Wholesale Poultry Products Act. Even after

Republican Richard Nixon became president in 1969, the Nader crusade marched on with the Federal Coal Mine Health and Safety Act. Nader got credit from the press for all of it. He even took aim at hot dogs and baby food, prompting changes in their ingredients.

As for the Corvair, publicity from the Ribicoff hearing caused sales to drop 56 percent in 1966, to just 104,000 cars. The next year they plunged another 75 percent, to just 27,000. The strong-willed engineer who had brought the Corvair to life fared better, however. On October 30, 1967, Ed Cole, who didn't have much use for Ralph Nader but had been outraged at his company's spying, became president and chief operating officer of General Motors. The only higher position was chairman and chief executive officer. That job went to Jim Roche.

On December 12, 1969, exactly ten years, two months, and one week after Ed Cole and his Corvair made the cover of *Time*, Ralph Nader made the cover himself under the headline "The Consumer Revolt." Beneath the art deco rendering of his portrait appeared the tail end of a Corvair, driving into history. The last Corvair had been built seven months earlier, ending a decadelong production run of a total of 1,786,243 Corvair cars, trucks, and vans. "To many Americans, Nader, at 35, has become something of a folk hero," *Time* wrote, "a symbol of constructive protest against the status quo."[45] Protest, constructive or otherwise, was becoming common in the late 1960s. Gone were the days of *Executive Suite* when William Holden could play a heroic business executive as the embodiment of American idealism in a hit movie.

During the tail-finned Fifties Americans had trusted, by and large, government officials, clergy, educators, and corporate executives. But that was before Vietnam, urban riots, campus unrest, and before the Chevy Corvair. Mistrust of authority became the new ambient attitude.

For Ralph Nader, mistrust was even becoming a business in the form of Nader's Raiders. A force of more than 100 law, engineering, and medical graduates joined his not-for-profit affiliates and accepted minimal pay to investigate government agencies and corporations. Nader's Raiders ushered in the era of influential NGOs, nongovernmental organizations, with a public-policy agenda and ample funding to pursue it.

Nader's budding empire of NGOs included the Center for the Study of Responsive Law, the Center for Auto Safety, and a network of "Public Interest Research Groups." It got a financial infusion in August 1970, when Nader and GM settled the invasion-of-privacy lawsuit he had filed nearly four years earlier. GM paid Nader $425,000 (equivalent to $2.5 million in 2012), just a fraction of what he had sought. One dismayed fan told the *Wall Street Journal* Nader was a "rat fink" for settling with GM, but Nader, characteristically, was undeterred.[46]

He desposited his windfall in First National City Bank—itself an organization he was investigating—and announced he would use the money to fund further investigations of GM and other corporations and organizations. A few months later, his raiders discovered that Justice Department lawyers had recommended breaking up General Motors and Ford on antitrust grounds, but that their recommendation had been quashed. The revelation caused a minor stir, but it soon faded. Japanese car companies would soon do to Detroit what the Justice Department didn't.

After all the drama, the federal government wound up officially exonerating the Corvair. "The handling and stability performance of the 1960–63 Corvair does not result in an abnormal potential for loss of control or rollover," a government study panel concluded in July 1972.[47] GM said tersely that the report "confirms our position." A furious Nader condemned the report as a whitewash, not without reason. Even some GM attorneys who defended the Corvair in court had developed qualms about arguing the company's case. But officially, Ralph Nader had lost his battle with the Chevy Corvair.

By this time Ed Cole was espousing a cause dear to virtually every Nader supporter, ironically. Cole was leading the quest to put catalytic converters on cars to eliminate most of the smog-producing chemicals spewing out of automotive tailpipes.

Catalytic converters required cars to use unleaded gasoline. But for decades oil companies, using a process developed and patented by GM, had added lead to their gasoline to boost the octane level and reduce the clunking sounds of engine "knock." The oil companies mounted a furious defense of leaded gas, as did Ford, Chrysler, and even most executives at GM. The Nixon White House secretly suggested keeping leaded gasoline around a few more years.

But Cole stood firm, just as he had done in defense of the Corvair, defying both his GM colleagues and the White House. "He had the balls to do it," a Ford executive recalled decades later.[48] Catalytic converters were installed on Detroit's cars from the mid-1970s onward. Ed Cole, however, would be remembered as the "Father of the Corvair" instead of the "Father of Clean Air."

The Corvair's story was Greek tragedy, with a car created by Ed Cole's genius and undone by his hubris. In October 1974 the two great protagonists, Cole and Nader, met face-to-face to debate. Cole had just retired from GM and was free from his corporate handlers, who had forbidden any public confrontation with Nader. Television's Phil Donahue came to Detroit from Chicago to broadcast the event.

Nader attacked the Corvair, while Cole defended his car. When Nader decried factory work as "inhuman," an irritated Cole replied, "IT ISN'T INHUMAN!" At one point Nader sniffed that he spent only $5,000 a year on living expenses, "compared to the millions of Ed Cole." But backstage, when it was over, Nader handed Cole a compliment. "You got the lead out of gasoline," he quipped, shaking hands with the retired industrialist. "Now how about getting the lead out of GM?"[49]

Though he officially lost the Corvair battle, Nader won the war in every conceivable sense. His victory launched America's consumer movement. It also revolutionized the country's regulatory climate. The most lasting legacy, however, was the Corvair's impact on American law. For decades, product-liability lawsuits had been a tangential presence in American law and business. To collect any money, a plaintiff had to prove that a manufacturing defect caused a product flaw that, in turn, caused harm. It was a narrow definition of what was defective. The Corvair changed that.

In the mid-1960s, courts started accepting proof of inherent design defects, like the Corvair's heavy rear end, as reason enough for plaintiffs to prevail. The American Law Institute persuaded every state to revamp its tort laws along those lines. "Public policy demands that the burden of accidental injuries caused by products intended for consumption be placed upon those who market them," the Institute declared, "and be treated as a cost of production against which liability insurance can be obtained."[50]

Thus the Corvair established lawsuits as a disciplinary device to deter manufacturers from producing unsafe products. As damage collection got easier, lawsuits got more common. It was a "collision of two sets of cultures," Marshall Shapo, a Northwestern University law professor, would write. "What may be called a justice culture and a market culture."[51]

In the 1970s another small car, the Ford Pinto, picked up where the Corvair left off. When three Indiana girls burned to death after their Pinto was struck by a truck, prosecutors charged Ford with criminal negligence for making a gas tank prone to explode in rear-end collisions. The company was acquitted, but Ford wound up paying millions in civil penalties for Pinto accidents. And its corporate image was tarnished for years.

The evolution of liability law that got its early impetus from the Corvair reached its ultimate expression not in cars but in coffee, of all things. In 1994, thirty-five years after the Corvair first appeared, a jury in New Mexico heard the case of an eighty-one-year-old woman who had been scalded by coffee purchased at a McDonald's drive-through window. The coffee had been heated to precisely the temperature, 180 degrees, that McDonald's had specified for optimum taste. But that was 20 degrees hotter than the coffee in other restaurants.

The injured woman was hospitalized for seven days to receive skin grafts for third-degree burns. She was just one of 700 McDonald's customers, it turned out, who had suffered coffee burns. The jury, which had seemed skeptical of her claim initially, awarded her $2.9 million. Whatever the merits of her case, however, the woman became the poster-granny symbol of a legal system gone awry. Public opinion sided with McDonald's. The jury's award was reduced on appeal, with the case later settled for an undisclosed amount.

A year after the McDonald's trial, a book by Philip Howard called *The Death of Common Sense* denounced trivial lawsuits and became a best seller. America's "justice culture" began to swing back to "market culture," but the pendulum never would swing back to where it was before the Corvair. The car had launched one of the growth industries of the late twentieth century, lawsuits. "The Model T put Americans on the road," Robert Marlow, a Corvair collector from New Jersey, wrote in 1996. "The Corvair put us in the hands of lawyers."[52] Some of Marlow's fellow collectors display their sentiments on the license plates of their lovingly restored cars, plates that say "RALPH WHO," "F RALPH," and simply but cleverly, "NADIR."[53]

These people likely didn't vote for Ralph Nader when he ran for president, at age sixty-six, in the year 2000. He got 95,000 votes in the state of Florida, which George W. Bush won by about 1,800 votes. Bush lost the national popular vote to Al Gore but prevailed in the Electoral College, when the U.S. Supreme Court upheld his tissue-thin victory in Florida.

Had Nader not been on the ballot, Gore surely would have gotten most of his votes. And had it not been for the Corvair, Nader wouldn't have been on the ballot. Thirty years after its demise, Ed Cole's flawed car was still shaping American life. It can safely be said, at any speed, that the Chevy Corvair's legacy helped make George W. Bush president of the United States.

In 1908 Henry Ford developed the Model T in a room in this building on a Detroit side street. Today the neighborhood around Piquette Street is a scene of urban blight, but Ford's old factory houses a successful commercial laundry on the first floor and a small Model T museum on the upper floors. *(John Stoll)*

The Abundant Faith Cathedral, which today sits across the street from Ford's original Piquette Street factory. *(John Stoll)*

Quickly outgrowing the Piquette Street factory, Ford commissioned a new industrial complex on the outskirts of Detroit. The Highland Park plant looked like this in 1913, the year Henry Ford introduced the revolutionary moving assembly line. In early 1914 he followed that with the $5 day, which gave rise to the American middle class. (*Detroit Public Library*)

Ford's 1912 Model T pickup, a far cry from Ford's lavish F-150 King Ranch truck a century later. In 1912, the chassis and the truck bed were purchased separately, and bolted together by the dealer or the customer, like a giant Lego toy. (*Ford Motor Company*)

Henry Ford and his son, Edsel, with a 1921 Model T Ford, had 60 percent of the U.S. car market, but Henry's refusal to modernize his car spelled opportunity for General Motors, which viewed automobiles as devices for social, as well as physical, mobility. *(Ford Motor Company)*

The power plant at Ford's manufacturing complex in Highland Park, Michigan, in 1920, when the company's dominance was at its peak. *(Ford Motor Company)*

James Couzens, the brilliant, hard-nosed finance chief who helped Henry Ford build his empire. Some historians credit Couzens, not Henry Ford, with conceiving the $5 day. But Ford and Couzens eventually grew apart, as typically happened with Ford and his closest business associates. *(Detroit Public Library)*

Alfred Sloan Jr., the executive who dominated General Motors from the 1920s to the 1950s. His declaration that GM would build "a car for every purse and purpose" led to the company's hierarchy of brands, which transformed the automobile into America's foremost status symbol. *(General Motors Heritage Center)*

The 1927 LaSalle, the first car designed by the legendary Harley Earl. The low, sleek lines stood in sharp contrast to the boxy shapes of most cars of the day, and made the LaSalle America's first "yuppie" car. *(General Motors Heritage Center)*

Harley Earl, who ran GM's design studios for thirty years, in the 1950s, when his tail fins ruled America's roads. *(General Motors Heritage Center)*

Earl with his 1951 LeSabre concept car. Note both the rear fins and the protruding cones on the front grille, which became standard on Cadillacs and were nicknamed "Dagmars," after a well-endowed television starlet of the day. *(General Motors Heritage Center)*

The original Chevrolet Corvette was unveiled in January 1953 at GM's Motorama exhibition in New York as an experimental model, the EX-122. Years later, *Sports Illustrated* described the car as "an albino toadfish with ill-fitting false teeth." *(General Motors Heritage Center)*

Zora Arkus-Duntov, around the time of his retirement in 1974. The Russian immigrant engineer's steadfast devotion to the Corvette kept the car alive at a time when GM's bosses wanted to kill it, and eventually made it an American icon. *(General Motors Heritage Center)*

The 1963 (CK) Corvette, a black hardtop and a red convertible, traditionally the car's two most popular colors. *(Collier Collection)*

The 1963 Corvette Stingray, with its distinctive split rear window, sparked conflict between Duntov and design chief Bill Mitchell, the successor to Earl. Mitchell liked the window's design, but Duntov thought it impaired visibility. The split window lasted just one year. *(General Motors Heritage Center)*

Virgil Exner, Chrysler's design chief in the mid-1950s, pushed tail fins to new heights, and launched Detroit's tail fin war. Nothing captured America's exuberant sky's-the-limit ethos of the late 1950s as much as tail fins. *(Detroit Public Library)*

Chrysler's 1955 models, the first to feature Exner's designs, sparked a surge in sales. This is the 1955 Plymouth Belvedere Sport Coupe, with America's biggest tail fins to date. Bigger were to come. (*Chrysler Historical Collection*)

The Dodge display at the 1957 Detroit Auto Show. Thanks to Exner, Chrysler's tail fins had grown taller since 1955, but their success would soon stir industry-giant General Motors to fight back. (*Detroit Public Library*)

Chuck Jordan in the 1970s, about fifteen years after he designed the 1959 Cadillacs, which had the tallest tail fins ever. Jordan eventually became GM's vice president for design. *(General Motors Heritage Center)*

The 1959 Cadillac Series 62 convertible, the epitome of fin-dom and the ultimate automotive expression of America's fascination with the space age. After 1959 fins began declining in size, disappearing for good in 1965. *(General Motors Heritage Center)*

The 1959 Cadillac Eldorado, a car George Jetson could love. The side panels resemble the fuselage of a jet engine. The red tail lights mounted astride each rear tail fin were nicknamed "gonads." *(General Motors Heritage Center)*

Ferdinand Porsche, the inventor of the Volkswagen Beetle, shown in his later years, probably the late 1940s. Porsche had met Adolf Hitler before he came to power, and Hitler's admiration for "the professor" helped Porsche get the contract to develop Germany's "people's car." *(Volkswagen archives)*

When the Grateful Dead's Jerry Garcia died in 1995, Volkswagen ran this ad in *Rolling Stone* and other magazines. The Beetle was born in Hitler's Germany but the car and its derivative, the Microbus, become symbols of America's 1960s hippie counter-culture. *(Volkswagen archives)*

Jerry Garcia. 1942-1995.

"Lemon" was one of Volkswagen's most daring Beetle ads from the early 1960s. The text tells how this car was rejected by a company quality inspector until a minor flaw could be fixed. VW's advertising, created by agency Doyle Dane Bernbach (now DDB Worldwide), defied convention with self-deprecating humor, giving Hitler's car its "cute" image in America. *(Volkswagen archives)*

Lemon.

American Gothic was evoked in this whimsical Beetle ad from Doyle Dane Bernbach from the early 1970s. The ad featured a real-life couple who lived in the Missouri Ozarks. *(Volkswagen archives)*

"It was the only thing to do after the mule died."

The headline on this 1963 VW Microbus ad would have drawn feminist ire just a few years later. The text, however, describes an independent-minded woman, at least in the context of the day. *(Volkswagen archives)*

Do you have the right kind of wife for it?

Heinz Nordhoff, the man who saved Volkswagen and, therefore, the Beetle, circa the late 1940s, when the company's revival was underway. Nordhoff's need for U.S. dollars to purchase modern production machinery led VW to begin exporting Beetles to America in 1950. *(Volkswagen archives)*

Ed Cole, then general manager of Chevrolet and later president of General Motors, made the cover of *Time* in October 1959, when his revolutionary rear-engine Corvair debuted. Cole dismissed suggestions that the Corvair was unstable, leading an unknown lawyer named Ralph Nader to pillory the car in his 1965 book, *Unsafe at Any Speed. (Courtesy of Time Inc.)*

Cole shown with his pet project, the Corvair, in 1960. The Corvair got 29 miles a gallon, remarkable for its day, but it was prone to spinning out around curves. *(General Motors Heritage Center)*

The Corvair as it appears in a GM photo montage from 1960. *(General Motors Heritage Center)*

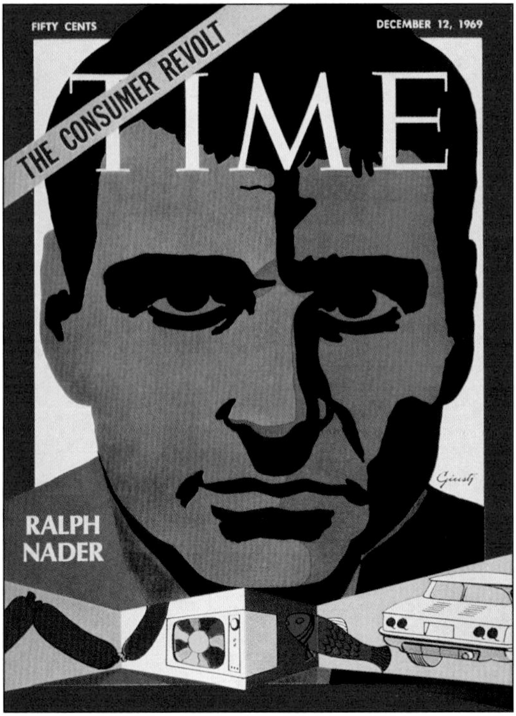
FIFTY CENTS
DECEMBER 12, 1969
THE CONSUMER REVOLT
TIME
RALPH
NADER

In December 1969, ten years and two months after Ed Cole made the cover of *Time,* his arch-nemesis, Ralph Nader, landed there. Note the Corvair in the lower right corner, driving into oblivion. GM killed the Corvair that year. *(Courtesy of Time Inc.)*

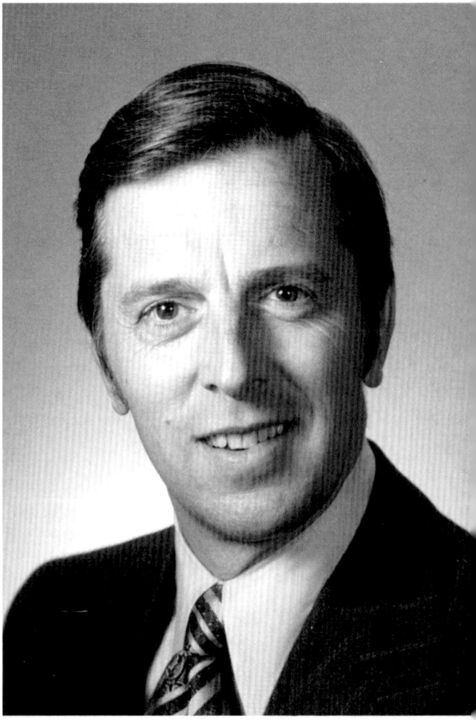

Harold "Hal" Sperlich, who played a leading role in developing both the Ford Mustang and the Chrysler minivans, shown circa the late 1970s. The Mustang captured America's baby boomers when they were just starting to drive, and the minivan lured them again during their child-rearing years. *(Chrysler Historical Collection)*

6

TURNING A "LIBRARIAN" INTO A "SEXPOT": THE YOUTH BOOM, THE SIXTIES, AND THE MAKING OF THE MUSTANG

The Kennedy years were so exciting. We were all over the demographic data back then, and it was all about youth.

—Hal Sperlich, Ford executive on the Mustang project[1]

The median Mustanger is 31 years old, compared to age 42 for the median purchaser of the regular Ford car.

—Ford press release, March 2, 1966

The car that almost killed Ford, the Edsel, had been conceived and developed with little market research. Simply deeming the car to be a good idea, the company built it, only to find that few people wanted it. But after the Edsel's demise in 1959, Ford was determined not to repeat that mistake. By the early 1960s the company had a twenty-person staff devoted to market research.

"We have experts who watch for every change in the customer's pulse-beat," boasted Lee Iacocca, a hard-charging young executive on a meteoric career path at Ford. "For a long time now we have been aware that an unprecedented youth boom was in the making."[2] Ford's market researchers documented this phenomenon with enough statistics to clog a computer.

Demographers projected the number of Americans between ages fifteen and nineteen was to increase by 41 percent during the 1960s. The number of twenty- to twenty-four-year-olds would climb even more, by 54 percent. The ranks of families earning more than $10,000 a year would jump 156 percent between 1960 and 1975. The number making between $5,000 and $10,000 would rise 27 percent.[3]

Only 21 percent of America's "housing units," as Ford's reports termed them, had two or more cars, but that percentage would surge. The number of women drivers increased 53 percent between 1956 and 1964, while the number of male drivers rose just 6 percent. College enrollment would double during the 1960s, and would increase by another 25 percent during the not too distant 1970s. College grads were just 19 percent of the adult population, but they accounted for 46 percent of new-car sales.[4]

Evidence of the youth boom was obvious even to nondemographers. "Now it's Pepsi—for those who think young," advised a 1961 advertising slogan, implying a connection between soda and brainwaves. *TV Guide* carried instructions for a new kids' dance: "Feet in prize-fighter position. Hips swivel from side to side as if rubbing the back with a towel. Twist feet as if putting out a cigarette."[5] It was called the Twist.

Little state colleges expanded into full-fledged universities to accommodate the hordes of baby boomers banging down the doors. One was the teachers college in the central Illinois town of Normal, so named because teachers colleges traditionally had been called "normal schools." The college was renamed Illinois State University in 1964. (The town, however, wasn't renamed.)

That May President Lyndon Johnson traveled to a college right in Detroit's backyard, the University of Michigan in Ann Arbor, to give a landmark speech. "I have come today from the turmoil of your capital to the tranquility of your campus," Johnson intoned, "to talk about the future of your country . . . The Great Society rests on abundance and liberty for all."[6] His words would ring with irony a few years hence, when campuses became scenes of riots instead of tranquility.

Just a month before Johnson proclaimed the Great Society, Ford introduced a new car called the Mustang, aimed at the youth market its researchers had assiduously identified. Unlike the ill-fated Edsel, its sales accelerated like a sports car peeling away from a stoplight. In New York a diner posted a sign in its window: "Our Hotcakes Are Selling Like Mustangs."[7]

The Mustang was perfect for its day. It had a sporty style inspired by little British roadsters. It was modestly priced, under $2,500 for basic versions but available with upgrades that produced big profits for Ford. And though it was compact, the Mustang was big enough for a family of four. While its styling was fresh and youthful, however, the Mustang *wasn't* an altogether new vehicle. The car's basic engine and mechanical underpinnings were carried over from the boring old Ford Falcon. Seymour Marshak, a former college professor who became Ford's head of market research, used a vivid analogy to describe how Ford transformed its ugly duckling into a swan.

"You can take a girl, put her hair in a bun, add horn-rimmed glasses and low-heeled shoes, flatten out her chest and her behind, and you've got a school librarian," he proudly told the *Detroit Free Press* after the Mustang had become a runaway hit. "Take the same girl in upswept hair, contact lenses, spike heels, fill out her figure top and bottom—and you've got a sexpot! We did much the same thing with a car."[8]

Had Marshak uttered those words a decade or two later he would have been hauled down to the human resources department and sentenced to weeks of sensitivity training unless he was fired outright, which might have been a kinder fate. But this was 1964 and women hadn't been liberated, Vietnam hadn't escalated, and the inner cities hadn't exploded. And Americans absolutely loved the Ford Mustang, which uniquely captured the youth ethos of its day.

In November 1960 John Fitzgerald Kennedy, at age forty-three, became the youngest man and the first Catholic ever to be elected president. The same month another man whose name ended in a vowel, Lee Iacocca, became the head of Ford Motor Company's flagship Ford Division. At just thirty-six years old, Iacocca was the youngest man ever to hold that post.

"It was energizing when Iacocca came into the Ford Division," recalled one of his even younger protégés, Harold K. "Hal" Sperlich, nearly half a century later. Sperlich was just thirty-one years old in 1960 when he got his first management job, overseeing "special studies," which would lead to a major role in making the Mustang. "It was like: here's Kennedy, here's Iacocca, and here comes the whooole youth market."[9]

The Kennedy years ended tragically on November 22, 1963, but the sense that youth was coming to the fore continued. In February 1964 the Beatles appeared on the *Ed Sullivan Show*. Girls screamed themselves hoarse and boys started growing long hair, causing parents to scream themselves hoarse, too. Two weeks later brash young Cassius Clay knocked out Sonny Liston to win boxing's world heavyweight championship, causing a sensation. Soon after, he said he was a Muslim and changed his name to Muhammad Ali, causing a bigger sensation.

Just as the Irish-Catholic Kennedy had broken down ethnic and religious barriers in politics, the Italian-Catholic Iacocca was doing the same thing in corporate America. His given name was Lido, for the Italian beach where his parents had honeymooned. To those, and they were many back in 1964, who stumbled over his name, *Time* explained that it "rhymes with try-a-COKE-ah." [10]

Iacocca was born in Allentown, Pennsylvania, where his immigrant parents lived. He studied industrial engineering at nearby Lehigh University, and then got a master's degree in mechanical engineering at Princeton. After joining Ford he moved into sales, figuring it was a better place to get noticed than the short-sleeved, slide rule–toting ranks of the engineers.

After ten years of successful but anonymous toil Iacocca was an assistant sales manager in Ford's Philadelphia zone. He hit on the idea of selling Ford's new 1956 models on an installment plan for payments of $56 a month. His "$56 in '56" promotion vaulted the Philadelphia office to the top of Ford's sales charts and jump-started his career. Iacocca was promoted to Ford's new glass-and-steel corporate headquarters, the "Glass House," as it was nicknamed, in Dearborn, just outside Detroit.

He came under the tutelage of another wunderkind, Robert S. McNamara. McNamara was one of the ten financial "Whiz Kids" who had joined Ford as a team out of the air force in 1946. During World War II he and his fellow number crunchers had overhauled the Pentagon's antiquated procurement system, enabling the surge in the production of tanks, planes, and guns that had muscled America and its allies to victory.

When the Whiz Kids arrived in Dearborn, Ford's financial systems were so primitive that the company paid its bills by stacking them up, weighing them, and writing checks based on dollars per pound of paper. McNamara and his colleagues replicated their success at the Pentagon, installing modern financial cost controls and management-information systems that saved the company.

McNamara was a no-frills man. He wore rimless glasses, parted his hair down the middle, and eschewed living in Detroit's nouveau riche northwest suburbs favored by the princes of America's car kingdom. He lived instead in the more ascetic and academic environs of Ann Arbor.

His signature product at Ford was the compact Falcon, launched in 1960 and derisively termed the "Plain Mac." It had all the flair of a McDonald's hamburger. Ford managers mocked him by whispering the words "super low cost, zero personality," leaving it unclear whether they meant the Falcon or McNamara himself. Usually it was both.

But the inexpensive and reliable Falcon was a hit. Ford sold 417,174 of them during the car's first year, a record for any new car up to that time. Thanks to its success, McNamara was named president of Ford Motor on November 9, 1960, the day after Kennedy was elected president. Iacocca, in turn, succeeded McNamara as head of the Ford Division.

Iacocca was the yin to McNamara's yang. His penchant for the blunt, colorful quip led *Newsweek* to describe him as "direct as the thrust of a piston."[11] Indeed, Iacocca once explained that he had no trouble supervising men many years his senior because "all of a sudden a guy is face-to-face with the reality of his mortgage payments."[12]

When a flustered PR man once told Iacocca that a critical newspaper article had appeared adjacent to a company advertisement only by happenstance, he snapped: "Are you telling me we've been fucked by juxtaposition?"[13] His gift for colorful nonstop gab led the *Wall Street Journal* to dub him the "Motor City's most famous motor mouth."[14] Reporters who interviewed him learned that the only way to interrupt his long-winded monologues was to jump in with a new question when Iacocca paused to puff his ever-present cigar.

Surprisingly, the anal-analytic finance man and the talkative supersalesman enjoyed rapport. Iacocca admired his boss's wizardry with numbers. McNamara stood awed by his young protégé's ability to push his ideas through Ford's corporate bureaucracy. He once even directed Iacocca to put his proposals in writing instead of presenting them in person, as he found the young man's personal pleas rather too persuasive.[15] The two men, in their new roles, might have made a terrific team, or perhaps just as easily a disaster. It would be neither, as things turned out.

Shortly after McNamara and Iacocca were promoted, President-elect Kennedy offered McNamara the post of Secretary of Defense. After just weeks as Ford's president, McNamara returned to the Pentagon. Iacocca was deemed too young to become president of Ford right away, much to his dismay. But he gained sudden latitude to pursue ideas that would have caused the straitlaced McNamara to squirm.

One idea was to have Ford sponsor folk music performances on college campuses, eventually leading to the ABC television series *Hootenanny*. Iacocca himself was more attuned to Frank Sinatra than to Pete Seeger or Joan Baez, but he saw that the kids were listening to different music, and figured that sponsoring it would burnish Ford's image among youth, which he deemed critical. "Young people are trend-setters in cars, clothes, and many other commodities," Ford's marketing researchers dryly noted. "They also generate a great deal of influence in the make and kind of car their parents buy."[16]

Iacocca also resumed sponsorship of auto-racing teams to enhance the image of Ford's cars. The Big Three had "voluntarily" withdrawn from racing a few years earlier to head off potential government regulation of their marketing, but Iacocca wouldn't be cowed. "More people watch automobile racing than baseball and football put together," he said.[17] At one point he even hatched a plan to have Ford buy Ferrari. The thought of selling a national icon to an American company, even one with an Italian-American near the top, proved too much for the Italians, and the idea died. But Lee Iacocca continued to think big.

Iacocca quietly launched the project that produced the Mustang in early 1961, shortly after taking the helm at the Ford Division. Despite the Falcon's successful first year, the public's attraction to "no frills" cars was wearing thin. The U.S. car market was witnessing a "bucket seat explosion,"[18] as Ford's marketing department termed it, but most Ford cars were as ugly as buckets, never mind the seats. At times, company executives even caught flak in their own homes. The preteen children of Donald Frey, Iacocca's director of product planning, told their father that Ford's cars were "dullsville," and Frey agreed.[19]

Frey, who had a Ph.D. in metallurgy and spoke both Russian and French, was an example of Iacocca's ability to command high-powered talent (Frey would eventually leave Ford to become CEO of Bell & Howell). Another example was Sperlich, who was both high-powered and high-strung.

Sometimes Sperlich would work in his office all night and then convene crack-of-dawn meetings of his staff in the men's room, where he washed and shaved while fine-tuning that day's presentations to the bosses.[20] Iacocca once said Sperlich approached product development as if it were hand-to-hand combat.[21] Sperlich deemed it a compliment.

Sperlich's business trips to Europe made him a fan of little two-seat roadsters, the MGB, the Triumph TR4, and the Sunbeam Alpine. They were small, stylish, and fun but utterly impractical. Two-seaters accounted for just 1 percent of American car sales. That was why McNamara had ordered a backseat for the Thunderbird, which boosted its sales even though sports car purists cried foul.

The Thunderbird, however, had a price tag that topped $3,500, nearly double that of most American cars. What Iacocca wanted, he told Sperlich and Frey, was a "poor man's T-bird," a lightweight, sporty four-passenger car that could be sold, profitably, for $2,500.[22] Frey and Sperlich determined they couldn't afford the cost of developing a whole new vehicle. Doing all the engineering from scratch and buying new production machinery would have sent the cost of development out of sight, and pushed the price of the car beyond Iacocca's $2,500 ceiling.

Since Iacocca wanted a low-priced Thunderbird, Frey and Sperlich studied using the chassis and other key components of the Thunderbird as the new car's building blocks. They concluded it was impractical. The "Bird" had a heavy, expensive chassis that couldn't be adapted to produce a lighter, cheaper car. It was a puzzle without a solution until Sperlich floated another idea: putting a roadster-style body, modified to include room for a rear seat, atop the chassis of the Falcon. The Falcon's chassis was lightweight and inexpensive to build, befitting Bob McNamara's no-frills car. It also was reliable and proportionately correct, or at least close enough.

The more Sperlich studied the idea, the more convinced he became it would work, and would be the *only way* the new car could meet Iacocca's decreed price target. After drawing up the engineering specifications, Frey and Sperlich asked Ford's designers to style a new sporty top for the Falcon.

On August 16, 1962, CEO Henry Ford II, Iacocca, and a

cohort of other executives gathered in the cloistered courtyard of Ford's Styling Center to review clay models of seven different designs for the new car. A different Ford styling team had developed each design. Six of the teams had been working for months. The seventh was thrown into the fray just for insurance, as it were, and had only a couple of weeks to do its design.

That last team's leader was Joseph Oros, who had started his design career under Harley Earl at GM before moving to Ford and designing the four-door Thunderbird. His design featured a long hood created by pushing the Falcon's passenger compartment backward nine inches, and moving the front wheels forward.

The rear "deck lid" that covered the trunk was short, but still allowed an ample trunk and a fairly roomy rear seat. The two doors were long, providing a stretched appearance and allowing rear-seat passengers to get in and out easily. The side panels sported stylish, rectangular "scoops" that looked like backward renderings of the letter "C." To Iacocca, it seemed like the model was in motion, even though it was standing still.[23] Henry Ford liked it, too, and the two men declared the last-minute Oros design the winner. Taking their cue, the other executives on the scene chimed agreement.

Using the Falcon's chassis as the underpinning for the new car was "like putting falsies on grandma," quipped one of Iacocca's colleagues.[24] Which was pretty much the same as saying Ford was transforming a school librarian into a sexpot, only more concise. Before the Sixties were over, Robert McNamara would be widely reviled for sending thousands of young Americans to fight and die in Vietnam. Yet McNamara's Falcon was the basis for the car that captured America's youth culture of the 1960s more than any other car, even the Beetle, because the Mustang was so mainstream. Lee Iacocca was the father of the Mustang, but Robert McNamara was its grandfather.

The design decision was a major hurdle for the new car, but not the last. It still faced the financial "program review" by Ford's all-important Product Planning Committee. Henry II himself chaired it. Maybe Henry liked the styling, but if he didn't like the numbers the car would be killed.

Program reviews throughout Detroit were mind numbing. They consisted of dozens of detailed cost projections for every part and for the machinery to produce them, forecasts of sales volume, profit margins, and overall profits. The only thing worse than sitting through the reviews was preparing them, which in this case was left to Sperlich.

He proposed an initial investment of just $50 million, barely one-tenth the amount Ford had lost on the Edsel. The amount was so small because many components would come straight out of the Falcon, not just the chassis but also the base six-cylinder engine, the transmission, and even the heater. After months of work, Sperlich put the all the information on glass slides, the ancient ancestors of PowerPoint used for formal presentations to top management.[25]

Because he was just a junior manager Sperlich wasn't allowed in the meeting itself. He sat in the projection room instead and watched silently while his slides, with all their facts, figures, and forecasts, were presented and discussed. The meeting's atmosphere was subdued, reflecting none of the euphoria that the Mustang later would create. None of the twenty-odd executives on the committee signaled an opinion until they sensed what Henry II would say.

When the presentation was finished Henry uttered his approval, quickly and matter-of-factly. He walked out of the meeting with Iacocca just as Sperlich emerged from the projection room. Still mindful of the ill-fated Edsel, Henry was nervous. "You've got your car," he said to Iacocca, Sperlich, and Frey. "It had better work."[26]

The next round of wrangling pitted the designers against the engineers. They waged running "battles of the inch" over the car's final specifications.[27]

The car's radiator cap wouldn't fit under the designers' low hood, forcing the engineers, after some wrangling, to design a new, recessed radiator cap. The angle of the accelerator pedal gave Ford's test drivers "ankle fatigue." The engineers moved the pedal forward seven-tenths of an inch and pushed the front seat back a half inch, providing a different pedal angle. Henry Ford insisted on an additional inch of backseat legroom even though Iacocca argued that stretching the car would compromise the styling. To nobody's surprise, Henry prevailed.[28]

He proved more pliable, though, on the sensitive issue of the new car's name. Ford's advertising agency suggested hundreds of possibilities. One was Torino—the Italian spelling of Turin, Italy's car capital—to emphasize the new car's European flair. Other ideas included Bronco, Puma, Cheetah, and Colt. Henry II liked Thunderbird II, "but nobody else seemed to," he said later with evident petulance.[29]

Iacocca believed that recycling an old name would shortchange the car. Mustang was added to the list to honor the legendary P-51 Mustang fighter plane of World War II. Iacocca figured young people had few memories of the war, but he loved the name anyway. "It had the excitement of the wide-open spaces and was as American as hell," one Ford advertising executive wrote.[30] Ford took the fighter plane name but decided to put a pony logo on the car.

Amid all the progress, disaster struck. Sperlich discovered he had miscalculated some of the Mustang's costs. The car wouldn't meet the profit projections promised to the Product Planning Committee. It was a major faux pas, so serious that he didn't even tell Iacocca about it. As Sperlich wrestled with what to do, company executives decided to increase the initial production plans for the car. The additional volume increased the profit projections. The relieved Sperlich was off the hook.[31]

A much less serious but widely publicized crisis struck in mid-March 1964, just after the first Mustang rolled off the assembly line, but a month before the car's public debut. Walter Buhl Ford III, a nephew of chairman Henry Ford II, spotted a hot-red new Mustang sitting in his mother's Grosse Pointe garage and couldn't resist taking it for a spin. The twenty-year-old "Buhlie," described by *Time* as "something of a legendary cut up around Grosse Pointe," dropped the Mustang with the parking valet at Detroit's Sheraton-Cadillac hotel—just as an editor from the *Detroit Free Press* walked by.[32]

The editor summoned a staff photographer to snap a photo, breaking the strict secrecy surrounding the Mustang and scoring a scoop for the "Freep." The unintended publicity, as it turned out, whetted the public's appetite for the car, prompting speculation that Buhlie's blooper was a Ford publicity stunt. The company, however, honestly denied any such plan.

On the night of April 16, 1964, Ford blitzed the airwaves by buying the 9:30 to 10 p.m. slot (Eastern time) on all three television networks to tout the Mustang. The company also bought ads in 2,600 newspapers, including special ads for the "women's pages," as they were then known, to celebrate the Mustang's Tiffany Design Award, the first ever given to a commercial product. The official introduction came the next day, April 17, at the New York World's Fair.

"Americans will have to be deaf, dumb, and blind to avoid the name," declared *Newsweek*.[33] Archrival *Time* likened the Mustang to "a high-strung pony dancing to get started on its morning run."[34] The simultaneous and similar cover stories prompted backslapping in the Ford PR department, which had cajoled both sets of editors into heralding the new car unbeknownst to each other.

The Mustang didn't evoke praise from every publication. "The basic engine for the Mustang is the 170-cubic-inch Falcon six," noted *Car and Driver*, "a piece of machinery about as exciting as a dish of baby food." The magazine added: "The hood lands on the grille with a fit that reminds us of the lid on one of our mother's more experienced sauce-pans."[35] But the public didn't seem to mind. One dealer in Texas had fifteen willing buyers for the only Mustang he had in stock. He auctioned the car to the highest bidder, who spent the night in the car to make sure no one else got it.[36]

When a Mustang was displayed at a stock car race in Alabama, 9,000 fans surged over the retaining wall for a look, delaying the race almost an hour. In Seattle a cement-truck driver passing a Ford dealership turned to gawk at the Mustang, lost control of his truck, and crashed into the showroom. In Detroit, a man parked his vehicle in the doorway of a crowded dealership and ran in, blocking anybody else from entering. It took the dealer a half hour to find the man and make him move his car.[37]

The first Mustang that Ford built was sold not in the United States but in Newfoundland, Canada. Stanley Tucker, an airline pilot in the provincial capital of St. John's, bought the only Mustang the local Ford dealer had in stock. For weeks his was the only Mustang in the province. Children who saw him driving would wave and shout "Mustang!"[38]

That was a frequent occurrence in the States, too, where kids whose fathers brought home the first Mustang on the block would remember the day decades later. "It turned mom from a thirty-three-year-old mother into a free-spirited co-ed," recalled John Hitchcock, whose father, a Ford purchasing manager, delighted his wife with a dark-green Mustang convertible.[39] Janette Hitchcock never went back to driving station wagons. Ford targeted women like her with an ad showing an attractive young housewife loading groceries into her Mustang. The headline read, "Sweetheart of the Supermarket Set."[40]

While many of Volkswagen's Beetle ads used humor, Ford's Mustang advertising didn't, at least not on purpose. One sales brochure titled *The Woman's Angle* included eight exercises, created for Ford by Elizabeth Arden, that busy women could do while riding as a passenger in a Mustang. "Place hands in front of body and open and close them repeatedly," said one set of instructions. "Shake vigorously. Then relax hands and with palms down make a figure 8."[41] All with a straight face, presumably.

The advertising aimed at men, meanwhile, was hormonal instead of cerebral. "A car to make weak men strong, strong men invincible," declared one ad that showed a smiling man sitting in front of his Mustang. Another Mustang ad said, "Desmond traded in his Persian kitten for an heiress named Olga." Still another read: "Wolfgang used to give harpsichord recitals for a few close friends. Then he bought a Mustang . . . Being a Mustanger brought out the wolf in Wolfgang."[42] Thank goodness his name wasn't Dick.

What most surprised Ford executives was the Mustang's appeal across generations. The median age of Mustang buyers was thirty-one. But one buyer in six was between forty-five and fifty-four, many of them people like Jack Ready Sr. Ready had a tough time growing up in the Depression, and he saw heavy action in World War II as a side-gunner in big bombers flying over Germany. In

1964 he was a forty-five-year-old school principal in Westport, Connecticut, and a straitlaced veteran who still sported his military haircut. True to form, he had driven Detroit's lumbering but practical station wagons for years.

So the Ready clan was stunned when, right after the Mustang went on sale, Jack Sr. announced he was getting one. It was a convertible in British racing green with a black interior, a white top, the top-of-the-line 289-cubic-inch V8 engine, and a floor-mounted automatic transmission. No one in the family had seen such a car before. "Buying the Mustang was completely antithetical to anything he had ever done," recalled his son Jack Jr., who was twelve at the time. "It came completely out of the blue."[43] The family gave Jack Sr. an ascot tie to wear, and he went cruising in the Mustang with the ascot on and the top down. Jack Jr. felt special when he got old enough to drive the Mustang on dates. Nearly fifty years later, he still keeps the car's original purchase papers.

Some buyers were even older. Mildred Griffith of Oconomowoc, Wisconsin, a Milwaukee exurb, was sixty-three years old and many times a grandmother when she bought her bright-yellow Mustang with a black interior and floor-mounted automatic transmission in the spring of 1964. This was right about the time, as it happened, when "The Little Old Lady from Pasadena" by Jan & Dean was climbing up the hit charts with the refrain "Go, granny, go, granny, go, granny, go!"

Grandma Griffith, who looked much younger than her years, seemed to take the song to heart. Even though her car just had the base engine, she routinely zipped her Mustang past her husband's Cadillac on the Milwaukee-area freeways, much to the delight of her grandchildren. One of them, four-year-old Jack, spent endless hours sitting behind the wheel of his grandmother's parked

Mustang, pretending to drive. Three of his older cousins later inherited the car, each in turn, before it gave out.[44] The Mustang, it seemed, could traverse a new feature on the American landscape: the generation gap.

In the 1960s car companies almost always launched new models in the fall, the beginning of Detroit's new "model year." But the Mustang's debut in the spring of 1964 would create an asterisk of history, like the one for Roger Maris's home-run record in 1961. Ford designated the cars as 1965 models, but they would be known forever as the "'64-1/2s."

The car was a little heavier and longer than Iacocca had wanted. But at $2,368, well under the average new car of its day, the Mustang was priced below the $2,500 maximum Iacocca had decreed for the base model. Even that version came with floor carpeting and bucket seats, features not found on economy cars of the day. Sperlich called the freebies "Robin Hood" features. Previously reserved for wealthy buyers of luxury cars, they now were being given to the common folks on a midmarket car. But not all the upgrades were free.

The Mustang was "a new generation of Ford for a new breed of Americans," declared an early Ford TV commercial, "who want the elegance of a European touring car and, until now, had to settle for basic transportation . . . With an unexpected variety of options, Mustang is the one car that's designed to be designed by you."[45] Indeed, the extensive list of extra-cost optional equipment held the key to the Mustang's popularity and profitability alike. The car's base six-cylinder Falcon engine produced a paltry 101 horsepower. But buyers could upgrade to one of three eight-cylinder engines: a 164-horsepower V8 for an extra $116, a 210-horsepower V8 for an extra $181.70, and the 289-cubic-inch, 271-horsepower "high performance" V8 that cost an extra $437.80 and included a special sports handling package.

All the Mustang's transmissions featured floor-mounted gear shifting. The standard transmission was a three-speed stick shift. But depending on their engine, Mustang buyers could opt for an automatic transmission or a four-speed manual, with prices ranging from $115.90 to $189.60. Power brakes cost an extra $43.20, power steering an extra $86.30, air-conditioning $283.20, and a push-button AM radio $58.50. There were nearly forty optional-equipment upgrades in all, and the beauty for Ford was that these options *sold*.

The Mustang's launch came amid a booming economy. Just two months before its debut, the Dow Jones Industrial Average topped 800 for the first time ever. Tycoons such as Harold Geneen of International Telephone & Telegraph and Jimmy Ling of Ling-Temco-Vought were buying companies and creating multi-industry conglomerates, thus creating some of the Nifty Fifty stocks of the market's "Go-Go Sixties."

Americans had money to spend. Mustang buyers shelled out an average of $500 apiece on options, adding more than 21 percent to the price of each car, the equivalent of spending an extra $6,000 on a $30,000 car today. "The Mustang is a multi-car," *Automobile Quarterly* magazine, the self-styled "Connoisseur's Publication of Motoring," marveled in late 1964. "It is a Family Car suited for young marrieds with two children; a Convertible with or without an automatic top; a Hardtop; a Rally Car with a big package of heavy-duty accessories. The Mustang is a Sports Car if you accept its definition as a vehicle docile enough to drive in city traffic day in and day out but sufficiently spirited to race on weekends."[46]

Ford sold more than 263,000 Mustangs between its launch date and the end of 1964, and nearly 525,000 the next year, well beyond the company's most optimistic projections. The Mustang factory couldn't keep up. In short order, Ford dedicated a second assembly plant to producing the car, and then a third. The term "pony car" soon entered the automotive lexicon to define small, sporty, and fun-to-drive cars. The Mustang was the original.

In the fall of 1964, the official start of the 1965 model year, Ford augmented its hardtop and convertible Mustangs with a fastback version. The car featured swept-back rear glass, louvered ventilation slots, and fake bucket seats in the back. (It was a single rear seat, but looked like two bucket seats.) Carroll Shelby, a Texan who traded in his racing career to develop high-performance cars, customized the fastbacks for racing by ditching the rear seat to reduce the car's weight, lowering and stiffening the suspension system, and slapping on oversize tires and large front disc brakes. He also installed a special four-barrel carburetor and made other engine modifications.

The resulting Mustang GT-350 packed 306 horsepower and ran from 0 to 60 miles an hour in 7 seconds, 30 percent faster than the fastest ordinary Mustang. The price was $4,547, more than $2,000 above the entry-level car. Shelby Mustangs were en route to becoming car collectors' icons, like Picassos for art aficionados but louder and faster.

In 1966 Ford sold a record 549,436 Mustangs. That February, total Mustang production passed 1 million cars. And Stanley Tucker, the Canadian pilot who had bought Mustang No. 001 less than two years earlier, bought Mustang No. 1,000,001. His new Mustang was a convertible, which, as anyone who had experienced "summer" in Newfoundland knew, marked him as an optimist. Captain Tucker would later return his first Mustang to Ford as a keepsake.

The company used the occasion to issue a new raft of market-research statistics. Some 25 percent of Mustang buyers were less than twenty-five years old, Ford said, compared to only 3 percent of other Ford buyers. Women represented 42 percent of Mustang buyers, versus only 31 percent of Ford's overall buyer base. "A whopping 35 percent are single," a Ford press release stated, compared to just 9 percent of other Ford buyers.[47]

By this time the Mustang was being celebrated in music and the movies. The signature song about the car was Wilson Pickett's "Mustang Sally," which became one of the top hits of 1966. Pickett sang:

> *I bought you a brand new Mustang,*
> *A nineteen sixty-five, huh!*

In 1968 actor Steve McQueen careened one of Carroll Shelby's Mustang GT fastbacks through the streets of San Francisco in *Bullitt*, riding on the edge of disaster while pursuing two hit men in a Dodge Charger R/T Magnum. The chase scene made movie history and burnished the Mustang's mystique.

Many Mustangs from the mid-Sixties were kept by their owners, or by their owners' children, as heirlooms. In 1972 an Arkansas man named Jeff Dwire bought a used 1967 Mustang convertible. His purchase would have been of little note, except that twenty years later his stepson, Arkansas governor Bill Clinton, ran for president of the United States. On August 19, 1992, some of Clinton's friends threw a birthday party for the forty-six-year-old candidate with a Sixties theme. The man who would become America's first baby-boomer president a few months later cruised into the party in the '67 Mustang. Two years later, as president, he drove the car around the Charlotte Speedway in North Carolina at the Mustang's thirtieth anniversary celebration.

When the *Mary Tyler Moore Show* debuted on television

in 1970, the opening scene showed the fictional Mary Richards cruising the freeways of Minneapolis–St. Paul in her new white Mustang. Ms. Moore's screen persona had evolved from the ditzy housewife she had portrayed on the *Dick Van Dyke Show*. She had become an independent, if not always self-confident, career woman. Her Mustang was part of the package.

But in 1970 Ford sold fewer than 159,000 Mustangs, less than one-third of the number sold in 1966. Sales had been sliding, in fact, since 1967, because the Mustang was becoming a different car. The 1969 hardtop model had ballooned to 2,838 pounds, 16 percent heavier than the first Mustang, and enough to change the car's driving dynamics. Ford added the extra chassis weight to accommodate a bigger optional engine: a 320-horsepower brute that displaced a massive 390 cubic inches, 30 percent more than the biggest engine on the original car. Because of the extra weight and the car's expanding exterior dimensions, Mustangs equipped with smaller engines felt sluggish and lethargic. The more-muscular Mustang was better for boy racers, but worse for everybody else.

"Within a few years of its introduction, the Mustang was no longer a sleek horse," Iacocca would write later. "It was more like a fat pig."[48] Its broad appeal as a "multi-car," offering everything from hot looks at a basic price to bona fide hot-rod performance, was fading.

The bloating of the Mustang was a competitive response. By the late Sixties, Ford was getting outgunned in the horsepower war among Detroit's Big Three. The 1967 Plymouth Barracuda packed 280 horsepower, up from 180 just three years earlier, and was gaining sales. More ominously, after being caught flat-footed by the Mustang's success, General Motors rolled out a new model in 1966, the Chevrolet Camaro, that was available with a V8 that outpowered the Mustang.

GM's Pontiac Division, meanwhile, was enhancing its once-dowdy image with another high-horsepower car that could never be labeled a "sexpot" or anything that smacked of femininity. The new Pontiac wasn't a versatile, version-for-everyone automobile like the original Mustang had been. It was instead particularly suited to the latter half of the Sixties when almost everything in America—music, movies, the civil rights movement, campus life—took on a harsher, tougher edge. The new Pontiac seemed to snarl instead of smile. Instead of a "pony car" it was a "muscle car," as enthusiasts termed it, built to go faster and to growl louder than anything else on the road. It was unabashedly rebellious, like the man who created it.

7

THE BRIEF BUT GLORIOUS REIGN OF
JOHN Z. DELOREAN AND THE PONTIAC GTO

This car was built in honor of Almighty God, in memory of my
dad, and of my fellow hometown veterans who did not have the
chance to live these memories.

—Sign on a restored Pontiac GTO

The parking lot is full at the stadium of the Kane County Cou-
gars, a minor-league baseball team in Geneva, Illinois, even though
there isn't a game. The Cruisin' Tigers are here. They're gathering
for their annual Indian Uprising convention, an allusion to Chief
Pontiac, an eighteenth-century leader of the Ottawa tribe. Tigers
and Indians might seem like mixed images, but not to the Cruisin'
Tigers. The chief was the namesake for Pontiac cars, and Pontiac
used tigers to tout its most famous model, the GTO. The Cruisin'
Tigers are among the largest local chapters of the GTO Association
of America.

This Indian Uprising occurs as the first decade of the twenty-
first century is nearing an end, but the Tigers are more focused on
the mid-twentieth century. That's when the Pontiac GTO outran

almost anything on America's roads, thanks to horsepower close to a Corvette's but at a much lower price. Nostalgia hangs heavily here in suburban Chicago's humid August air. The sound of Randy & the Rainbows circa 1963 blares over the loudspeaker:

> *Oh, Denise, shooby doo,*
> *I'm in love with you, Denise, shooby doo . . .*

Many of the 200 or so GTOs on display have stuffed-tiger toys sitting on the engines or the roofs, and others have tiger tails (fake, one presumes) dangling outside the hood. The tiger accoutrements are inspired by the GTO's original tiger-themed advertising campaign: "For the man who wouldn't mind riding a tiger if someone'd only put wheels on it," one early ad for the car declared.[1]

The theme grew from there, sometimes to the point of absurdity. In 1966 Pontiac's advertising agency tried to film a real tiger behind the wheel of a GTO. But after being lured into the car with forty pounds of red meat the frightened animal went berserk, chewing up the dashboard and the steering wheel, and clawing long gashes in the seats.[2] The king of the jungle, alas, didn't want to be king of the road.

Just as memorable, though in a different way, was a magazine ad showing a woman in a tiger-print bikini leaning over the trunk of a GTO . . . with a tiger tail dangling from the bikini's bottom.[3] GM's senior management was not amused. The GTO's tiger-themed advertising was soon tamed.

The tiger advertising was the brainchild of a young Detroit ad-agency executive named Jim Wangers. And here at the Indian Uprising, more than forty years after his tiger ads first appeared, Wangers is hanging out with the Cruisin' Tigers and basking in their collective adulation. At age eighty-one he has just written a new book titled, with an advertising man's subtlety, *Pontiac PIZAZZ!*. People here line up to meet him, waiting to hear words of wisdom from the guru. "Have you read my new book?" seems to be his mantra.

The license plates on some of the GTOs are like message boards: "HE GONE," "GRRRR!," "KEITHS 68," and so on. The plates aren't the only prized features. A Cruisin' Tiger named Bill Nawrot points to his 1972 GTO's "side splitters," exhaust pipes angled out the side of the car instead of pointed out the back like normal exhausts. "The only year the side splitters were standard was 1972," explains the friendly Nawrot, a fifty-eight-year-old phone company employee, like a sommelier explaining the unique attributes of a 1982 Bordeaux.

In other years the "side splitters" had been extra-cost options, like the heavy-duty "metallic" brake package and the prized "Tri-Power" carburetor, which consisted of three two-barrel carburetors lined up over the engine instead of the standard four-barrel carb. More barrels meant more horsepower, not to mention more profits for Pontiac. The GTO's optional-equipment list was "as long as your arm and twice as hairy," Pontiac boasted.[4]

The mystique and the throaty roar of the GTO still resonate with the Cruisin' Tigers, mostly men in their late fifties and early sixties, some paunchy, others gray-haired, others nearly bald. Their vintage GTOs are like potent elixirs that vividly evoke the flat-bellied, full-haired, testosterone-driven days of their past. The Cruisin' Tigers haven't discovered the fountain of youth. But they have discovered something almost as good: the fountain of makes-you-feel-like-youth.

The GTO also marked the career zenith of one of Detroit's most charismatic yet enigmatic figures, John Z. DeLorean, who gained fame for his gull-winged sports cars, the starlets he escorted on both arms, and his ignominious end. But before all that, in the late 1960s the six-foot-five, ruggedly handsome DeLorean cruised the corridors of General Motors, the pillar of America's corporate establishment, with an air of defiance.

His signature car, the GTO, was all about rebellion. Not the antimaterialist, aesthetic, hippie sort of rebellion whose converts were driving around in Volkswagens. Hard-core GTO drivers didn't want to raise anybody's consciousness. They just wanted to raise hell. The GTO was launched in 1964, the same year as the Mustang, but the two cars captured different dynamics in a uniquely turbulent decade. Just as the early Sixties were about the Beatles, civil rights, and the Mustang, the late Sixties were about the Rolling Stones, race riots, and the GTO.

Events that seemed unthinkable just a few years earlier started happening all the time. In April 1966 *Time* asked ominously on its cover, "Is God Dead?" Four months later an ex-marine named Charles Whitman climbed the bell tower at the University of Texas at Austin, killing thirteen people and wounding thirty-one with high-powered rifles before himself being shot by police. That same year the Rolling Stones released "19th Nervous Breakdown," a song about alienated youth that seemed to fit the times.

The next year, 1967, brought the Summer of Love to San Francisco, but not to Detroit. The U.S. Army was dispatched to quell a race riot in which forty-three people died, and the dark smoke billowing over the city inspired Gordon Lightfoot to write a ballad called "Black Day in July." Warren Beatty and Faye Dunaway romanticized rebellion in *Bonnie and Clyde*. Nineteen sixty-eight brought the horrific assassinations of Martin Luther King and Bobby Kennedy. Police and protesters waged war outside the Democratic National Convention in Chicago.

In this atmosphere, setting young kids loose on the streets

in outrageously overpowered cars qualified as good, clean fun. Automotive-emissions regulations didn't exist, but they were coming. Cheap gasoline did exist, but it was going. People were breaking rules, rejecting norms, pushing limits, and embracing excess. As it happened, those were the same things that John De-Lorean did to create the Pontiac GTO.

Like Lee Iacocca, John DeLorean was the son of immigrants. His father, a factory worker at Ford, was hard-drinking and abusive. Eventually he left his wife, who raised their four sons alone. John excelled in music as a child, and won a scholarship to the Lawrence Institute of Technology in suburban Detroit to study engineering. In 1950, after getting his degree, he started working at Chrysler, but soon moved to Packard Motor for what seemed a better opportunity. It wasn't. By 1956 Packard was in financial distress, due in part to its ill-timed merger with ailing Studebaker two years earlier, and shut its doors on August 15. A month later DeLorean landed an engineering job at GM's Pontiac Division. At age thirty-one, he had left two automotive also-rans and had graduated to the world's leading car company.

DeLorean was restless and inventive. He developed more than a dozen patents and innovations, including turn signals that could be used when drivers just changed lanes as opposed to turning corners. In 1961 he became Pontiac's chief engineer at just thirty-six years old, unusually young for such a post in Detroit at the time. But young leadership was just what Pontiac needed.

For years Pontiac had been regarded as an "old man's" car. But Pontiac had transformed its once-stodgy reputation by embracing stock car racing, enjoying success until General Motors banned corporate support of racing. The bosses at headquarters worried that supporting racing might tag the company as promoting reckless driving, and prompt unwanted government regulations. The racing ban presented a special blow to Pontiac.

Unlike the other GM divisions—Chevrolet, Oldsmobile, Buick, and Cadillac—Pontiac had built its entire image around racing and performance. Pontiac's "Wide Track" marketing campaign promoted cars with a low-slung and assertive look that, in truth, did more to enhance sales than to improve performance. Without racing, Pontiac executives feared, the division would lose the image that had attracted customers to Pontiac instead of Chevy or Ford.

In the early 1960s DeLorean hosted Saturday-morning hot-rod sessions for Pontiac's engineers at GM's Proving Ground in Milford, Michigan, forty miles from Detroit. The engineers fueled themselves with hot coffee and fueled their cars with high-test gasoline. Decades later corporate consultants would call this sort of pursuit a "team-building exercise," a term that would have made DeLorean and his colleagues gag. They wanted to build fast cars, not touchy-feely teams.

The Saturday test-track regulars included a couple of DeLorean's key engineers, a chassis expert named Bill Collins and an engine guru, Russ Gee. On one morning, in the spring of 1963, the group put a prototype of the compact 1964 Pontiac Tempest coupe on a lift. Collins and Gee were looking at it from underneath. The Tempest had a 326-cubic-inch engine, big for a car of its size. But it occurred to Collins that Pontiac's 389-cubic-inch engine used on the full-sized Bonneville model might slide easily into the Tempest. The bigger engine would require heavier front-end springs, he mused, but the Tempest's engine mounts would suffice.[5]

Putting a big engine in a small car wasn't an original idea. Moonshiners in Appalachia did it in the 1930s to outrun revenuers. More recently, hot-rodders in Southern California had been doing it to have fun. Car companies had dabbled with the idea, but not seriously. The Chevrolet Corvette was an exception, but it was a premium-priced two-seater with a heavy-duty sports car suspension, too expensive and too impractical for everyday buyers.

A week later, at the next Saturday-morning session, Collins and Gee put the 389 in the Tempest. DeLorean took it out on the track. He later described the sensation of driving the lightweight car with a monstrous engine as "electrifying."[6] Over the next few weeks DeLorean and his colleagues kept tinkering, tuning the car's suspension system, adding high-performance tires, and making other adjustments. The Pontiac boys were having fun. And they were developing a car that could burnish Pontiac's high-performance reputation despite the corporate ban on racing.

There was, however, a major bump in this road. General Motors had a corporate rule against building cars with more than one cubic inch of engine displacement for every ten pounds of vehicle weight. The Corvette, with its sports-car suspension, got some leeway from this safety rule, adopted by the company to make sure no engine would be too powerful for a car to handle. The Tempest, however, weighed only about 3,400 pounds, making the 389 engine nearly fifty cubic inches too big. It might fit into a Tempest but it wouldn't fit into GM's rules, which the company's powerful Engineering Policy Committee enforced. Even the chief engineer of Pontiac couldn't ignore it.

DeLorean wasn't yet a high-living, jet-setting legend. Pictures from GM corporate events in the early Sixties show him dressed in the loose-fitting dark suits and narrow, dark ties that gave GM executives sartorial parity with the Soviet Politburo. But underneath the conservative corporate facade, DeLorean's inner rebel was emerging.

After studying the Engineering Policy Committee's rules,

DeLorean found a loophole. The committee only reviewed the specifications for *new* cars. Its mandate didn't cover optional equipment on *existing* cars. He proposed offering the 389 in an extra-cost optional-equipment package on the high-end version of the existing Pontiac Tempest, which was dubbed the Tempest LeMans. DeLorean had found a way to use the bureaucratic rules that defined the Engineering Policy Committee's purview to keep the corporate bureaucracy off his back. It was a brilliant piece of corporate jujitsu.

It wasn't as brazen, however, compared to what DeLorean did after that. He needed a name for his new option package to signal its high performance. And there wasn't any automotive brand that shouted "high performance" more than Ferrari. At the time the Italian company was making a limited-production model called the 250 GTO. But the sales were so small that Ferrari had neglected to lock up the legal rights to the name. DeLorean had yet another loophole to exploit. He decided to call his car the Pontiac Tempest LeMans with the GTO package. It was wordy, but it was catchy, too.

The initials GTO "stand for Gran Turismo Omologato, which roughly translated means a grand touring car ready to go," explained a Pontiac publicist named Solon E. Phinney.[7] More literally the name meant a grand touring car that had been homologated, or sanctioned, for racing.

"The sporty-car, purist crowd is so offended by the pirating of a Ferrari name that they can do little but stamp their collective feet and utter lady-like oaths," reported *Car and Driver.*[8] It's a safe bet, however, that most young hot-rodders cared little about the controversy. Many didn't even know what GTO really stood for. That didn't matter, however. Before long American kids, playing on those initials, would nickname the car the "Goat."

DeLorean still faced opposition from Pontiac's sales executives, who figured that whatever the car was named, dealers wouldn't want a teenage hot-rod on the showroom floors. But DeLorean's boss, Pontiac general manager Elliott "Pete" Estes, assuaged the sales guys by asking them to sell just 5,000 of the GTO-equipped cars. Dealers snapped them up within days, found them easy to sell, and ordered more. The Pontiac sales people were surprised and delighted.

When GM's corporate mandarins learned what had happened they were annoyed. But DeLorean's car had passed the point of no return. Canceling the car would rile Pontiac dealers, something to avoid. Besides, selling a souped-up Tempest would be profitable. Pontiac priced the GTO option at $295, on top of the car's $2,491 base price, but GM didn't have to bear the costs of retooling a factory for a totally new car.

It was remarkable. All this planning and finagling had occurred between the spring and fall of 1963, so the "GTO option" car would be ready to launch as a 1964 model. John DeLorean liked to go fast, on the test track and in the office. During those six months he had broken the rules and flouted authority, presaging the behavior that would lead him to destruction in later years. But against all odds, his car had made it to market. And ironically, by pulling Pontiac out of organized auto-racing but then acquiescing to the GTO, GM's bosses had stopped promoting racing on tracks . . . and started promoting it on America's streets.

The 389-cubic-inch V8 produced a monstrous 325 horsepower, compared to just 140 horsepower for the six-cylinder engine in the standard Tempest, even though both cars shared the same body. The Tempest LeMans with the GTO option, as a result, was a wolf in sheep's clothing. Ken Crocie, a young hot-rodder from Kendall

Park, New Jersey, who bought his car with $200 down, was happily surprised when his first insurance bill arrived. Crocie got the standard 10 percent discount for compact cars, as did most other early GTO buyers.[9]

America's insurance companies didn't know they were insuring hot rods instead of meek compact cars. And it would take them a few years to catch on, partly because Pontiac initially kept a low profile for the car. Pontiac's marketing maestros feared the corporate higher-ups might get cold feet. They confined GTO advertising to auto-enthusiast publications, where the extra horsepower would resonate with potential customers without drawing unwanted attention. Pontiac's product catalog for 1964 didn't even mention the GTO option, and a brochure for dealers carried the muted headline: "Performance Option for LeMans Sports Coupe and Convertible."[10]

But the low profile wouldn't last. Wangers, eager to promote the car (he was an adman, after all), approached the editors of *Car and Driver* with an idea. He suggested pitting Pontiac's GTO against the Ferrari GTO on a track, figuring that whatever happened, the mere mention of Pontiac in the same article as Ferrari would be a no-lose proposition. The magazine's editors loved the idea. In late 1963 they rented the Daytona International Speedway for the week between Christmas and New Year's to conduct their test.

Wangers shipped two Tempest GTOs down to Daytona, but unbeknownst to the editors, one of the cars was a ringer. Wangers had replaced its regular 389 engine with a bigger and more powerful 421-cubic-inch engine, which wasn't even available on the car. The adman wanted to rig the race. It turned out it wasn't necessary.

The March 1964 issue of *Car and Driver* carried a dramatic

cover touting the comparison between the "Tempest GTO" and the Ferrari GTO. But inside the magazine, the editors confessed that they really hadn't obtained a Ferrari for the test. They simply made assumptions about what the Ferrari *would have* done. After clocking a Wangers GTO-on-steroids accelerating from 0 to 60 miles an hour in a neck-snapping 4.6 seconds, the magazine passed judgment. "The Pontiac will beat the Ferrari in a drag race . . . ," the editors wrote. "Pontiac, God love 'em, went the hairy-chested route . . ."[11] The GTO, it seemed, invited references to body hair. The whole comparison was blatantly bogus, but that didn't matter. The publicity impact for Pontiac (and the magazine) was enormous.

Around the same time, the GTO received another PR boost from, of all things, a car-crazy high-school kid. His name was John "Bucky" Wilkin. He lived in Nashville, where his mother, Marijohn, was a songwriter. One afternoon, sitting in physics class, he sketched a song about the laws of motion. Well, at least the motion of the hot new Pontiac. His mother liked the lyrics and used her music-industry connections to set up a recording session with Bucky and a group of freelance "session musicians." Afterward the group was named Ronny and the Daytonas and the record was released.

Many people assumed the song came from the Beach Boys and that its title was "Little G.T.O.," though it was simply "G.T.O." Misconceptions or not, the song jumped to No. 4 on the hit charts, thanks to its catchy tune and clever lyrics:

> *Little GTO, you're really lookin' fine*
> *Three deuces and a four-speed, and a 389*

The "three deuces" were the two-barrel carburetors, which cost an extra $92. They created a rich gasoline-to-air mixture that boosted the GTO to 348 horsepower from the standard 325. The "four-speed," an extra $188, was the floor-mounted, four-speed manual transmission. The "389," of course, was the engine. The lyrics also extolled the car's tachometer, a display-needle that tracked the engine's revolutions per minute.

If the song seemed like the perfect sales pitch, it was thanks in no small part to the irrepressible Wangers. Before he recorded the song, the young Wilkin had called Pontiac for advice about the lyrics and had been referred to the adman, who suggested a few tweaks. "What we had was a 2-minute and 20-second commercial for our new Pontiac," Wangers would later write.[12]

Indeed, the last verse included the line "Gonna save all my money, and buy a GTO," which was an outright invitation to buy the car. Pontiac sold 32,450 Tempest GTOs in 1964, far outstripping the initial target of 5,000 cars that the sales people had resisted.

That was just the beginning. The next year sales more than doubled to 75,352 cars. The new "vertically stacked" dual headlights and continuing buzz in the motor press fueled the sales momentum. "Ferocious GTO," declared a headline in *Motor Trend*, whose editors advised ordering heavy-duty brakes because "a car of the GTO's potential could use extra stopping power."[13]

In July 1965 DeLorean, at age forty, moved up to the top job at Pontiac, succeeding Estes. The GTO's unexpected success made him the natural choice. DeLorean seemed attuned to what young people wanted, and had a knack for doing things that made a car feel special. A month after he took the helm, Pontiac released a "cold air induction kit," sold under the catchy name Ram Air, that dealers or owners could install. The kit transformed the distinctive air-intake scoops on the GTO's hood from ornamental items to functional devices that sucked additional air into the engine and gave the GTO a little extra kick.

In September, with Pontiac preparing to launch its 1966-model cars, DeLorean dropped the pretense of having a GTO option on the Tempest LeMans and named the car, officially, the GTO. Pontiac dressed a man up in a tiger suit, dubbed him the "Mystery Tiger," and had him drive a car called the "GeeTO Tiger" at drag strips around the country. GTO sales for 1966 hit yet another high of nearly 100,000 cars.

The number was less than 20 percent of the Mustang's sales that year, but that was beside the point. The GTO wasn't for "the sweetheart of the supermarket set." It was for speed freaks. Besides selling well, it boosted Pontiac's image and solidified its position as the third-largest automotive nameplate in America, after Chevrolet, GM's flagship, and crosstown rival Ford.

"Our growth was coming primarily out of the hide of Chevrolet, and this had Chevy dealers in a dither," DeLorean would boast later, like a boy who had beaten his big brother.[14] But DeLorean couldn't afford to gloat too much. He faced new challenges, courtesy of the stuffed shirts at corporate headquarters. Late that summer, as the 1966 model year wound down, the GM board of directors met in the company's New York office on 5th Avenue and 59th Street. Just off the first-floor lobby was a large display room where each GM brand took turns showing its cars. That month it was Pontiac's turn, and chairman Jim Roche, the same man who had apologized publicly to Ralph Nader, walked into a display room that had been converted into a veritable jungle.

"There were growling sound effects, jungle music, tiger tails, tiger rugs, tiger heads everywhere . . . ," as Wangers later described the scene.[15] The conservative and reserved Roche, a financial man with no marketing background, was shocked. He did some roaring himself. He passed the word down the line to DeLorean: kill the GTO's tiger advertising, immediately. That wasn't all. Next, headquarters told Pontiac to drop the "three deuces" Tri-Power carburetor from the 1967 GTOs. In the wake of the Corvair disaster, Roche wanted to position GM on the side of safety.

DeLorean's engineers quickly retuned the GTO's four-barrel carburetor to match the Tri-Power's horsepower. But a clear conundrum was developing. The GTO was popular because it was the meanest street-legal car in America, with an image to match. But DeLorean had to downplay that image to mollify his cautious corporate bosses. It was time for a softer approach.

With the tiger banished, the next advertising campaign used a slightly dyslexic play on the GTO's initials, and labeled the car "The Great One." A sales brochure for the 1967 model gushed: "Enough of the [GTO's] essence may be captured in words to create within the heart of the initiated an undying devotion to The Great One."[16] Pontiac took out splashy, two-page color ads in newspapers showing the GTO's extra-cost options: a reclining front passenger seat, fake-wood steering wheel, and a tachometer mounted on the car's hood (instead of the dashboard) for onlookers to admire. The caption: "Now you know what makes The Great One great."[17]

By this time, however, other car companies knew, too, or at least were catching on. Pontiac's competitors, not just Ford and Dodge but also GM's Chevrolet and Oldsmobile divisions, had been surprised by the GTO's success. But they were rushing to get in on the muscle car act.

The first serious competition surfaced in 1966 when Dodge made a 425-horsepower engine available on two of its models, the Charger and the Coronet. The engine was the Street Hemi, named for the hemispherical shape on the ends of its eight combustion cylinders. The Street Hemi was so brutish that Dodge sold only 468 of them in 1966, though it sold 37,344 Chargers overall. But Dodge had outmuscled the GTO. Chevrolet, meanwhile, launched the 1967 Chevelle SS (for Super Sport) 396 with a "high-compression" engine that cranked out 375 horsepower, slightly more than the GTO's 360. The Mustang went muscular with an optional "big-block" V8, with 320 horsepower. Chevy responded to the Mustang with the sleek and sporty Camaro, available with the SS 396 engine. This was war, Detroit-style.

Muscle cars produced muscular profits, with prices $1,000 or more higher than basic sedans and coupes, due to features that ranged from the functional (heavy-duty suspensions and brakes) to the cosmetic (rally wheels and racing stripes). Other competitors joined the fray, including the Plymouth Barracuda and the Oldsmobile 442. By 1968 even lowly American Motors, where George Romney had popularized the compact car a decade earlier, responded with the two-seat, 315-horsepower AMX fastback.

By the late 1960s, the muscle car proliferation was producing a boom in street racing. Races often were spontaneous affairs in which cruising kids spotted potential rivals, exchanged knowing glances, met up at a stoplight, and zoomed into the night with tires squealing, rubber burning, and exhaust pipes roaring. It was happening everywhere.

In Greenview, Illinois, a farm town of fewer than 1,000 souls near the state capital of Springfield, the local racing strip was a deserted country road where the county sheriff was unlikely to venture. In New Brunswick, New Jersey, home to Rutgers University, the strip was Livingston Avenue, where the stoplights marked convenient starting points for late-night showdowns.

In Chicago, Fulton Street and Clybourn Avenue were both

wide streets that ran through industrial parks with little late-night traffic, making them perfect drag strips. When the races got out of hand, the city's fire department hosed down both streets on Friday and Saturday nights to make them too wet for racing.

Ground zero for street racing, though, was Detroit's Woodward Avenue. The street began downtown and sliced on a straight, diagonal line through city neighborhoods and then increasingly affluent suburbs before ending in the city of Pontiac, where GM's Pontiac Division was headquartered. It had four lanes in each direction, which was handy because some of the races were four cars across, starting at a stoplight and ending when the winner pulled ahead of the others. Regular racers had noms de guerre, such as Peanuts, Cheater, or Stripes. Sometimes money was at stake, other times girls. Night after night, "Woodward was like a social affair," George Poynter, a Woodward regular as a teen, recalled decades later. He bought the first of his four GTOs in 1964 from a buddy shipping out to Vietnam, and met his first wife while cruising Woodward in one of them.

Just off Woodward, the Royal Pontiac dealership specialized in modifying GTOs to hit nearly 110 miles an hour. As Poynter put it, "you could feel the sensation in the seat of your pants." Or maybe in the front of your pants. A few years later, in 1973, such scenes would be memorialized in the movie *American Graffiti*.[18]

The median age of GTO buyers was under twenty-six, well below the median age of forty-three for all new-car buyers.[19] But it wasn't just kids who were racing. The street sport also attracted some adults. In 1967, Wangers got word that a well-known local racer had switched to a Plymouth Street Hemi GTX, and was itching to race a GTO. The challenge couldn't go unmet. He enlisted a former Mystery Tiger driver to drive a GTO modified by Royal Pontiac.

"The night of the race a huge crowd had assembled, at least two hundred cars," Wangers later wrote. "The Hemi GTX was a jet-black coupe, and was parked right in the center of the crowd. We pulled in alongside of them with our white 'Goat.'" The contestants and the crowd then cruised out to a lightly traveled stretch of the Edsel Ford Freeway east of Detroit. After the "Go" signal, the GTO "literally ran away from the Hemi . . ."[20] Predictably, the killjoys at GM headquarters frowned on "victories" of this sort. Later that year they issued more advertising strictures, banning any suggestion of aggressive driving. But DeLorean and Wangers wouldn't be deterred.

They responded with a two-page full-color ad showing a dark-green 1968 GTO with two young men inside parked behind a Woodward Avenue road sign. "The Great One by Pontiac," declared the caption. "You know the rest of the story."[21] The ad implied that the GTO was waiting to wipe out some sucker in a Chevelle SS, an Olds 442, or a Plymouth Barracuda. The suits at headquarters killed the ad.

A few months later Pontiac rented a billboard on Woodward, showing a GTO under the headline: "To Woodward With Love From Pontiac." Local town officials complained that even that modest statement promoted street racing, at least subliminally. The corporate axe fell again.

By now DeLorean was causing consternation around the company for more than his advertising tactics. He was sporting hair down to his collar, double-breasted suits and, horror of horrors, blue shirts. GM's view of shirts was sort of like Henry Ford's attitude toward cars: employees could wear any color they wanted as long as they wore white. DeLorean wasn't exactly a hippie, but he worked hard to be hip (not that it took much effort) amid GM's cloistered, country-club, corporate conformity. But many people found working for him energizing, and followed him as though he were the pied piper. Besides, he also produced results.

In 1968 the GTO got restyled with new, curvy lines and was named *Motor Trend*'s Car of the Year. The new model came with a rubber-composite front bumper painted the same color as the car, instead of a traditional chrome bumper that was easily damaged, even by minor bumps. Pontiac promoted it with TV commercials that showed a man taking a hammer to the bumper, which re-mained unscathed. This time, nobody at headquarters complained.

The GTO's real problem, though, was that the competition was getting tougher. In 1968 Plymouth launched a new muscle car called the Road Runner, named for the cartoon character that always outran and outwitted Wile E. Coyote. The car's horn was even tuned to mimic the bird's trademark "beep beep" call. The Road Runner was mostly gimmickry, except it had a 383-cubic-inch V8, almost as big as the GTO's. Its base price was $2,800, sev-eral hundred dollars less than the GTO. The car magazines fawned over the new Plymouth. Its sales accelerated like its namesake.

Pontiac's response was schizophrenic. DeLorean created a committee to recommend how to price the GTO below the Road Runner. The suggestions included substituting a bench seat for the front bucket seats and using a smaller engine, 350 cubic inches instead of the GTO's standard 400. DeLorean was irate. "It's a 400-cubic-inch world," he snapped.[22]

Pontiac lurched in the opposite direction, and the result was a new, more expensive version of the GTO. Its garish hot-orange color was matched by brightly colored decals and a rear-end spoiler wide enough that, as some automotive writers noted, one could fry an egg on its surface. The new GTO's name, decreed by DeLorean himself, was the Judge. It came from the punch line to a recurring skit on television's *Laugh-In* comedy show: "Order in de court, order in de court. Here come de Judge." Pontiac's advertising declared: "The Judge can be bought."

In actual fact, though, the Judge wasn't being bought. Pontiac dealers sold just over 6,800 Judges in 1969, out of nearly 73,000 GTOs. That number, in turn, was well below the 84,000 Road Runners sold that year. "The Great One—the Pontiac GTO—is still kingpin," Pontiac's advertising boasted. But the sales numbers said otherwise.

It was a problem, however, with which DeLorean wouldn't have to deal. In mid-1969 General Motors announced it was promoting the forty-four-year-old DeLorean to run Chevrolet, its largest division and the traditional stepping-stone to the presidency of the corporation. DeLorean's mandate was to reverse a multiyear Chevy sales slump that had been caused, in no small part, by his own success at Pontiac.

To mark DeLorean's departure, an artist who had designed the decals for the Judge did a full-color caricature sketch that summed up the DeLorean years at Pontiac, and indeed the last few years in the country at large. It showed a Judge sitting between two hostile groups: the Hippies and the Establishment. DeLorean, clad in a black turtleneck and beads, was one of the Hippies, who were bedecked with flowers, peace symbols, and various arrangements of facial hair. The Establishment was led by a dour-looking Jim Roche, wearing a striped suit and sticking out his tongue toward DeLorean in a classic Bronx cheer.[23]

DeLorean reveled in his hip, rebellious image. He divorced his wife of fifteen years, Betty, and in the late summer of 1969 strolled into Chevrolet's new-model press preview with his new wife. She was California-blond Kelly Harmon, the daughter of football legend Tom Harmon and less than half DeLorean's age. The new Mrs. De-Lorean got more attention from the journalists than the new Chevys.

DeLorean overhauled his wardrobe with tight-fitting Italian suits, flared pants, and wide, brightly colored ties. "He'd sit in a meeting brushing his hair, dressed like a floozy and calling us the Establishment," lamented Bill Mitchell, the design chief who had once called Picasso a "queer." [24]

DeLorean also started lifting weights, shed excess pounds, and had plastic surgery to sharpen his jawline. When he got the facelift, he went AWOL for a couple weeks, and told colleagues he needed the surgery to repair injuries he had suffered in a driving accident. His explanation drew snickers on GM's jungle-drum gossip grapevine. But the DeLorean party was just winding up, even as the muscle car craze and the assorted other crazes of the late 1960s were winding down.

"The excitement of the 1960s vaporized in the '70s," DeLorean wrote years later, "and with it the innovative leadership and electrifying youth image of the Pontiac cars." [25] It was tantamount to saying, "Without me at the helm, Pontiac went brain dead." It was partly true, but then there was another side to that story.

Car insurance companies were starting to use mainframe computers to track accident records. They finally figured out that young muscle car drivers were cracking up their cars like a cook cracking eggs. Out went compact-car discounts, to be replaced by high-horsepower surcharges. By 1969 GTO owners in some cities had to pay $1,000 a year to insure their $3,600 cars. Even worse, the high-performance Pontiac persona DeLorean had created with the GTO was passing from hip to has-been.

The Woodstock music festival that August marked the high point of the Sixties. In December, the last month of the decade, the Rolling Stones held a concert at the Altamont Speedway in northern California. As the crowd got unruly a young black man started to charge the stage, where the security force, a gang of local Hells Angels, intercepted him. In the ensuing scuffle the young man was stabbed to death.

The tragic incident presaged other events of 1970. That spring protests against the Vietnam War took a deadly turn on college campuses, and Ohio National Guardsmen shot and killed four students at Kent State University. That summer student radicals blew up the Army Mathematics Research Center at the University of Wisconsin in Madison, killing a young scientist. That fall Janis Joplin became the latest, but not the last, rock star to die of a drug overdose.

The GTO was about rebellion when rebellion was about harmless hijinks. But it was getting difficult to separate the sight of kids spinning out in their cars from the spectacle of a country that seemed out of control. Amid all the craziness—Hells Angels providing "security," National Guardsmen killing kids, and students blowing up buildings—signs of retreat from rebellion emerged.

In the spring of 1970, just before they broke up, the Beatles released *Let It Be.* The album's title song was a hymn to finding comfort amid chaos. Later that year George Harrison's first solo album, *All Things Must Pass,* featured an overtly spiritual song, "My Sweet Lord," which became a surprise hit. The next year's most popular movies included *The French Connection* and *Dirty Harry,* in which cops were cast as heroes. Clint Eastwood's Detective Harry Callahan—"I know what you're thinking, punk"—oozed scorn for the self-indulgence of the Sixties. Crowds loved the movie, even if critics didn't.

The GTO might have survived this backlash. But the car seemed sacrilegious to the new secular religion of environmentalism, which sprouted in the Sixties and flowered on April 22,

1970, the first Earth Day. Devotion to clean air didn't much square with the GTO's guttural exhaust pipes and fuel economy of 11 miles a gallon. In the late 1960s GTO drivers had been viewed, at worst, as a bunch of harmless, hormonal kids. A few years later they were becoming rebels without a clue. GTO sales plunged nearly 45 percent in 1970, to just over 40,000 cars. Worse was to come.

The environmental movement's first big triumph was outlawing lead in gasoline, with a big assist from Ed Cole's catalytic converter. There was a phase-in period, with low-lead gasoline allowed for a few years before unleaded gas was required. But the path was clear.

Less lead, however, meant less horsepower. Not until the mid-1980s would automotive engineers figure out how to make high-horsepower engines run on lead-free gas. Meanwhile, Detroit scrambled to convince car buyers that muscle cars still had muscle. "Now you might think these engines would perform like turtles on the low-lead gas they use," teased a Pontiac sales brochure for the 1971 GTO. "Quite the contrary."[26] The truth was different. The GTO's most powerful engine produced only 310 horsepower in 1971, down from the 360 of prior years. That year GTO sales tumbled by another 75 percent, to just 10,500 cars.

By 1973 the GTO only mustered 250 horsepower. Pontiac dropped the car as a distinct model, and returned the GTO to its original status as an optional-equipment package on the LeMans. There wasn't much purpose in a muscle car without muscle, and a mere 4,800 GTO-equipped LeManses sold that year.

That fall the Arab oil embargo produced America's first oil crisis. For a while gasoline was so scarce that it was rationed. In many cities cars with license plates ending in even numbers could only buy gas on Tuesdays, Thursdays, and Saturdays, while odd-numbered plates had the other days. Another fuel-economy measure, the 55-mile-an-hour speed limit, derisively dubbed the "double nickel," kicked in.

It was at this point that Bill Collins, the engineer who had first proposed putting a big engine in the little Pontiac Tempest, suggested in a Pontiac management meeting that the GTO be discontinued. "What's the point?" he asked.[27] Collins didn't get any argument. The 1974 GTO would be the last one for thirty years.

Over its eleven-year run Pontiac sold just 514,793 GTOs, fewer than the Mustang's sales in 1965 or 1966 alone. But the car's influence as the original muscle car far outweighed its numbers. The GTO streaked through America's automotive and cultural atmosphere like a meteor, only to crash and burn when the atmosphere changed. John DeLorean would crash and burn, too, but that would take longer.

DeLorean's marriage to Kelly Harmon lasted three years. As the two became estranged, DeLorean started dating Ursula Andress, Raquel Welch, and other Hollywood starlets. He appeared in gossip tabloids as often as car magazines. On Thursday nights he would commandeer a General Motors jet from Detroit to Los Angeles, where a GM junior executive would meet him with keys to a company car and to a hotel room in Beverly Hills or Bel Air. He would party through the weekend and fly back to Detroit Monday night, showing up in the office on Tuesday morning. On Thursday nights it was back out to Hollywood again.[28]

His bosses tolerated this flight pattern because DeLorean still produced results. He eliminated layers of management, reorganized engineering to designate individual project managers responsible for each of Chevy's models, slashed inventory, and installed computerized financial controls. On September 18, 1971,

BusinessWeek touted him on its cover with the headline: "John Z. DeLorean: A Swinger Tries to Cure Chevy's Ills."

In 1972, under DeLorean's leadership Chevrolet became the first automotive nameplate on earth to sell more than 3 million vehicles in a single year. It was a major milestone, and in October of that year DeLorean was promoted yet again: to group vice president in charge of GM's entire car and truck business. The rebel who had defied corporate headquarters was now a headquarters man himself.

But now instead of developing cars, he spent his days listening to presentations with groups of old men whom he regarded as brainless bureaucrats. They in turn deemed him arrogant and weird. His latest statement of rebellion, wearing cowboy boots to the office, didn't help.

In April 1973, after just six months in his new job, DeLorean walked out of General Motors altogether, quitting his $650,000-a-year job (equivalent to $3.5 million today) and gaining pop culture luster as "the man who fired GM." In truth, the bosses were happy to see him go. As parting gifts the company gave him a Cadillac dealership, which was a money machine in those days, and arranged for him to run the National Alliance of Businessmen, an organization of socially conscious executives.

A month after he left, the forty-eight-year-old DeLorean married his third wife: Cristina Ferrare, a twenty-two-year-old Italian model. His star status remained undimmed, thanks not only to his glamorous new wife but also his well-publicized high-living lifestyle. It included an apartment on Park Avenue, a 434-acre estate in New Jersey's horse country, and a farm in California.

All the while, he yearned to return to developing cars. Two years later, in 1975, he started a new car company, DeLorean Motor, luring Bill Collins and a handful of other former GM acolytes to join him. DeLorean announced his company would build an "ethical sports car," a concept that he never really defined but that sounded good anyway. Comedian Johnny Carson invested $500,000, and DeLorean added about $700,000 of his own money. All that, however, was just a pittance of what it would take to start a car company. The bulk of the required capital, $120 million, came from the British government. In return, DeLorean agreed to build his assembly plant in Northern Ireland, where the government was eager to get people building anything besides bombs.

Putting the plant in Ireland meant that DeLorean Motor would be located everywhere, and thus nowhere. His company's headquarters was in New York. The engineering offices were moved to England, as another sop to the British government. The sales office was in Los Angeles. Meanwhile, at the assembly plant in a suburb of Belfast, Protestant and Catholic workers entered through different doors to comply with local custom.

The far-flung empire was great for DeLorean's jet-setting lifestyle. It wasn't so great for coordination and control. Meanwhile Collins, who had done the basic engineering sketches for the new car, walked out in a dispute over money and his role in the fledgling enterprise.

The first car was supposed to debut in 1979, but glitches of all sorts delayed initial production until 1981, a full six years after DeLorean formed the company. When the car finally appeared it was striking and stylish, just like DeLorean himself. Dubbed the DMC-12 but usually just called "the DeLorean," it was a sleek, stainless-steel roadster with gull-wing doors, a louvered back window, and a French-built V6 engine. It wasn't clear which of those features made the car "ethical."

The engine was mounted in the rear, just where the Corvair's had been, but DeLorean's car had better suspension and wasn't prone to rear-end spinouts. It had other problems, however. The quality was lousy, thanks to ill-fitting parts. The price was $26,000, astronomical for its day and $8,000 more than the Corvette.

The price didn't deter Johnny Carson. He backed his investment by buying a DMC-12 in L.A. and then got busted for driving it while drunk in 1982. The incident produced a fresh burst of publicity, albeit not entirely welcome, for the car. But there was no getting around the problems: the DMC-12 was a high-priced, low-quality car being built in a war zone.

The one thing that wasn't suffering was DeLorean's lifestyle, thanks to his salary of $475,000 a year and another $1,000 a week in expenses. The California ranch, with its hot tub for eight, needed maintaining, after all. "John would sooner be sterilized than go second class," one of his growing number of critics grumbled.[29]

The fledgling company couldn't cope with it all. After building only 8,563 cars DeLorean Motor entered bankruptcy in February 1982. The factory in Belfast was shuttered. DeLorean's investors, including British taxpayers, lost almost all their money.

DeLorean himself was getting pinched, too. Cristina sold four sable coats for $15,000 apiece. The couple put some Louis XV clocks on the auction block at Sotheby's, and DeLorean sold his small shareholdings in the New York Yankees and San Diego Chargers.[30] But DeLorean still had dreams of raising money to revive his company. Normal people would have tried an IPO, or junk bonds. DeLorean tried selling cocaine. In October 1982 the FBI filmed him in a Los Angeles hotel room trying to close a deal, and arrested him. Two years later he went on trial, but won a miraculous acquittal after his lawyers argued he was a victim of entrapment.

In 1986 he went on trial again, charged with financial fraud and embezzlement in the collapse of DeLorean Motor. His lawyers argued the company's failure didn't prove theft. DeLorean, wearing his cowboy boots in court, won another acquittal, defying the odds again.

In between his two courtroom victories DeLorean experienced another miracle of sorts: the DMC-12 became larger in death than it ever had been in life. One of the hit movies of 1985 was *Back to the Future*. Michael J. Fox played teenager Marty McFly, who travels into his parents' past in a DMC-12 that was converted into a time machine. Critics scorned the movie's heavy-handed product placements—for Pepsi, Nike, and Calvin Klein—but most agreed that the DeLorean car really fit its part. The price of used DeLoreans—there being no other kind—soared. Even in disgrace, John DeLorean still worked marketing magic.

He would live twenty more years, occasionally popping into the headlines, as when he and Cristina divorced and DeLorean got married again, for the fourth time. He dodged creditors constantly before dying at age eighty in a New Jersey hospital, a testament to hubris run amok.

During his Icarus-like career DeLorean created two automotive time machines. One of them, the DMC-12, was for the movies. The other, the Pontiac GTO, allowed real people to feel as though they were traveling back in time. GM tried to tap their nostalgia by relaunching the GTO in 2004. But the car, imported from Australia, flopped. Hard-core GTO fans still want the American original, as any Cruisin' Tiger can attest.

"It makes me feel seventeen again," explains one of them, John Skwirblies, at an Indian Uprising, describing the appeal of his "sundance orange" 1972 hardtop with a truncated "ducktail spoiler" rising subtly from the rear end. The license plate, fittingly, reads "DUCKTAIL." Behind the wheel, Skwirblies isn't just another suburban salesman with a house and a family. "I like showing off and I like the attention," he says. "My car is my getaway."[31]

8

OHIO GOZAIMASU: GODZILLA, MR. THUNDER, AND HOW A LITTLE JAPANESE CAR BECAME AMERICA'S BIG ICHIBAN

Two kids, a cat or a dog, cable TV, a health club, a home in the suburbs and a Honda in the driveway.

—The *Washington Post*,
describing the American Dream
in the late twentieth century[1]

On August 20, 1979, eighteen-year-old Brad Alty, fresh out of high school in Mechanicsburg, Ohio, headed to his first day at his new job when disaster struck. The car he was driving, a 1970 AMC Gremlin, broke down.

Such breakdowns weren't unusual during the Seventies. Detroit, long America's symbol of economic might, had become its showcase for industrial shoddiness. GM, Ford, and Chrysler rushed to downsize their cars for a new era of clean air and high-priced gas. But the quick lurch from muscle cars to econo-cars overwhelmed their engineers and hurt the quality of Detroit's cars.

As a result, Chevrolets, Fords, and Plymouths rattled, rusted, and rolled over. And those were the good ones.

The decade's most infamous car was the Ford Pinto, with its gas tank that made it prone to explode when the car was hit from the rear. While the Pinto's defects could be lethal, the Gremlin's shortcomings were loveable, at least if you didn't own one. The car's pug-ugly, chopped-off shape was first sketched out by an American Motors designer on the back of a Northwest Airlines airsickness bag.[2] Its truncated tail spawned a popular joke: "What happened to the rest of your car?"

The Gremlin debuted on April Fool's Day 1970 and died in 1980. The car's life spanned the decade of Watergate, defeat in Vietnam, two oil shocks, the Iranian hostage crisis, inflation, stagflation, and national "malaise," as President Jimmy Carter's "crisis of confidence" message was memorably dubbed. Not to mention bell-bottom pants, *The Brady Bunch*, and disco dancing.

By the time Brad Alty's mother arrived at his broken-down Gremlin and ferried him to work, he was two and a half hours late. The young man asked his mom to wait outside, assuming he'd be fired. But his new bosses put him to work anyway, sweeping up dirt and painting yellow lines on the factory floor. A few weeks later Alty and the other new workers started building motorcycles.

They built just three to five a day at first, and at the end of the day each bike would be disassembled, piece by piece, for a painstaking inspection. Alty thought it was mindless. "I thought I had made a mistake going to work there," he recalled decades later. "It was like, 'What the heck am I doing here?'"[3]

What he was doing, unknowingly, was helping launch an American revival in the 1980s that would be as striking as its doldrums of the 1970s. During the new decade the United States would beat inflation, recover its national pride, and win the Cold War. Along the way Americans started building good cars again.

GM, Ford, and Chrysler, however, didn't lead the way. Instead it was Honda, a third-tier Japanese car company when it built the Ohio motorcycle plant where Brad Alty worked. A few years later it would build a new automobile assembly plant right next door, and Alty would be transferred there to help build the Honda Accord. It didn't have tail fins, Dagmars, gonads, or even, at first, a high-horsepower engine. It was plain, simple, and reliable.

From the mid-Eighties onward, most Accords sold in America were built there, instead of in Japan. The success of a Japanese car built in America signaled the globalization of the U.S. economy in a remarkably visible way. Within two decades, tens of thousands of Americans would work in the company's expanding network of U.S. factories. Other foreign car companies—Japanese, German, and Korean—followed Honda and opened their own plants in the United States. Their arrival would spark culture clashes ranging from the comic to the tragic.

Globalization brought opportunity to some people, including Brad Alty, but it also cost other people their jobs. Many of them worked in the factories of Detroit, which in the 1970s were becoming scenes of increasingly disruptive labor-management strife. Honda worried that it might have the same problem but was willing to take that risk, along with many others. Its decision to build a factory in America was audacious. The pace of its U.S. expansion was almost reckless. Amazingly, it all worked.

Thirty years later Alty still worked at Honda. But everything else about his life had changed, thanks to the crazy dreams of a little man that flowered amid the gray desperation of postwar Japan.

• • •

Soichiro Honda was born in 1906 in a rural region southwest of Tokyo. He didn't see his first car, a Model T Ford, until he was eight years old. The sight and smells of the machine fascinated the boy. He started hanging around garage shops, where on many days he got the soot-streaked face that prompted schoolmates to call him "Black-Nosed Weasel." He left school early to become a self-taught engineer, just like Henry Ford. There would be plenty of other parallels between the careers of the two men.

During World War II young Honda helped manage a company that made piston rings for military vehicles, and learned the ins and outs of factory production. In 1948 he started his own firm, Honda Giken Kogyo, to build motorcycles. Soichiro Honda was forty-two years old and a late bloomer, just like Henry Ford.

The company's first products were rudimentary bikes with little motors that had been used to power Japanese army radios during the war. Later it started producing more conventional, but very small, motorbikes that were known as "Bata-Batas" for their distinct sound. It would be years before Honda made engines that hummed like sewing machines.

Soichiro named his various early motorcycle models in alphabetical order: the A-Type, the Dream D, the Cub F, and so on, as Henry Ford did with his first cars. Also like Ford, Honda enlisted a partner to handle his company's financial and administrative affairs, so he himself could focus on engineering. The Japanese version of Ford's James Couzens was Takeo Fujisawa.

Soichiro's and Fujisawa's personalities were yin and yang. Fujisawa was an introverted accountant. He spent nights at home listening to Wagner's operas and rarely drove a car himself. Soichiro was a Japanese party animal. He once drove a car off a bridge in Tokyo with two geishas inside. Fortunately, nobody died. "Takeo Fujisawa was the stalk that supported the gay flower Soichiro," a

Japanese journalist later wrote.[4] The two men, however, had one personality trait in common: an explosive temper. Honda's employees gave them nicknames: "Godzilla" for Fujisawa and "Mr. Thunder" for Honda. At least it was better than Black-Nosed Weasel.

By the early 1950s Honda was expanding rapidly due to Japan's postwar economic recovery. One young job applicant who was being interviewed by a company official had the temerity to ask, "When did you begin working for Honda, sir?" The interviewer replied: "Yesterday."[5] Honda went public on the Tokyo Stock Exchange in 1954, by which time it was the largest motorcycle manufacturer in Japan. But soon the hectic growth caused a raft of quality glitches, and the company's sales plunged.

At this point Soichiro Honda did something that would set the tone at his company for the next three decades. He declared that Honda Motor would respond to its lapses in quality by pursuing a new corporate goal: winning the Tourist Trophy at the annual races on England's Isle of Man, the most prestigious motorcycle races in the world. It was as if the Chicago Cubs decided they would learn to play better baseball by winning the World Series. Soichiro's motivational strategy was laughably bold.

But in 1961, just five years after he set his seemingly impossible goal, Honda stunned motorcycle fans by sweeping the top five positions in the races at the Isle of Man. By that time Soichiro was emboldened enough to pursue other dreams, including selling Honda motorcycles in America.

In 1959 Honda entered the U.S. market with a little motorcycle called the Super Cub 50. The little bike wasn't at all like most motorcycles of its day. It featured a step-through design that made it easy to get on and off, even for women. It also had a clean "four-stroke" engine that didn't require mixing gasoline with oil, unlike the two-stroke, lawnmower-type engines on many other motor-cycles.

The Super Cub was a hit, due to clever advertising designed to undermine the stereotype that motorcycles were only for Hells Angels. "Honda has more fresh ideas than boys around a bikini," declared one ad. "You meet the nicest people on a Honda."[6] The last line became the company's advertising slogan, and it caught on. The company had "transformed the motorcycle from a rather raffish curiosity to something approaching a household appliance," observed *Car and Driver*.[7]

In 1964, the same year that the song "G.T.O." was propelling sales of the Pontiac GTO, Honda got a similar sales boost from a song called "Little Honda," which extolled the virtues of its cute little motorcycles. The song was written by the Beach Boys but popularized by an impromptu group called, conveniently, the Hondells, and the lyrics could have been a Honda commercial:

> *It's not a big motorcycle,*
> *Just a groovy little motorbike . . .*

That year, Honda sold a couple hundred *thousand* motorcycles in the United States, up from just a couple hundred five years earlier. Meanwhile, against all odds, Soichiro Honda decided to start manufacturing cars.

Honda unveiled its first car, the S360, at the Tokyo Motor Show in October 1962, but its debut was inauspicious. The Japanese government's economic planners had concluded that their country already had enough automakers, and were preparing an edict that would prevent more companies from entering the car business. Honda rushed out the S360 to beat the deadline, and the slipshod results were evident. Japanese journalists derided the S360 as a motorcycle on four wheels.

To make matters worse, Soichiro Honda and Fujisawa feuded over marketing strategy. The founder wanted to portray the S360 as a hot rod. But Fujisawa wanted it positioned as practical, everyday transportation. Each man insisted that the company's display at the motor show should reflect his vision of what the car should be.

Their disagreement created a dilemma for a young employee named Tetsuo Chino, who was assigned to develop the display. After much agonizing, Chino prepared two different displays at opposite ends of the Honda's exhibit area. "One showed the car as a sports car and the other showed it as an executives' car," Chino would confess years later. "Mr. Honda and Mr. Fujisawa visited the show separately and I carefully took each one to his own corner of the booth."[8] Chino's strategy worked. Honda Motor launched its first car, and young Chino launched his career.

The next year, 1963, Honda introduced the S500, which was like the S360 but with a slightly bigger engine. Honda Research and Development, an organization with special status in the company, developed both cars. Honda R&D, as it was known, was funded with an earmarked portion of Honda's annual corporate revenue. The lab was Soichiro Honda's idea, and his way of protecting the company's product-development programs from Fujisawa's periodic cost-cutting campaigns. Honda R&D became a proving ground, both for new engines and also for Honda's most promising young engineers. They worked directly under Honda-san himself, who spent most of his time in the labs instead of in the president's office.

Among the promising young recruits in the early 1960s was Soichiro Irimajiri (eerie-muh-jeerie), a graduate of the elite University of Tokyo. In one episode the young engineer changed the design of a piston on a motorcycle engine, causing Honda to lose a race. An enraged Soichiro demanded that Irimajiri resign. "I don't like college men," Mr. Thunder snapped. "They use only their heads," as opposed to their instincts. When the young man refused to quit, Soichiro insisted that he apologize to every one of his colleagues at Honda R&D, and accompanied him along the way. Irimajiri never forgot the humiliation, and never made the mistake again.[9] Honda R&D became the company's Marine Corps. Its elite engineers referred to other Honda employees as "civilians."

In this demanding atmosphere, only Fujisawa could stand up to the company's strong-willed, temperamental founder. He did so at critical junctures, most notably at the end of the 1960s, when safety problems plagued Honda's cars.

A disaster occurred in July 1968 at the French Grand Prix in Rouen. A Honda car spun out of control and slammed into a wall. The driver couldn't escape before the flames engulfed him. The Honda engineers on the scene were horrified. They weren't sure why the car spun out, but they suspected the fire was caused by the car's air-cooled engine.

Soichiro Honda, like Ferdinand Porsche and Ed Cole before him, liked air-cooled engines because they eliminated the need for liquid coolant and the apparatus required to handle it. Soichiro insisted on using them in most of Honda's early automobiles, including its Formula One cars, where the high speeds of racing make the heat stresses on engines especially severe. Honda's engineers had doubts about this, but Soichiro was the boss.

A year later another safety issue hit closer to home. A Japanese man was killed in an accident involving a Honda minicar, the air-cooled N360 model. The man's family sued Soichiro personally, charging him with willful negligence in designing a car that lacked stability in taking turns. It was similar to the complaints against the Corvair, another air-cooled car. The N360's engine was in the front, unlike the Corvair's, but the perception that Honda's cars were wobbly was widespread. *Car and Driver* called them "evil-handling."[10] Spinouts. Fires. Fatal accidents. Lawsuits. And on top of all that, Japanese prosecutors launched a criminal investigation of the company.

Without telling Soichiro, Fujisawa began a series of quiet conversations with the engineers in Honda R&D. They confessed qualms about some of the company's engineering designs, and especially about tying Honda's future to air-cooled engines. Safety issues aside, they said, air-cooled engines posed a strategic problem. Tighter automotive-emissions regulations loomed in the United States, and the engineers believed air-cooled engines couldn't comply with the new rules. The same conclusion later led Volkswagen to discontinue the Beetle. But Honda's young engineers couldn't raise these issues themselves.

In late 1969 Fujisawa sat down with the founder, one on one, and described the engineers' concerns. The conversation got testy. Fujisawa put the matter to Soichiro directly. "Which path are you planning to take, Honda-san?" he said. "Are you the president, or an engineer?" Fujisawa's message was that Soichiro, as the company's president, couldn't put his pet engineering projects ahead of the company's interests. Honda had to focus on improving the safety of its cars.[11]

After some awkward pauses and audible sighs, Soichiro Honda relented. He agreed to allow his engineers to develop water-cooled engines for Honda's future cars.

Fujisawa had intervened just in time. In July 1971 a special investigator for the Tokyo Public Prosecutors office announced that while the N360 had inherent stability problems, there was no hard proof that the car's design had been the primary cause of accidents.

A month later the prosecutor said nobody at Honda would face criminal charges. The personal lawsuit filed against Soichiro, likewise, went nowhere. It was hardly a thundering vindication for either the man or the company, but it was enough.[12] Meanwhile, Soichiro's agreement to let his young engineers pursue their ideas was about to produce a breakthrough.

In late 1972 the engineers at Honda R&D unveiled a revolutionary new water-cooled engine called the CVCC, for "Compound Vortex Controlled Combustion." The CVCC had an extra "precombustion" chamber that made the air-fuel mixture inside the engine burn especially cleanly. The result was more fuel economy and less harmful engine emissions.

The CVCC's impact on Honda, both the man and the company, was profound. In 1973, the year after Honda introduced the engine, Soichiro retired as president and chief executive of Honda Motor. The press hailed him as a Japanese business maverick in a land of corporate conformity, and a man who had defied seemingly impossible odds to start the only successful car company launched since World War II. And the CVCC was hailed as a prime example of the engineering genius that Soichiro had fostered at the company that bore his name. His previous opposition to water-cooled engines was unknown outside Honda, and he had blessed the project in the end.

As for the company, when America's new emissions law took effect in 1975, Honda made the CVCC optional on the Honda Civic, a subcompact model, which also had a much better suspension system than Honda's earlier cars. The Civic CVCC got 39 miles a gallon and, even more impressive, it was the only car that met the new emissions requirements *without* a bulky catalytic converter device, even using leaded gasoline.

The CVCC was the automotive equivalent of a little company being started in a Silicon Valley garage and revolutionizing the computer industry. Amid all America's ills of the Seventies, little upstart companies were producing innovations that couldn't be matched by bigger competitors hampered by corporate tunnel vision. It was a trend that would revitalize the American economy in time.

The Civic came in just two colors, one more than the Model T. But the colors were yellow and orange, and both were bright enough to cure hangovers in a Japanese pachinko parlor. The garish hues didn't matter, though. The CVCC propelled Honda's U.S. sales to 100,000 cars in 1975, compared to fewer than 5,000 in 1970. "If it were the standard car in America most of our automotive problems would disappear," wrote *Road & Track*.[13]

A year later, in 1976, Honda launched a new model, the Accord. It was bigger than the Civic, though nearly three feet shorter than GM's compact Chevy Nova. The Accord also had just 68 horsepower to the Nova's 110. But those numbers didn't tell the whole story. The Accord was 60 percent lighter than the Nova but roomier inside, thanks to its front-wheel-drive design that eliminated the need for a drive shaft. The Accord's large windows and low dashboard added a sense of spaciousness. Its four-cylinder engine had an overhead camshaft that produced a high-revving action with sewing-machine smoothness, which would become Honda's hallmark. The five-speed manual transmission was so slick that it seemed coated with Teflon. And it provided broad gear spacing that improved the car's acceleration, and also boosted its fuel economy. The little car was fun to drive.

Chevy's rear-wheel-drive Nova, in contrast, was strictly low-tech. Its six-cylinder engine had old-fashioned internal pushrods instead of an overhead cam. It had to work harder and use more fuel. The three-speed automatic transmission couldn't compete with the performance of a five-speed, and the windows were small. The Nova felt cramped, sluggish, and slow. The Accord, while smaller, felt roomy, agile, and peppy.

In 1978 Honda's U.S. sales nearly tripled, compared with three years earlier, to 275,000 cars. That March, Brock Yates, a senior editor at *Car and Driver* and one of America's most influential car critics, made a surprising revelation. One of his NASCAR-driver pals not only owned a Honda Accord, but loved it. "He has fallen in love with some midget-sized Honshu shoebox?" Yates wrote, with mock incredulity. "Sure. And tell me Dick Butkus wears wedgies and that Anita Bryant's got a girlfriend." [14]

Yates also confessed that he himself owned an Accord and was equally enamored. From a guy who worshipped at the altar of high horsepower, this was the automotive equivalent of Nixon going to China. "A wide body of customers exists for a car that embodies proper integration of form and function, a car that *works* . . . ," he explained. [15] How revolutionary was that?

In just five years Honda had evolved from making clunkers to producing cars the critics loved. It was as if chimpanzees had evolved from apes into men in a couple of generations. Accord owners took to waving and honking in mutual admiration when they passed each other, just like Beetle owners a decade earlier.

The story might have ended there, with Honda just another Japanese car company cashing in on Detroit's lapses of the 1970s. But Honda's next audacious move was unfolding. The little company believed its expansion prospects were limited in its home market, where Toyota and Nissan dominated the car business. Honda's leaders also figured trade tensions would limit their ability to export cars to America. The only solution, as unlikely as it seemed at the time, would be to build cars in America. They needed a factory, and they had just the man for the job.

Shige Yoshida grew up during World War II, joined Honda in the early 1960s, and went to America a decade later. He was a low-key, methodical manager who played a major role in expanding the company's U.S. sales operations. But in 1977, quite unexpectedly, Yoshida got a special assignment: to find the best location for a factory in America, which was no small task. America was a big place, with wide differences in state laws and local customs.

Yoshida traveled the country, visiting California, Tennessee, Kentucky, and even Las Vegas, where he concluded that the local workforce had certain talents, but not for building cars. Months of research led him to rural central Ohio, where Honda could find cheap land, good roads, and an ample supply of labor among local farm families, whose children seemed to have a strong work ethic. It was the ideal place, he concluded, for Honda's new venture.

When Yoshida returned to Japan to present his recommendation to Honda's board of directors, he decided, out of courtesy, to inform Soichiro Honda. The founder had retired four years earlier, but he was still a frequent visitor to Honda's engineering labs, where he often tinkered. After the two men exchanged pleasantries, Yoshida told Soichiro about his decision to recommend Ohio and explained why. He asked, "Do you approve, Honda-san?"

Mr. Honda became Mr. Thunder again. "Why are you asking me to approve?" he snapped. "I know nothing about America. This is for you to decide."[16] Even in retirement, Honda-san remained formidable and feisty.

Yoshida got a better reception from Honda's board, which approved his recommendation. On October 11, 1977, Governor James A. Rhodes convened a press conference in Columbus, with

Yoshida and other Honda executives on hand, to announce the company would build a motorcycle factory in Marysville, a village thirty miles northwest of the capital. While the plant would be small, with fewer than 100 employees at the outset, it was a coup for the governor. Ohio had lost to Pennsylvania a year earlier when Volkswagen had opened an auto-assembly plant, the first foreign car factory in the United States. VW's decision had been a setback for Rhodes, whose pride in bringing jobs to Ohio was encapsulated in his longtime political slogan: "Jobs and Progress."

A reporter asked whether Honda might one day expand and build a car factory in Marysville, and the governor replied: "I wouldn't be surprised. You know these Japs are pretty smart." When the audience gasped Rhodes quickly realized his faux pas and added, "Of course, you know that by 'Japs,' I mean Jobs and Progress." The crowd broke into relieved laughter.[17]

The press conference kicked off Rhodes's charm offensive with Honda. When the governor discovered that Soichiro Honda, like many retirees, liked golf, he hosted him at the Muirfield Village golf club near Columbus, designed by native son Jack Nicklaus. They were the proverbial odd couple. Rhodes spoke no Japanese and Soichiro spoke no English. And the American politician stood nearly a foot taller than the diminutive Japanese businessman. But there was a favorable omen about their relationship. The Japanese term for "good morning" was *ohayo gozaimasu*, with *ohayo* pronounced just like "Ohio."

Even building a small factory in Ohio was a major step for Honda. As a little car company, just the sixth- or seventh-largest in Japan, Honda couldn't bear the financial burden if a major investment in America went awry. Starting with a motorcycle factory, as opposed to a full-fledged auto-assembly plant, was Honda's version of a test drive. The company moved cautiously in other ways, too.

Yoshida screened each and every job applicant for attention to detail. He asked them to write their first names on a nametag when they arrived, and then to place the tag on their left shoulder for their interview. Some would put the tag on their right shoulder, and others forgot to wear it at all. Yoshida crossed them off the list.

He told the first crop of managers he hired that the factory employees would be called "associates" instead of workers. The managers, for their part, had to wear the same white jumpsuit uniforms as the "associates," instead of the shirt and tie that signaled managerial superiority in American factories. What's more, managers weren't assigned special parking places close to the factory, much less the heated parking garages that the bosses got in Detroit. At Honda, parking would be first come, first served, regardless of rank. And there would be no executive dining room, no separate white-collar bathrooms, and no separate locker room for the managers to change into their work clothes. The rules surprised Honda's American managers, who complained that the company was ignoring American customs, but Yoshida pushed back. Such practices were important, he explained, to show all employees they were part of a common purpose.

That stance stood in contrast to Detroit, where cross-purposes were the rule. By the late 1970s American auto plants had become combat zones in a destructive cold war between labor and management. Just a couple hours' drive from Marysville, GM's massive assembly plant in Lordstown, Ohio, was gaining national notoriety for the "blue-collar blues," a potent blend of worker alienation and sabotage. The Chevrolet Vegas built there "regularly roll off the line with slit upholstery, scratched paint, dented bodies, bent gearshift levers, cut ignition wires and loose or missing bolts," reported *Time*.[18]

Meanwhile, Volkswagen was stumbling at its factory near Pittsburgh, where it was building the Rabbit, successor to the venerable Beetle. The Germans had staffed the factory with veteran managers from Detroit, and then recognized the United Auto Workers union. Before long the same labor-management strife that plagued Lordstown and other Big Three factories afflicted the Volkswagen plant. It was little wonder, then, that Yoshida proceeded cautiously. At first he hired just sixty-four people, associates and managers, who would become known in Honda history as the "Original 64." The little plant at Marysville started production on September 10, 1979.

Just a few days later, Yoshida received a fax from headquarters in Tokyo, directing him to make immediate plans to build an automobile assembly plant next to the motorcycle factory. He was stunned. The decision to begin building a car factory seemed reckless. He still didn't know whether his approach to hiring and labor relations would work. Unbeknownst to him, that wasn't the only problem with moving quickly.

Back at Honda's headquarters in Tokyo, another problem had surfaced. The company's initial financial analysis showed that building a car factory would be too risky to justify the investment. The analysis presented a dilemma for the man charged with preparing it, Tetsuo Chino, the same guy who, nearly two decades earlier, had created different displays at the Tokyo Motor Show to satisfy Messrs. Honda and Fujisawa. Now he needed different numbers instead of different displays. For days he was stymied before he tweaked some of the plan's assumptions to "dress up the figures just a bit," as he later would confess.[19] Once again, with a little fudging, Chino solved a problem. Honda's board of directors approved the project.

With the final decision made, Honda's Marysville car factory would be built in just two years, one year faster than the company's contractor had proposed. "It was all about speed," said one of the Honda executives in Ohio. "Crazy speed."[20]

In May 1982, just months before the new assembly plant was scheduled to begin production, Honda sent associates from Ohio to Japan, to train at the company's car factories there. The ever-methodical Yoshida held a special prep session for the employees selected to go, including Brad Alty. Yoshida advised them on how to deal with jet lag, how to eat raw fish (or politely avoid it), and how to cope with differences in customs and currency.

When Yoshida invited questions, an associate popped up and asked, "Where is the airport here in Columbus, and how do we get there?" Yoshida was incredulous. It had never occurred to him that, for many of the young Americans, this trip would be not just their first to Japan but also the first airplane ride of their lives.[21]

By that standard Alty was a veteran traveler. He had flown once before to California, when he was twelve. He did have one special problem, however, for which Yoshida couldn't provide any advice. He was supposed to be married in a few weeks, when he would still be in Japan, and the arrangements were all in place. Alty begged his fiancée to postpone the wedding, and she reluctantly agreed. Once he arrived in Japan he called her every night, Tokyo time, and talked for two hours. Then Alty got a $400 phone bill, and the couple agreed to shorter, and fewer, phone calls.

When Alty got to Japan, his biggest shock wasn't sushi or jet lag, but the crowds. The only crowds in central Ohio were at Ohio State football games. Ohio didn't have trains, either, which presented another adjustment problem for Alty and his American cohorts. He and a couple colleagues took a weekend trip and downed too much *biru* on the train, falling asleep and not waking up till the end of the line. It took them hours to double back.[22]

Alty spent two months in Japan. He and the other Ohioans visited the homes of Honda workers, toured Mount Fuji, and watched a baseball game with the local professional team, the Seibu Lions. The young Ohioan drank it all in, never figuring that more trips to Japan would lie in his future. He returned home on July 24 and was married just a week later. During his absence, however, a shocking incident had laid bare the depth of anti-Japanese sentiment in America.

On June 19, 1982, a Chinese-American named Vincent Chin was enjoying his bachelor party in Detroit, at the Fancy Pants strip club on Woodward Avenue, the same street where kids had raced their GTOs fifteen years earlier. A group of local autoworkers was also in the club. After a few beers they started throwing taunts at Chin and his friends. "It's because of you little mother-fuckers that we're out of work,"[23] snarled one man, a supervisor at a nearby Chrysler factory, apparently in the mistaken belief that Chin was Japanese.

As the scene got uglier, the Chin party departed and took refuge at a nearby fast-food restaurant. It was hardly a safe haven. The Chrysler supervisor and his stepson, a laid-off Chrysler worker, tracked Chin down, dragged him outside, and beat him with a baseball bat. As he lost consciousness, Chin whispered to a friend, "It isn't fair."[24] He slipped into a coma and died four days later, at age twenty-seven, at Henry Ford Hospital.

The tragedy made national and international headlines, and prompted soul-searching in America, including Detroit, where the deep recession of the early 1980s had cost thousands of people their jobs. The tenor of the times worried many of Honda's American employees. Alty even got grief from his own brother for "working for the Japanese." When he took weekend trips up to Michigan, where his father owned property, the young man would remove his Honda employee parking ID from his car's windshield. Just driving his Honda in Michigan was risky enough, he reasoned. Labeling himself a Honda employee would be downright stupid.

Despite the anti-Japanese backlash, however, there was no turning back for Honda. On November 10, 1982, the new Marysville assembly line gave birth to a boxy gray four-door Accord sedan. It was the first Japanese car made in America.

Honda hoped building cars in Ohio with American workers would defuse trade tensions, and make Americans more willing to buy Japanese cars. Many Americans were wary at first. Detroit's problem-plagued cars had given American automobiles such a bad reputation that, throughout the early 1980s, many Honda customers insisted on getting Accords from Japan. The most sophisticated shoppers even learned the codes for the vehicle identification numbers that designated Japanese-made cars.

Some Honda dealers, likewise, were suspicious of the first American Accords. In suburban Chicago, one dealer noticed a recurring rattle in the dashboard of an Accord, and asked his service manager to make the car as good as the ones from Japan. The manager began doing the repair work, only to find that problematic car actually *was* from Japan.[25]

It would take a couple of years, and hundreds of similar incidents, before the doubts of dealers and customers went away. But in 1985 Honda was named number one in customer satisfaction in the J.D. Power ratings, the gold standard for car quality. General Motors, Ford, and Chrysler were embarrassed.

By then the Marysville plant had hired more workers to run on two shifts. A new spurt of "crazy speed" expansion was about to begin. In 1985 Honda opened a new factory fifty miles from Marysville in the village of Anna, a hamlet so small that a curious cow once wandered into the farmhouse that served as the construction office.[26] The Anna factory made key components—engines, suspensions, and the like—so the Accords made in Marysville were soon assembled with American-made parts, instead of parts shipped from Japan.

Later that year Honda opened a second assembly line at Marysville. Production had expanded so rapidly that one line wouldn't suffice. Line Two, like Line One, was U-shaped. The assembly area looked like a giant double horseshoe. And the welding robots could handle different cars, giving Honda unprecedented flexibility to build two different models, the Accord and the Civic, in the same plant. The expansion and the new flexible automation set a manufacturing milestone.

In the fall of 1985 the Marysville plant started building the 1986-model Accord, a revamped version. It was longer and wider than its predecessor, with a new engine designed with multiple air-intake valves for each cylinder. It was the equivalent of adding extra nostrils to the human nose. The "multivalve" design boosted horsepower 30 percent with little sacrifice in gas mileage.

Model changeovers at GM, Ford, and Chrysler required assembly plants to shut down for two or three months so new machinery could be installed and the plants reconfigured. But Honda decided to make a "rolling model change" without any shutdown at all. After months of preparation and planning, the first 1986-model Accord started down the beginning of Line One while the last 1985 model was rolled off the line's end. It was a quantum leap in manufacturing efficiency that Detroit had never deemed possible.

Honda's triumphs were the work of "amateurs," as a reporter from Detroit exclaimed while visiting the Marysville factory.[27] He was right. Some of the managers were midcareer attorneys who had gotten bored with billable hours and bond issues. Many of the associates were farm kids, who didn't have set ideas about what "couldn't be done."

The physical work was far from easy. The pace on Honda's assembly lines sometimes approximated an aerobic workout, causing some associates to lose ten or twenty pounds after a few months on the job. And factory discipline was strict. The associates couldn't drink soda, smoke, or snack on the assembly line. At the Big Three, car seats stained with Mountain Dew or cigarette burns sometimes wound up in showrooms. And Honda workers weren't allowed half a dozen unexcused absences before being fired, like their unionized counterparts in Detroit.

Honda's employees were paid about the same as those in Detroit, but they were expected to use their brains as well as their bodies. They suggested ways to improve factory efficiency, with the promise that nobody would get laid off as a result. Associates who spotted a problem were empowered to stop the assembly line so the issue could be addressed immediately. The concept was more than many of them could grasp at the outset. "Early on we were a little confused because we expected to be told what to do," one early employee confessed.[28]

The specter of American workers thriving under Japanese management brought the Marysville plant attention from two formidable forces in America: the United Auto Workers union and Hollywood. Since organizing the Big Three in the 1930s and 1940s, the UAW had represented *every* autoworker in *every* American car factory for more than four decades. The union launched an organizing drive in Marysville, and scheduled a showdown election for early 1986.

It never happened. In the weeks leading up to the election, the union realized it faced certain defeat. UAW leaders asked that the election be postponed "temporarily," but they never renewed their request for a vote. It was a major setback to the union's power, from which it would never recover.

Not long after, a movie called *Gung Ho* opened in American theaters. It depicted the comic culture clashes that resulted when a Japanese company took over a failing American car factory. The workers in the movie had to sing the company song and do calisthenics before their shifts.

As it happened, the movie's stereotypes were partly true. There wasn't any company song at Marysville, but managers did encourage associates to limber up before their shifts to the tune of music piped into the factory. The exercise routine was voluntary, but participation got a boost when Honda brought in the Ohio State cheerleaders to demonstrate their favorite stretches, a performance that managed to cross cultural boundaries.

Some of the factory language was weird, too. The "associates" learned "counter-measures," which was Honda-speak for fixing glitches on the factory floor. Employees also learned about the "three joys," which sounded vaguely like an Oriental noodle dish. It referred instead to the three major aspects of the car business: producing them, selling them, and buying them. Joyfully, of course.

Honda's Ohio employees knew this sort of pedagogy drew snickers. Some of them even went to see *Gung Ho*, and thought it was funny. But they themselves were having the last laugh. "Crazy speed" meant constant expansion for Honda, and more opportunities for promotion. With the new assembly line the company built 324,000 cars at Marysville in 1987, nearly 100,000 more than the year before. The next year the factory started making a new version of the Accord, the coupe, which wasn't built at Honda's factories in Japan. Honda started exporting Accord coupes from America to Japan, an improbable effort that brought new responsibilities to

Brad Alty. The export program made him a frequent flier. He made more than a dozen trips to Japan in less than two years to make sure the Accord coupes met quality standards.

By 1988 Honda had more than 6,500 employees in Ohio, a far cry from the days of the Original 64. That same year, after losses totaling hundreds of millions of dollars, Volkswagen closed the doors to its factory in Pennsylvania, the one that Jim Rhodes had tried but failed to bring to Ohio. Some 2,500 people lost their jobs.

Multiple factory startups. Dressed-up numbers. "Crazy speed." Rolling model changes. Besides all that, Honda dominated the Formula One racing circuit. Its engines rolled up victory after victory in the most demanding auto races in the world. Honda's decadelong 1980s adventure was a corporate sojourn composed of hundreds of personal stories, in which individual improbabilities produced collective implausibility. The leaders were people who had survived the privations of postwar Japan and the occasional wrath of Soichiro Honda.

One of them was Soichiro Irimajiri, the same man who, two decades earlier, had been forced by Mr. Honda to apologize, one by one, to each of his colleagues at Honda R&D. After that humiliating start his career fared better, and it was "Mr. Iri," as the Americans called him, who presided over Honda's burst of American manufacturing expansion in the 1980s. He lectured at the Harvard Business School, where his speech was titled "The Racing Spirit." Audacious philosophy, as well as innovative engineering, drove Honda's expansion.

Another key leader was Tetsuo Chino, the expert at arranging auto-show displays to satisfy feuding bosses and dressing up numbers to justify building factories. He ran Honda's American holding company, managing the organization he had helped to create.

But perhaps the most remarkable leader of Honda's American adventure was Hiroyuki Yoshino, whose personal story was far more compelling than Honda's corporate success. In 1945, in the immediate aftermath of World War II, Yoshino was eight years old and stranded with his family in Manchuria, where the occupying Japanese army had stationed his father. To get back to Japan the family had to make an arduous trek over hundreds of miles to reach the Manchurian coast. Yoshino's infant brother didn't survive the trip. After they made it back to Japan more disaster struck, when their home was destroyed in an earthquake. "I realized then that people could survive in an open field," Yoshino later recalled.[29]

Yoshino was in charge of Honda's American manufacturing operations on January 4, 1990, the day that automakers announced their U.S. sales for the prior year. Honda had sold 362,707 Accords in 1989, beating the second-place Ford Taurus by nearly 15,000 cars. For the first time ever America's new *ichiban* car (it meant "number one" in Japanese, as Marysville's associates knew) was a foreign nameplate, the Honda Accord.

In just twenty years Honda had gone from making clunkers to building American factories to leaping ahead of GM, Ford, and Chrysler on the world's biggest automotive scoreboard, the U.S. sales rankings. The unlikely victory came as a surprise, even within Honda. Nearly 60 percent of the Accords sold in America that year were made in Marysville. The associates celebrated with a little ceremony on the assembly line led by Yoshino, who thanked everyone for their hard work. Eight years later he would be named president of Honda Motor, the latest successor to Soichiro Honda himself, completing his improbable personal journey.

The Accord's victory rattled Detroit. "We're not too happy

about it," conceded a Ford spokesman, "but at least we have No. 2 and No. 3."[30] A Chrysler executive warned xenophobically: "We are having our pants removed an inch at a time by a carefully orchestrated, totally committed economic aggressor."[31] In truth, had Japan's auto industry been "carefully orchestrated" by the government, the Accord never would have existed. Honda had defied the Japanese authorities to enter the car business. But dispassionate observers saw symbolic portents in the Accord's ascent. "It was an embarrassing year for the American auto industry," wrote the *Washington Post*.[32] The *Los Angeles Times* deemed the development "yet another sign of Japan's increasing dominance over Detroit's troubled auto makers."[33]

The Accord's sales triumph signaled the unmistakable globalization of the world's economy, and the blurring of industrial distinctions that had once seemed clear. What was a foreign car, anyway: a Honda built in Ohio or a Ford made in Mexico or a Chrysler built in Canada, as many were? Or what about the Dodge Stealth, made in Japan by Mitsubishi and sold in the United States under an American brand?

Such questions were relevant well beyond the auto industry in the new global economy. Was flying JAL any less patriotic than flying United, and did it make any difference that JAL used Boeing planes? Was single-estate Sumatran less "American" if brewed in a Braun instead of a Mr. Coffee, where all-American hero Joe DiMaggio served as the TV pitchman?

America itself had installed democratic capitalism in Japan after World War II. It had done the same in Germany, where in 1989 the Berlin Wall fell, freeing East Germans from tyranny. It also freed them from driving the Trabant, a flimsy Communist car that made the Corvair seem safe, and gave them access to Opels, made by a

German car company owned by General Motors in Detroit. Globalization wasn't a one-way street.

Honda's success in Ohio prompted it to build more factories in America and launched a flood of foreign automotive investment in the United States and Canada. Toyota, Nissan, Mazda, Mitsubishi, Subaru, BMW, Mercedes-Benz, and Hyundai all built American assembly plants. Even Volkswagen would return. In the process, America's diet as well as its economy got globalized. You could find sushi in Ohio, *bibimbap* in Alabama, and spaetzle in South Carolina, right alongside fried okra and biscuits with gravy. And maybe Alka-Seltzer, too, if you were lucky.

In 1989 Soichiro Honda became the first Asian inducted into the Automotive Hall of Fame in Dearborn, Michigan. Right next door, the first Accord built in America was enshrined at the Henry Ford Museum, alongside such American icons as the Corvette, the Mustang, and the GTO. The Accord would never be a flashy hot rod like those cars. It appealed to people who wanted safe, reliable, and enjoyable driving, not to people who saw their car as an extension of their personality, or of anything else.

Thirty years after starting at Honda, Brad Alty would be a senior manager at Marysville, overseeing the work of a couple thousand associates. Instead of a Gremlin he was driving a Honda SUV built in Alabama. By then Honda would be a much bigger and more conventional company, with more corporate bureaucracy, less daring, and slower corporate speed. But the Accord, which evolved into a midsized sedan with a high-horsepower engine, would still be among the top-selling and most respected cars on American roads. "The Accord has just kind of followed the baby boomers as they have grown up," a Honda spokesman explained.[34]

In truth, however, it followed only some boomers. Others

required more room than the Accord or any other sedan could provide. In the mid-1980s they discovered a new type of vehicle from Detroit, which finally was mustering its response to the Japanese challenge. The new vehicle would become ubiquitous on American roads, and a potent symbol in American politics.

9

THE CHRYSLER MINIVANS:
BABY BOOMERS BECOME SOCCER MOMS AND
A, UM, DRIVING FORCE IN AMERICAN POLITICS

I'm telling you, you never get pulled over in a minivan. You could have a hooker strapped to your hood, doing Mach 5—the cops'll go, "Eh, let him go. He's suffered enough, for God sakes."

—Craig Shoemaker, comedian[1]

During the 1980s and 1990s, many thirtysomething baby boomers made a discovery. Having children of their own actually meant making compromises, even sacrifices. Like spending their evenings bathing kids in the bathtub instead of lounging on the deck in the hot tub. Or choosing between a hot car and a family car.

For new parents, sports cars were out of the question when their lives and their automobiles were filled with Big Wheels, soccer balls, and Barbie dolls. Cramming the little cupcakes into the back of any sedan, be it a Honda Accord, a Chevy, or a Ford, worked for a year or two. But after that, when legs and arms got longer, kid-stuffing transformed car trips into hell on wheels. Young parents needed something bigger, even if that meant something more boring. In the tug-of-war between fun and functionality in American cars, functionality was about to assert itself again, even more dramatically than it did with the Honda Accord.

When the baby boomers themselves grew up in the 1950s and 1960s, "practical family transportation" meant the station wagon, words that surfaced mercifully repressed memories in the young adults of the 1980s. In station wagons, the annual long-distance trek to grandma's house or Camp Cutup amounted to anger-management boot camp for both parents and children.

The kids were crammed into the rear seat between the adults in front and the luggage in back. Billy hit Bobby, Bobby passed gas, and Susie shrieked at being in the backseat with "the boys." Bathroom breaks provided only the briefest respite. And often as not, the Buick Estate or Ford Country Squire—station wagon names that evoked a bucolic unreality—blew a fuel pump and broke down on the Pennsylvania Turnpike.

The new solution to this family-transportation dilemma came from the unlikeliest of origins: a car company that had barely escaped death. In 1979 and 1980 the Chrysler Corporation was bankrupt in everything but name. It avoided outright liquidation only because of a controversial government rescue. Chrysler's leaders were corporate castoffs, including Lee Iacocca and Hal Sperlich, who had been fired from Ford by Henry Ford II in a brusque and humiliating fashion.

Iacocca and Sperlich were the same two men who had collaborated in 1964 to create the Ford Mustang. Going to work for lowly, brink-of-bankruptcy Chrysler was neither man's first choice, but it turned out to be a stroke of luck. By 1983 the two men and their cohorts had engineered remarkable personal and corporate comebacks. Against all odds, they had restored Chrysler to profitability. Their tools were Sperlich's cars, Iacocca's salesmanship, and an

American economy rebounding from the deepest recession, until that time, since the Second World War.

Meanwhile many of the teenage baby boomers who had been wowed by the Mustang in the 1960s had completed personal journeys of their own. They had gone to college, grown up, gotten haircuts, taken showers, found jobs, gotten married, and started families. Not always in that order, of course.

Thus the stage was set for Iacocca and Sperlich to capture the mood of America's largest generation once again, exactly twenty years after their stunning success with the Mustang, at a time when many boomers were settling into family life. The two men responded not with a new car but with a totally new type of vehicle. Like the Mustang, this one would help define the lifestyle of a generation, or at least the lifestyles of baby boomers who were into painting the nursery instead of painting the town. All this would signal a shift in America's love affair with the automobile from sleek cars to tall and burly trucks, a change that would resonate with America's restless national psyche in the last two decades of the twentieth century.

The minivan's equivalent of Neanderthal man—that is, an evolutionary dead end—was a lumbering, ungainly vehicle from the 1930s called the Stout Scarab. It earned its name because it resembled a giant, rolling beetle, with a rounded body that had neither a hood nor a trunk. A Detroit inventor named William Stout envisioned the Scarab as the ideal vehicle for wealthy businessmen, literally an "office on wheels."

The Scarab had an engine mounted in the rear, over the drive wheels, which was a space-efficient design like that of the Volkswagen Beetle, then being developed in Germany. But the Scarab was three feet longer than the Beetle, so long that it had four windows on each side. Its cavernous interior held a couch, swivel chairs, and a removable table, features similar to what would be found in the Chrysler minivans half a century hence. The Scarab, however, had a price tag to match its size: $5,000, equivalent to more than $80,000 in 2012. The strange design, lofty price, and the onset of World War II killed the Scarab after only about a dozen were built.[2]

The next evolutionary creature was the Volkswagen Microbus, which spread and multiplied far more successfully than the Scarab. But it was too quirky, and too beloved by the hippies, to become a mainstream vehicle. Faring worse was Chevrolet's Greenbrier van of the early 1960s. It had the misfortune of being built on a Corvair chassis and died an early death in 1965, four years before the Corvair itself was relegated to history.

As the Greenbrier faded from memory, Hal Sperlich was a thirty-five-year-old junior executive at Ford, basking in the glow of the Mustang's success. He had been raised in Detroit in the 1930s and 1940s, at a time when local kids could aim for the pinnacle of American business without leaving their hometown. In 1951 he graduated from the University of Michigan, the American auto industry's unofficial finishing school, with a degree in mechanical engineering. He joined Ford Motor in 1957 and entered the ranks of the company's product planners.

The job of product planners was to develop the plans (or "programs," in car talk) for successful new vehicles. To do it, they had to reconcile the conflicting desires of the engineers, designers, and financial staffers. A bigger engine would make a car faster, but it also made it more expensive to build. A curve in a fender might make a car look better, but it also added to the complexity and cost of manufacturing. And so on.

Sperlich became one of the few Detroit product planners who

saw value in small cars. Proof positive, in his view, was the Mustang itself, a small car that offered stylish functionality at a low price. It had made huge profits for Ford. Likewise it made the careers of Lee Iacocca and Sperlich himself. After Iacocca became president of Ford in 1970, Sperlich followed him up the corporate ladder, becoming vice president for product development in 1972.

As the 1970s unfolded, Sperlich came to see small cars as the best bet for fighting Japanese imports, a quest he viewed as the equivalent of a military mission for Detroit's automakers. "Economically, World War II started on the decks of the battleship Missouri in Tokyo harbor," he once told an interviewer. "The Japanese . . . probably muttered under their breath, 'Unconditional surrender, my ass.'"[3] It was a common view among Detroit executives, though few would say it as bluntly as Sperlich. His all-night work sessions and crack-of-dawn staff meetings at Ford were legendary. He would argue for his ideas with a passion that often annoyed his colleagues, even those who admired his considerable talent. "I know all of Hal's speeches," said one executive who worked with him for decades at Ford and later at Chrysler. "He knows all of mine, which are shorter."[4]

But passion and intensity were the price of innovation at Ford, as Sperlich saw it. He had a point. Conservative corporate bureaucrats who specialized in analysis paralysis dominated the company. The financial staff, for example, projected the profits from any new vehicle by estimating its share of its particular market segment. There was logic to this exercise, unless it was applied to a new type of vehicle that didn't fit an existing segment. In that case, projecting even a 100 percent market share in a nonexistent segment meant zero sales, and thus no profits. For the financial staffers, saying no was always safer, careerwise, than saying yes. Ideas for innovative new vehicles often died on the drawing boards.

Even new ideas that survived the initial analysis faced a gauntlet before getting to market. One such vehicle in the early 1970s was a cut-down version of Ford's full-sized van, the Econoline, which typically was used as a commercial delivery truck. A prototype of the little van, code-named Nantucket and intended for everyday family use, was built in 1972. But a year later the Arab oil embargo sent car sales into a tailspin, and the Nantucket fell victim to Ford's corporate cost-cutting.

At the time Sperlich, intrigued with the idea of a small van, figured that even a smaller version of the Econoline would still be too big for everyday use by families. In the early 1970s he began to toy with a new design that he called the "Mini-Max." It would have minimal length, making it small enough to fit into the average garage. But it would offer maximum interior space, and thus provide an attractive alternative to traditional station wagons.

The only way to reconcile a small exterior with a big interior would be to build the van on a front-wheel-drive platform instead of the Econoline's rear-wheel-drive chassis. Front-wheel drive would eliminate the need for a heavy, space-consuming drive shaft, just like it did on the Honda Accord and other front-wheel-drive cars. As it happened, Ford used a front-wheel-drive platform to make its Fiesta subcompact car, which was built and sold in Europe. Sperlich himself had helped develop the Fiesta during his brief tour of duty at Ford of Europe, where he had become enamored of front-wheel-drive design.

But the Fiesta might as well have been on the moon instead of in Europe. At the time Ford of Europe was virtually a separate company from Ford's operations in the United States; the two operations jealously guarded their turf. The right platform for the Mini-Max, unfortunately, happened to be on the wrong continent. Iacocca and Sperlich wanted to bring a larger version of the Fiesta to the United States, but they were rebuffed. Sperlich's Mini-Max was caught in corporate limbo.

Increasingly, Sperlich himself was caught in limbo, too. His persistent badgering for the Mini-Max grew more and more annoying to chairman Henry Ford II. He flatly rejected the idea when Sperlich presented it to him in 1976.[5] Worse yet, behind-the-scenes warfare was breaking out between Henry II and Iacocca. Sperlich, being an "Iacocca guy," got caught in the crossfire. Not long after Henry II rejected the Mini-Max, he summoned Iacocca and ordered him to fire his young protégé. Iacocca protested, but to no avail. "Don't give me any bullshit," Henry said. "I don't like him."[6] Sperlich landed a job over at Chrysler. It was the weak link in Detroit's Big Three, but it was better than unemployment.

At that point the Mini-Max might well have died, along with Sperlich's once promising career at Ford. But then fate intervened on Sperlich's side in the unlikely figure of his nemesis, Henry Ford II.

On June 13, 1978, Henry Ford II called Lee Iacocca to his office and fired him with the words: "Well, sometimes you just don't like somebody."[7] At least Hank the Deuce gave a consistent explanation for his firings. He didn't need to say more. The Ford family controlled Ford Motor with super-voting shares, and Henry occasionally would remind people: "My name is on the building."

Iacocca was fifty-three years old and in the prime of his career. He had never worked for any company besides Ford since joining the company thirty-two years earlier, straight out of Lehigh University. He had been Henry's heir apparent, seemingly destined to be Ford's first CEO outside the Ford family. The firing, which made front-page news across America, left Iacocca shaken. He had no future plans.

At the same time, just a few miles from Ford headquarters, Chrysler was entering a full-scale crisis. For years the company's cars had been uninspired or unreliable, often both. Chrysler's management had been equally inept. Within months the company began to court Iacocca. Sperlich, who saw Chrysler's troubles mounting around him, urged his former mentor to come aboard. Five months after being fired Iacocca joined Chrysler as president, with the understanding that he would get the top job in about a year.

The day Chrysler hired Iacocca the company reported a loss of $160 million, its largest quarterly loss ever. The bad news was just beginning. The company had a' habit of building cars that hadn't been ordered, and then storing them in empty fields and parking lots around Detroit for months until dealers could be persuaded, or pushed, to take them. The practice was nicknamed, with exquisite euphemism, the "sales bank." It was the equivalent of a McDonald's store making hamburgers without any customers in sight, except that the unsold cars rusted instead of spoiled.

The sales bank produced losses totaling hundreds of millions of dollars, and it was just one example of the company's distress. "If I'd had the slightest idea of what lay ahead for me when I joined up with Chrysler," Iacocca later wrote, "I wouldn't have gone there for all the money in the world."[8] In September 1979 Iacocca was named chairman and CEO, three months ahead of schedule. By that time Chrysler was running out of cash. No banks would lend it money.

In desperation, Iacocca mounted a furious lobbying campaign for government help. He begged the U.S. government, and Canada's, too, to cosign loans from banks. He enlisted help from union officials and from politicians in places where Chrysler had factories. The prolonged negotiations sometimes took bizarre turns.

At one point Chrysler's young treasurer broke down and cried, convinced the rescue effort would collapse. At another juncture he summoned key bankers to tell them Chrysler was filing for bankruptcy, and waited for their faces to turn white before reminding them that it was April Fool's Day. When all seemed lost, Iacocca even tried to sell the company to his archenemy, Henry Ford II, but Henry rebuffed him. Iacocca then began a brutal cost-cutting drive that would slash Chrysler's workforce in half over the next four years.

In early June 1979 Chrysler ran out of cash and suspended payments to its parts suppliers, signaling an imminent corporate collapse.[9] But the company managed to hold on for a couple more weeks, and at the eleventh hour it met the dictums of the Chrysler Corporation Loan Guarantee Act, which gave the company access to fresh bank loans. Even then, however, fate seemed stacked against Chrysler.

On June 23, fire broke out in the New York office building where the loan documents were to be signed. At 2 a.m., after the flames had been extinguished, anguished Chrysler officials waded into the smoky, watery offices to retrieve the papers. They moved to a nearby building and spent the rest of the night sorting through thousands of documents so the closing could occur the next day.[10] Chrysler had survived. The nick-of-time rescue never would have happened without Iacocca's personal persuasiveness and his sheer determination. Ironically, when Henry Ford II fired his onetime protégé, he had unwittingly kept one of his competitors in business. Chrysler's future was anything but secure, but the company did have some hidden assets.

Soon after Sperlich had arrived at Chrysler, he had discovered that the company's European subsidiary, Simca, had developed a spiffy little front-wheel-drive subcompact, a car not unlike the

Ford Fiesta in Europe. Chrysler had to sell Simca to raise cash, but it retained the rights to produce and sell the little car in North America. In 1978, after several engineering modifications, Chrysler launched the Simca in the United States and Canada as the Dodge Omni and Plymouth Horizon, with prices starting around $2,500.

For once, Chrysler's timing was perfect. When the Iranian hostage crisis occurred a year later, gasoline prices soared for the second time in a decade. The small cars from General Motors and Ford still used rear-wheel-drive designs. They felt heavy and cramped and got mediocre gas mileage. In contrast, the Omni's and Horizon's front-wheel-drive platform featured a sturdy four-cylinder engine that sat sideways, in transverse fashion, under the hood, which saved weight, added space, and increased fuel economy. The cars got more than 30 miles a gallon. *Motor Trend* named them Cars of the Year. Amazingly, crippled little Chrysler had an advantage over its major rivals in one segment of the market.

Sperlich used the same vehicle architecture, a front-wheel-drive platform with a transverse-mounted engine, to develop two compact sedans, the Dodge Aries and Plymouth Reliant. Chrysler launched them in 1981 with a marketing campaign built around their internal code name, the K-cars. They had boxy styling and weren't exactly hot rods, but the K-cars could hold a family of five (albeit in cramped quarters), and topped 30 miles a gallon in highway driving. They were almost two and a half feet shorter than the models they replaced, and their weight was "reduced a startling 1,000 pounds, on average," as *Popular Science* wrote. The magazine added, "The fate of Chrysler may be riding on the fortunes of the new model line."[11] If so, that fate was fortunate. Chrysler sold more than 300,000 K-cars in their first year on the market, surprising even Sperlich.

The success of the K-cars allowed Sperlich to resume his quest for a small, family-sized van. Iacocca liked the idea, but many of Chrysler's engineers were dubious, and not without reason. For decades, each of Detroit's Big Three car companies had been organized with two separate engineering staffs, one for cars and one for trucks. They were different kinds of products, with different kinds of customers. Cars were for personal use. Trucks were aimed at business buyers: contractors, farmers, and companies with material to haul.

Chopped-down vans had never succeeded. They were in never-never land, too small for businesses but too big and ungainly for parents to ferry the kids around town. But Sperlich envisioned something different: a trucklike vehicle that would be built on a car chassis. It would thus be light, maneuverable, and easy to drive. The man who had built the Mustang on the platform of the Falcon wanted to build his new van on the platform of the K-car.

To convince his skeptical engineers, Sperlich wrote a satirical internal memo that he distributed just as K-car sales were taking off. The memo was from the "president" of a mythical car company that made only one kind of vehicle, small vans like the one Sperlich envisioned. The president had received a proposal from his engineering team to build a new, revolutionary kind of vehicle called a "sedan."

The president's memo expressed puzzlement at the proposal. Just why, he asked, did the engineers want to chop off the back end of their little van and replace it with something called a "trunk," which could hold a few suitcases but not any people, and not even the family dog? What was the purpose of making a less functional vehicle? Why did the engineers expect customers to buy this "sedan" instead of buying a convenient, versatile little van big enough to hold almost anything but small enough to fit in the family garage?[12]

The memo was pointed, passionate, and sarcastic, pretty much like Sperlich himself. But it made the point, and Chrysler's engineers set their skepticism aside. The project to make a van that looked like a truck but drove like a car, code-named the T115, moved quietly forward.

So did Chrysler's quest for survival, but in rather more public fashion. On March 21, 1983, Iacocca made the cover of *Time*, with his face mounted on the front of a K-car, in place of the car's grille. The headline read: "Detroit's Comeback Kid." Five months later, on August 12, Chrysler repaid its government-guaranteed loans in full, seven years before the loan payments were due. The government had received stock warrants in return for issuing the loan guarantees. The taxpayers made money on the deal.

The company celebrated with a ceremony at the Waldorf-Astoria in New York, with the nation's news media on hand. In the hotel's ballroom Iacocca posed before a giant four-foot-by-eight replica of the loan-repayment check mounted on a piece of cardboard. It was just "like Patton in front of the flag," a Chrysler PR man exulted, referring to the opening scene of the 1970 movie *Patton*.[13]

The whole event was pure, unabashed schmaltz, but the parallels were undeniable. Only five years earlier Iacocca had been fired, just like Gen. George Patton had been briefly sacked during World War II. But now, like the blustering World War II general, Iacocca had returned to lead his troops to improbable victory. And Chrysler was about to unveil a new vehicle that would roll over America's highways. It wasn't quite like Patton's tanks rolling over Europe, but it did get better gas mileage.

The May 1983 cover of *Car and Driver* might easily have been mistaken for the cover of *Sports Illustrated*. It showed five members of the Detroit Pistons basketball team, none of whom were small men, posing side by side in front of a new Chrysler that would be launched that fall. The vehicle had a short hood and a large passenger compartment that extended all the way to the back. In automotive jargon it was a "one-box" design, as opposed to the typical "three-box" configuration—hood, cabin, and trunk—on most cars. The headline read: "A Van for All Seasons." The article called it a "minivan."

"Picture a van that is three inches shorter, ten inches narrower and *fifteen inches lower* from road to roof than the next-smallest [van] on the market," the magazine wrote, "yet has enough room for the Detroit Pistons and their luggage and can be stuffed into your garage just like a normal automobile."[14] The minivans had a 2.2-liter, four-cylinder engine and a five-speed manual transmission that allowed them to get 24 miles a gallon in the city and more than 35 on the highway. Customers could buy a slightly larger four-cylinder engine and an automatic transmission, which reduced the fuel economy a bit. But the minivans still got great mileage, considering how many people and how much stuff they could hold.

The advantages were possible because Chrysler had the front-wheel-drive platform that Ford lacked, at least in America. "We started work on that thing 15 years ago at Ford," Sperlich, now president of Chrysler, told an interviewer. "We needed front-wheel drive to get the height down and still keep the cube shape. We couldn't get the O.K. at Ford because we couldn't guarantee getting our money back. At Chrysler, we had front-wheel drive and so we did it."[15]

The basic version of the minivan came with two rows of seats, which could hold five people, and was priced around $9,000. More upscale versions—with the bigger engine, an optional third-row seat, and kitschy fake wood-trim on the side—could cost up to $14,000. They were sold as the Plymouth Voyager and the Dodge Caravan, a name chosen because it stood for "car and van."

The vehicles weren't without flaws. When they were fully loaded with kids and all their attendant gear, the tiny four-cylinder engine could barely climb the hills of Pennsylvania, much less the mountains of Colorado. While the optional third-row seat could be removed to create extra cargo space, doing so was like pulling impacted wisdom teeth.

There were quality glitches, too, including a tendency for the sliding side door to stick. At the "Job One" ceremony at the factory in Windsor, Ontario, where the vans would be built, Iacocca drove the first minivan off the assembly line with a full complement of dignitaries: government officials, union officers, and executives. When the notables in the second row tried to climb out of the van, the sliding door wouldn't budge, leaving them trying to tug it open while the onlooking journalists started to snicker. A Chrysler PR man, sensing disaster, rushed over to blame the glitch on the van's childproof lock. In truth Chrysler's minivans didn't even have childproof locks, but the journalists didn't know that. A few moments later the door popped open, and a PR disaster was averted.[16]

But the drawbacks were more than offset by the minivans' pluses. Not only were they spacious and economical, they offered a high seating position, giving short people, especially women, a clear view of the road for the first time. They offered room for two, three, or even four children to spread out with ample space in between them, like little demilitarized zones. The uneasy truces among the kids in back meant more sanity for the parents in front. Many minivan buyers had never entered a Chrysler dealership before. The company's customer base had consisted largely of "Joe

Six-packs," as Chrysler executives called them, meaning blue-collar workers in their fifties or beyond. But minivan buyers tended to be young professionals in their thirties and early forties.

Car and Driver predicted the Voyager and Caravan would sell out. The magazine was right. Some buyers had to wait weeks or months to take delivery of their minivans, a shortage of supply that some dealers used to jack up the price. The magazine also forecast that the minivans would boost Chrysler's financial fortunes, and it was right about that, too. In February 1984, five months after the minivans debuted, Chrysler resumed paying dividends to its stockholders for the first time in five years. By February 1986 its stock had surged above $48 a share, a 1,500 percent increase since the dark days of 1980.[17]

Minivans led the company's profit parade because Sperlich had kept the development costs low by reusing the platform for the K-cars. All automotive platforms grow obsolete as styles and technology change, but the K-car chassis was still in its prime. It was the Mustang trick all over again, except that nobody would ever call a minivan a "sexpot."

Chrysler's minivans caught Ford, General Motors, and the Japanese car companies in their corporate blind spots. The two Detroit companies rushed out minivans of their own, the Ford Aerostar and Chevy Astro, but both were simply chopped-down versions of the companies' full-sized, rear-wheel-drive vans. They were sluggish and cramped. They drove like lumbering trucks, especially when pulling in and out of parking spaces, as Sperlich could have predicted.

A few years later GM belatedly launched a front-wheel-drive minivan called the Chevy Lumina. It wasn't exactly luminous. It had a bloated body and elongated front nose that made it look like a "plastic pachyderm," as one Chrysler executive snarked.[18] A couple years later GM gave the van a nose job to improve its looks, but it didn't help much.

Toyota's first minivan, meanwhile, was an egg-shaped creature called the Previa that had the engine positioned under the front seat. The position of the engine caused the chassis to be elevated and required short people to pull themselves up to enter the car, a big drawback for women. The Previa was a rare misfire for Toyota. The secret to the minivan's success wasn't style. It was basic functionality in an easy-to-drive package. Chrysler executives found themselves amazed by how their bigger competitors, time after time, managed to miss the mark.[19]

Throughout the 1980s and much of the 1990s, Chrysler dominated the minivan market, devoting a second factory and then a third to producing its revolutionary new vehicle. At one point Sperlich proposed a smaller vehicle, a mini-minivan, but the original minivans were selling so well that Chrysler saw no need for it.[20] Sperlich retired in 1988 with both the Mustang and the minivan on his resume. Either one would have been commendable; having orchestrated both was remarkable.

Neither Iacocca nor Sperlich was a baby boomer, but they had nailed the needs of the largest generation in American history twice, at critical junctures. The Mustang offered youthful excitement as boomers reached driving age, while the minivan provided convenient practicality when they were having children. But the minivan was about to do something that the Mustang never did. In the last decade of the twentieth century it became a symbol of a potent political force.

In late 1991 Marta Buser, a homemaker from upscale Overland Park, Kansas, told the *Kansas City Star* about the first time she made dinner for her kids in the family's minivan. It was a school day. She was coaching her daughter's soccer practice from 4 to 5 p.m. After that she had to get her son to soccer practice from 5 to 6:30.

Somewhere along the way she pulled out peanut butter, jelly, bread, potato chips, apples, and oranges, and the kids ate en route. Maybe it wasn't osso bucco with braised asparagus spears, but the children probably liked it better. It wouldn't be the last time they had dinner in the family minivan. "My kids eat in the car, they change clothes in the car, they do their homework in the car," Buser explained.[21] Presumably they could have done all those things in a station wagon, but only with lots of elbowing, shoving, and the wanton smearing of peanut butter and jelly all over the car seats.

By the early 1990s similar scenes played out out all across America, and many families developed their own particular, and peculiar, minivan rituals. In Detroit, where the minivan began, Edrie Robinson, the wife of a Chrysler attorney, spent more than a decade ferrying their four children to and from school, doctors appointments, and soccer games, all in a succession of minivans. They were Chrysler minivans, of course, which by the 1990s came in stretched versions with big V6 engines, leather seating, and sliding side doors that opened and closed with the touch of a button.

The Robinson children's seating arrangement inside the minivan was determined by the immutable pecking order of age. The older kids claimed the choice "captain's chairs" in the second row, relegating the little ones to the third-row bench seat in the far back. After a few years, however, the younger children got more assertive. Thus began protracted negotiations, and the occasional fight, over whose turn it was to take the captain's chairs.

Lefty, the Robinsons' yellow labrador retriever, played his own part in the ritual. He sat in the second row on the minivan's floor, but he also hopped up into the driver's seat whenever Edrie got out to pump gas. The kids took to buckling Lefty in with the seat belt and putting his paws on the steering wheel. When the guy at the cash register would ask Edrie which car was hers, she would shrug and reply: "The one over there. With the dog driving."[22]

When the kids (as opposed to the dog) started driving, they inherited their mother's minivans. One of the boys tarted up the van with soft green interior lighting, fake-flame decals on the front fenders, and a license plate that said: "MAN VAN." Edrie Robinson thought it looked like a suburban pimpmobile and refused to drive it. She went back to driving regular sedans instead.[23]

In 1995 a Denver woman ran for local office with the slogan "A soccer mom for city council." She won, and political consultants latched onto the term. Across America women like Edrie Robinson were labeled "soccer moms," and minivans were called "mommy-mobiles." The term presented something of a marketing dilemma for Chrysler. Company executives didn't want people to think minivans were only for women. In truth, many American men drove minivans, too. But Chrysler also wanted to market its minivans as the ideal vehicle for mothers raising active families.

It was a delicate balance that came close to veering over the edge. At one point the company drew up plans for a television commercial that showed an embryonic minivan growing inside a mother's womb. Before the commercial aired, however, wiser heads prevailed. The minivan-embryo commercial was aborted.[24]

By the mid-1990s minivan-driving soccer moms were becoming an identifiable force in American politics. At least as identified by those self-described trend definers, the nation's political pundits. So began the search for the soul and psyche of the soccer mom, in which political consultants commanded ridiculous sums to ascertain the obvious.

They divined that soccer moms were baby boomer women,

often well-educated, who had left their budding careers to be full-time homemakers. In the process, stay-at-home moms struggled to find fulfillment and respect in a society that increasingly worshipped career women. Joining the Junior League didn't cut it anymore. They had to juggle their kids' busy schedules and somehow stay connected with workaholic husbands who spent long days away from home, often working alongside professional women.

These mothers proceeded to manage their kids' development with the same energy that they had once devoted to their own, now-truncated careers. Soccer was just part of the package. It became the suburban white kids' sport, as opposed to Little League baseball for the blue-collar kids, or basketball for inner-city black children.

The stereotypical soccer mom had center-left leanings, especially on social issues. Political consultants and spinmeisters began advising their candidates on how to appeal to this group. Their typical advice: Be against global warming, but don't threaten anyone's right to drive a stretched minivan. Be for increased school funding, but not for higher taxes. Be tolerant of abortion, but express personal qualms about it. Most of all, don't be too strident about anything lest you offend sophisticated suburban sensibilities. Soccer moms in the swing states, after all, were more likely to be Presbyterians than Baptists.

With America's cities solidly Democratic and its rural regions reliably Republican, the politicos concluded that swing-vote suburban mothers would decide the 1996 presidential election. The *New York Times* sent a reporter to interview a soccer mom in Pasadena because "she and her comrades in the minivan brigades are politically hot," the *Times* article explained.[25]

Their collective political clout came as news to at least some of these women. "I have to go home and thaw something for dinner," a harried suburban homemaker told the *San Francisco Chronicle*. "I spend so much time going to soccer games that I don't think I can really be a political force."[26] But she and others like her were just that, as it turned out. Exit polls showed that in the 1996 presidential election, Bill Clinton dominated the soccer mom vote. Punditry prevailed.

Despite the collective power of soccer moms, however, women often resented the term. Many thought it implied they were banal, boring people who mindlessly overscheduled and overstressed their kids while forfeiting any intellectual life of their own. The image was pervasive. In Maine, a high-school girl who often hauled her friends to and from school events in the family's minivan told her dad she wanted to stop doing it so often. "I'm too young to be a soccer mom," she explained.[27]

Real soccer moms rebelled in a variety of ways against the stereotype. Some displayed a bumper sticker that said: "I May Drive a Minivan. But I Can Still Party Like a Rock Star!" A Louisiana woman named Laurel Smith, a mother of three and veteran minivan driver, started a website called MomsMinivan.com. She offered her opinions on various new minivan features, which multiplied with each passing year. Smith praised the drop-down video screens with DVD players that let the kids watch movies in the backseats. She also liked the third-row seats that folded easily into the floor to create extra cargo space, and the little tables that could be set up in the back to create portable game rooms or dining rooms. She started using the site to suggest car games for kids (e.g., "How to Catch a Cootie") and to recommend family-travel products. The most popular was the Yak Pack, a "vomit clean-up kit" that included a motion-sickness bag, odor neutralizer, and antimicrobial hand wipes.[28] (Don't leave home without it.) MomsMinivan

.com grew into a business, supported by advertising and product endorsements.

In 2001, meanwhile, a woman whose Dodge Caravan had 110,000 miles on the odometer and fossilized french fries wedged between the seat cushions published a book called *My Monastery Is a Minivan*. "Mothers who still imagine themselves wearing motorcycle boots and listening to Jimi Hendrix . . . poke fun at those of us driving these wimpy mommymobiles," wrote the author, Denise Roy. "Moms like this will only drive huge SUVs with smoked windows so they can put the kids in back and pretend they're still single and sexy. They don't fool me."[29] Nor many other people, one suspects.

In his late seventies, Hal Sperlich remained an avid minivan driver. Minivans remained unmatched in practicality and efficiency, he liked to remind people. Which was true, but that was just the problem. Even during the minivan boom years, between the mid-1980s and the mid-1990s, many Americans preferred showing off instead of showing up at the kids' soccer game. Young adults who wanted to flaunt their sophistication and success needed another vehicle for their aspirations. They found it.

10

THE BMW 3 SERIES:
THE RISE OF THE YUPPIES AND THE ROAD TO ARUGULA

There they are, the men with carefully wrinkled $800 sports jackets . . . the BMW cowboys . . . they're all here, grazing among the arugula.

—Restaurant review in the *Los Angeles Times*, October 1, 1989[1]

Soccer moms weren't the only Americans in the late 1980s and the 1990s more or less defined by their automobile. There were also the yuppies. The term stood for "young urban professionals," a new generation of adults with high-paying jobs in business, finance, medicine, law, and the like.

After the word "yuppie" was coined around 1980, it gave rise to such derivatives as buppies (black urban professionals) and guppies (gay urban professionals). There were more to come. Many yuppies enjoyed spendthrift lifestyles in the early years of their marriages because they were DINKs, which meant Double Income, No Kids. As time went by and one spouse traded his or her career (usually hers) for homemaking, some free-spending DINKs became ORCHIDs: One Recent Child, Hideously In Debt.

ORCHIDS sometimes became the SITCOMs: Single Income, Two Children, Oppressive Marriage. So went the alphabet-soup sociology of the day.

Yuppies, distinguished not only by their age and their occupations, were people who had to buy to live, just as sharks had to swim to breathe. But they couldn't buy just ordinary stuff. Theirs was a restless and creative materialism, a constant search to find the most distinctive and expensive version of just about anything. They favored $2 Dove Bars over 50-cent Eskimo Pies. Their beer was Anchor Steam instead of Budweiser. They chose Macallan single malt over J&B, Camembert over Kaukauna Club, Air Jordans over sneakers, Starbucks over Dunkin' Donuts, Perrier and San Pellegrino over tap water. And so on. By buying upscale versions of everything, they separated themselves from the unenlightened, the uneducated, and the unwashed. And they were proud of it.

Yuppie pride surfaced one night in the late 1980s in Ann Arbor, Michigan, when four young friends went out for a beer. After a couple rounds, one of them, a software developer, blurted out, "Listen, I get up in the morning, put on my French-cut suit, climb into my BMW, and drive to my high-tech job. Now, I ask you, what do I have in common with you people?"[2] Not much after that outburst, as it turned out.

Just being expensive wasn't enough to make something an object of yuppie desire. It had to be the right *kind* of expensive. Yuppies needed stuff different from what their parents had owned. In the case of cars, their parents had coveted Cadillacs as yuppies were growing up. But few things were more offensive to yuppie sensibilities than Cadillacs with vinyl roofs and wire wheels. "The sort of person who buys a BMW would rather be force fed Velveeta cubes than drive a Cadillac," the *Los Angeles Times* observed in 1987, when it reported on "The New Status Seekers in the 1980s."[3]

Yuppie gear had to provide benefits, real or imagined, beyond the comprehension of the hoi polloi but knowingly appreciated by the cognoscenti. In 1987 *BusinessWeek* reported on the booming sales of high-priced pet food. Many yuppies indulged their cats (Himalayans or Bengals, no doubt) with haute cuisine called Amoré and Fancy Feast instead of (sniff, sniff) Purina Cat Chow. It was what "you might expect from BMW aficionados," the magazine explained.[4]

BMW's image makers complained about the characterization, but it was like trying to punch holes in Jell-O (a substance, like Velveeta, that never passed through yuppie lips). So what if a few yuppies liked Saabs, Volvos, and Mercedes-Benzes? Nothing came close to defining the young affluents of the 1980s and 1990s like their BMWs. Crashing into a BMW, went one joke of the day, was "vehicular yuppie-cide." In 1985 a group of young San Franciscans threw a "Yuppie cotillion" that they publicized with the unauthorized use of the BMW logo, declaring that it stood for "Beauty. Money. Wealth." The suggested dress for men was black tie with Nikes.[5]

Such affectations defined yuppies as "bourgeois bohemians," in the words of David Brooks, author and *New York Times* columnist. He shortened the term to "Bobos." They had mastered the oxymoronic art of being downscale elitists, somehow blending the seemingly irreconcilable worlds of WASP and hippie cultures, of high-church music and grunge rock, to form the new American establishment. One Midwestern yuppie was so taken by the term that he even named his dog Bobo. (The dog was a dachshund, a breed developed in Germany, just like BMWs.)

"Bobos don't want gaudy possessions that make extravagant statements," Brooks wrote. "That would make it look like they are trying to impress. They want rare gadgets that have not been discovered by the masses but are cleverly designed to make life more convenient or unusual. . . . Nobody wants to talk about a diamond necklace over dinner, but it's charming to start a conversation about the host's African-inspired salad serving forks."[6]

For such people, BMWs were the perfect cars. They didn't have chrome, tail fins, or other gaudy ornamentation. They had instead high-performance engineering, making BMWs the automotive equivalent of high-performance clothing, like arch-supporting, ankle-reinforcing hiking boots or moisture-wicking, scent-eliminating underwear. Functional luxury was the hallmark of the BMW 3 Series, so called because it was a series of different models built on the same chassis but available with different engines and body styles: coupes, sedans, convertibles, and even a sporty little station wagon.

For several decades the 3 Series cars were the smallest and generally least expensive in BMW's American lineup, and thus the most affordable to the nouveau yup. BMWs conveyed "unpretentious exclusivity," company marketing executives often said, with no apparent hint of irony. In that respect the cars were like the $150 designer jeans that filled the ergonomically correct seats of many a Bimmer (pronounced "beemer," in case you didn't know but were afraid to ask).

Surprisingly, the yuppie-car image did no discernible damage to BMW in the United States, despite the fears of the bosses at Bayerische Motoren Werke (Bavarian Motor Works), the company's proper name. Between 1970 and 1986, BMW's U.S. sales jumped tenfold, to nearly 97,000 cars from just 9,800. Sales kept growing after that.

Notwithstanding the affectations of Bobo Bimmer owners, however, the cars themselves were terrific, at least from the late 1960s onward. Their functional-luxury features included small but smooth, high-revving engines, exceptionally precise steering, and road-hugging suspension that provided a hands-on, albeit butt-bouncing, driving experience. BMWs were the opposite of Detroit's gaudy Buicks, Cadillacs, and Lincolns, which wallowed through turns like a walrus on a skateboard. Even archrival Mercedes-Benz, as BMW fans saw it, built overweight cars for the insecure ignorati who prized hood-ornament prestige over pure performance.

BMW's corporate journey wasn't nearly as smooth as the engines that powered its cars. The company, launched during World War I to make aircraft engines for Kaiser Wilhelm's war machine, switched to making motorcycles in the 1920s and then started producing cars in the 1930s. In the late 1950s BMW almost went bankrupt.

Not until the early 1960s did the company gain corporate stability thanks to the intervention of two reclusive German half brothers. One of them was nearly blind and couldn't even drive a car. The other was the stepson of one of Hitler's most notorious henchmen. Nobody predicted then that, just twenty years later, BMWs would become the preferred personal transportation of America's young elite, some of whom unabashedly adorned their Bimmers with a bumper sticker that expressed the yuppie creed: "He Who Dies with the Most Toys Wins."

BMW marks its official corporate birth date as March 7, 1916, when Bayerische Flugzeugwerke was incorporated in Munich.[7] A year later the company adopted a corporate logo known as the "roundel," a circle with vivid blue and white quarter panels. It was patterned after the flag of the Free State of Bavaria, though for decades popular belief held that the logo depicted a white airplane propeller whirring against a blue sky.[8] Either explanation was more appealing, anyway, than a blue-eyed Bavarian stumbling out of a beer hall with his eyes buzzing.

Business boomed during the Great War. The company's engines developed a reputation for excellent reliability at high altitudes, no small matter when the Red Baron was doing dogfights against British biplanes. But after the war, demand collapsed. The victors imposed strict limits on German plane production.

That issue ended in 1922, when a fledgling company called BMW absorbed BFW by buying the factory and using it to build motorcycles. The first motorcycle that the company designed itself, the lightweight and fast R32, launched in 1923. Just six years later a BMW bike set a world motorcycle speed record of 134 miles an hour. From early on, speed was part of BMW's corporate ethos.

BMW's first car, launched in 1928, was a rebadged Austin Seven, built under license from a British automaker and sold in Germany as the BMW 3/15. But in the early 1930s BMW started building cars of its own—small, performance-oriented machines that got progressively better. In 1940 a BMW 328 Berlinetta won the Mille Miglia, a prestigious Italian road race that Alfa Romeo had dominated for years.

When World War II began, race organizers suspended the running of the Mille Miglia, and production of automobiles at BMW stopped. The company returned to its roots of making military aircraft engines, including the first jet engines to fly in combat. Fortunately, they were deployed too late to save Hitler's Reich.

BMW's World War II history wasn't pretty. As German casualties mounted and labor became scarce, the Nazis forced prisoners of war, concentration camp inmates, Jews, and others, to work in the company's factories. The slave-labor issue dogged BMW for decades, only subsiding when it joined other German companies, including Volkswagen and Daimler-Benz, in paying reparations to victims and their families.

After the war BMW lost its factory in Eisenach, which had the misfortune of being located just a few miles inside the Soviet occupation zone, later East Germany. When the plant resumed building cars it took the name Eisenach Motor Works, and the cars built there were EMWs. Germans also called them *minderwertig*, meaning shoddy merchandise, because that's what they were.

The other BMW factory was in Munich, in the American Zone, but it was pocked with bomb holes and the Soviet Union hauled off much of the machinery as war reparations. The factory took to making pots, pans, shovels, and bicycles. None of it evoked "Beauty, Money, Wealth," but it kept things going. BMW resumed production of motorcycles in 1947, and soon followed by building its first postwar cars.

Unfortunately, they were the wrong cars. The first, the BMW 501, was an ungainly, underpowered colossus, a German version of the Detroit land barges of the day. But most Germans who could buy cars, which wasn't many of them in the early postwar years, could only afford small cars. The company muddled along for a few years until 1954, when it made a strategic shift.

BMW licensed the rights from an Italian company to build the Isetta, a pug-nosed "bubble car" that had just one seat to accommodate a driver and passenger and a single, front-mounted door. Getting in was sort of like climbing in through the windshield of an ordinary car. The 9.5-horsepower engine packed less than half the power of a VW Beetle, and the car itself looked like it had rolled right out of an animated cartoon.

A few years later BMW launched a small car of its own design. It looked like a stretched Isetta. Sales improved but profits didn't because management squandered the revenue, leaving BMW broke by 1959. The company's major creditors pushed to have BMW absorbed by Daimler-Benz, the maker of Mercedes cars. Daimler was more than Germany's leading car company; it was the country's most prestigious company of any kind.

Before the deal could be consummated, however, a group of BMW shareholders managed to block it. BMW secured new financing, including a substantial additional investment from two of those shareholders who were half brothers, Herbert and Harald Quandt. Their family history had more twists and turns than a BMW test track.

Herbert was handicapped from birth with impaired vision. He was born in 1910 to industrialist Günther Quandt and his wife, Antonie, who died in a flu epidemic in 1918, when Herbert was eight. Three years later Günther married a blond, blue-eyed beauty named Magda, who at seventeen years old was only half his age. Within a year Magda gave him another son, Harald. But eight years later, in 1929, Günther and Magda divorced.

In 1931 Magda married again, this time to a man who was a rising star in German politics, Joseph Goebbels. A leading member of the National Socialist Party, he would become the Nazi Minister of Propaganda. Goebbels's boss was the party leader, Adolf Hitler, who served as a witness at his wedding to Magda. Ten-year-old Harald Quandt thus became Goebbels's stepson, destined for the Hitler Youth and then the Luftwaffe.

During the 1930s Goebbels, amid his growing duties and many extramarital affairs, kept Magda pregnant most of the time. The couple had six children of their own: a picturesque Nazi family whose haunting 1942 home movies, probably produced by Goebbels's propagandists, can still be viewed on the Internet today.[9] But the family's home life wasn't as perfect as the pictures.

Besides Goebbels's many dalliances, Magda was rumored to

have had some of her own. What's more, her stepfather, Richard Friedländer, was a Jew. Friedländer, by all accounts, was devoted to his family, including his stepdaughter. But he vanished in a Nazi concentration camp, receiving no assistance from "the First Lady of the Third Reich."

In April 1945, with the Russians closing in on Berlin, the Goebbels family, parents and children, retreated to Hitler's bunker. Everyone, that is, except Harald. He was languishing in an Allied prison camp. On April 28 Magda wrote a farewell letter to Harald, presumably without knowing his whereabouts. "We only have one goal left: loyalty to the Führer even in death," her letter said.[10] She meant it. On May 1 Joseph and Magda drugged and poisoned their six young children. They then committed suicide.

Günther Quandt, the father of both Herbert and Harald, survived the war. He had built an industrial conglomerate by providing the Hitler regime with ammunition, batteries, and other war matériel. Stories would emerge years later about prisoners who worked at the Quandt factories being mistreated, being whipped by the guards after their work shifts, or being threatened by vicious dogs.

The Allies arrested Günther Quandt in 1946, but he was released two years later without charges.[11] He died in 1954 and left Herbert and Harald control of his industrial empire, including significant shareholdings in both BMW and Daimler-Benz. It would have been natural, under the circumstances, for the two half brothers to support Daimler's effort to absorb ailing BMW in 1959. But lured by the chance to control a car company themselves, they opposed the transaction. Much corporate and legal maneuvering followed, keeping BMW independent. The Quandts began buying more BMW shares, and by the fall of 1960 they had acquired a majority stake.

Herbert and Harald were publicity shy, no surprise given the family history, and their descendants remain that way. But the brothers' determination and financial commitment saved BMW from certain corporate death. In time, the company would make the cars that defined the lifestyle of affluent young Americans, many of them the sons and daughters of the soldiers who had defeated Hitler's Germany.

Though virtually blind, Herbert Quandt had a vision for BMW. He wanted to return the company to its prewar roots, making technically advanced, small, sporty cars. His eyesight was so bad that he could only evaluate the designs of new models by rubbing his hands over them to feel their shapes and curves. To understand the machinery under the hood, Herbert would quiz the company's engineers.[12]

He took a more active role in BMW than Harald, but both brothers backed, conceptually and financially, the development of a new model called the BMW 1500, which the company unveiled at the Frankfurt Motor Show in the fall of 1961. The new car had a plain-vanilla "three-box" design: a squarish hood and square rear end, with a boxy four-door passenger compartment sandwiched in between. Beneath the bland exterior sat a smooth 1.5-liter four-cylinder engine with an overhead camshaft that controlled valve movement with more precision than Detroit's older pushrod design. It provided surprisingly peppy performance in the car's lightweight body. The 1500 had independent rear suspension and disc brakes, sophisticated features rare in American cars of the day. The car was engineered for the demands of the German autobahns, where the only speed limit was the driver's own daring.

The 1500 was an instant success. Within two years it revived the financial fortunes of the ailing company, and produced cash flow that allowed BMW to develop a series of successor models. In March 1966 came the 1600-2, so called because its engine had 1.6 liters of displacement instead of 1.5 liters, and two doors instead of four. BMW's engineering creativity didn't extend to its automotive nomenclature.

Despite its new success, however, BMW remained deeply in the shadow of Mercedes-Benz within Germany, and was barely known outside the country. In 1967 the company sold fewer than 3,700 cars in America, while Mercedes sold nearly 27,000.[13] But at least Herbert Quandt's back-to-basics strategy had been validated. Further proof came the next year, 1968, with the launch of a new model with two doors and a 2-liter engine. It was the BMW 2002.

The 2002 debuted at the height of America's muscle car era, when teenagers were cruising around in Pontiac GTOs and Firebirds with snarly styling, loud tailpipes, and 300-plus-horsepower engines. The automotive-enthusiast magazines—*Motor Trend*, *Road & Track*, and *Car and Driver*—encouraged the overdosing on high-horsepower V8s. It came as a shock, then, when the editor of *Car and Driver*, David E. Davis Jr., lampooned the muscle-car mentality and lavished praise on a car with just four cylinders, half as many as the American muscle machines.

In his review, headlined "Turn Your Hymnals to 2002," Davis wrote, "In the suburbs, Biff Everykid and Kevin Acne and Marvin Sweatsock will press their fathers to buy Firebirds with tachometers mounted out near the horizon somewhere and enough power to light the city of Seattle, totally indifferent to the fact that they could fit more friends into a BMW in greater comfort and stop better and go around corners better and get about 29 times better gas mileage."[14] It was as if a prophet had descended from the mountaintop, invoked divine revelation, and proclaimed a new religion.

Muscle cars still ruled American roads, but the impact of Davis's review was instant. The 2002 gained a small but cultlike group of followers. BMW began to emerge in America.

The company's success with the 2002 was a stroke of luck, and not just because Davis praised the car. BMW's managers had made a virtue out of necessity. They had wanted to export the 1600, which was basically the same car they sold in Germany, but its engine couldn't meet U.S. air-pollution standards. BMW's engineers stuffed a slightly larger 2-liter engine into the body of the 1600 and created the 2002. It was plan B, but it worked.

Never mind that the 114-horsepower engine provided barely a third as much power as American muscle cars. The car's light weight and tight construction made it nimble, agile, and fast. Its four-speed manual transmission could stay in second gear until the car hit 60 miles an hour, whereas most American cars groaned to upshift at 30. The 2002's high-revving engine wasn't drop-dead fast from a standing start. But it could cruise at 100 miles an hour seemingly without effort, and felt as though it could keep accelerating forever. The price was just $2,850, which was hundreds of dollars less than the GTO. Yes, there was a time when Americans could buy a BMW for less than a Pontiac.

As the 2002 gained popularity, BMW's future was being reshaped yet again. In 1967 Harald Quandt died in a plane crash in Italy at age forty-six, having lived a short life against the backdrop of tragic history. His death left Herbert as the sole patriarch of both the Quandt family and BMW. Three years later Herbert installed Eberhard von Kuenheim, just forty-one years old, as the company's new CEO. It was an unlikely choice.

Von Kuenheim had been born in 1928 to an aristocratic family in German East Prussia, territory that is now part of Poland. Tragedy marked his early life. His father died when he was a boy, and after the war ended his mother perished in a Soviet prison camp. But like many other German children in East Prussia, the teenaged von Kuenheim had been evacuated to western Germany in early 1945, just ahead of the advancing Russians.

After the war, in the new West Germany, he studied mechanical engineering and spent more than a decade working for a machine-tool company. In 1965 Herbert Quandt hired him as a strategic advisor on technology issues for the Quandt companies. Three years later he asked von Kuenheim to run an ailing machine-tool manufacturer that the Quandt family owned. Von Kuenheim revived the company's fortunes in just two years.

Impressed by the turnaround, in January 1970 Herbert made the young Prussian the new head of BMW, despite his lack of automotive experience. BMW was entering the 1970s in good financial health, in contrast to its dire condition a decade earlier, but von Kuenheim faced formidable challenges. For all its progress, BMW remained a quirky company focused on making small cars for a niche audience. It was overshadowed by Mercedes-Benz, a much larger company with a far broader product line. That the two companies would become head-to-head competitors seemed absurd.

The new CEO concluded BMW was a company with strong technical competence offset by woeful strategic impotence. Among its many shortcomings was the lack of a worldwide sales organization. "We were very provincial," von Kuenheim would recall years later. "Not a European company, not even a German company. It was a Bavarian company."[15]

Nonetheless von Kuenheim set an ambitious goal: making BMW the global leader in sporty luxury cars, a segment where he thought the competition was weak and the opportunity for growth strong. He also set a strategy of going upscale. BMW was a small company that didn't make many cars, he reasoned, so every car it made should command a premium price. To make his vision a reality, however, the onetime refugee needed to modernize BMW. And BMW needed to make a more modern, more sophisticated car.

In 1973 BMW celebrated the completion of its landmark new corporate headquarters in the heart of its hometown of Munich. The complex consisted of four round towers, clustered together, a striking design that evoked the company's high-tech four-cylinder engines. The new headquarters also symbolized BMW's new prosperity, although that remained a work-in-progress.

Developing new cars was glamorous, but von Kuenheim had to tend to a lot of unglamorous infrastructure work before his ambitions for the company could be realized. One early move was developing a company sales organization outside of Germany instead of relying on independent distributors, who took a hefty share of the profits that BMW otherwise could have poured into product development.

Unwinding their distribution rights took time, money, and lots of lawyers. It was a painstaking process that had to be handled country by country. In 1975, with the job completed in the United States and Canada, the company established BMW of North America. It was two decades after Volkswagen had set up its own American sales arm, but BMW's move proved timely. A year later, in 1976, BMW launched its new 3 Series sedans, successors to the acclaimed but aging 2002, in both countries. That was timely, too. The first baby boomers were hitting thirty, and had money to spend.

The first cars in the 3 Series lineup were two-door coupes and four-door sedans, all of which had four-cylinder engines, ranging between 1.6 liters and 2 liters. Convertibles, station wagons, and six-cylinder engines would enter the lineup, too, but not until later.

The new car offered more size and comfort than the hard-edged 2002, but it retained the sporty performance of its predecessor. One BMW ad said the 3 Series "Goes Like Schnell," a pun on the German word for "fast." The company also adopted an advertising tagline that it would keep for decades: "Ultimate Driving Machine."

As the two oil shocks of the Seventies sent gas prices soaring, almost every other automaker, even Volkswagen, was moving to front-wheel drive to reduce the weight of its cars and maximize fuel economy. But BMW stuck with rear-wheel drive. Its engineers insisted it provided the proper front-to-rear weight balance for agile automotive performance. "Front-wheel drive just does not feel like a BMW," one company executive explained, rather archly, some years later.[16]

BMW wasn't alone in adhering to rear-wheel-drive purism. Mercedes-Benz held on, too. BMW's growing prominence was making comparisons between the two companies obvious. Both made upscale, sophisticated luxury cars. But America's nouveau Bobos knew better. To them the Mercedes tristar hood ornament smacked of superficial ostentation, as opposed to the more subtle snootiness of BMW's front grille with its distinctive twin-kidney shape.

The cars felt different, too. Driving a BMW was like wearing a form-fitting, trim-cut Italian designer suit instead of sturdy but boxy threads from Brooks Brothers. Mercedes gave top priority to big cars, which made its smaller cars feel like cheaper versions of the real thing. But BMW's roots lay in small cars. Its bigger models, the 5 Series and 7 Series, evolved from its small cars, which remained at the core of the company. *"Eine Wurst, drei Grösse,"* the company's engineers would say, meaning "one sausage, three sizes."[17] That wouldn't become an advertising slogan, mercifully, but the point was clear.

The yup-noscenti, meanwhile, reveled in the intricate combinations of numbers and letters that formed the BMW model nomenclature. Heck, any dodo who had bologna on Wonder bread for lunch could say "Impala." But only Bobos who appreciated the texture of truffled risotto could rattle off "318i" or "325xiT," and understand what those alphanumeric combinations meant.

The first number in each car's name indicated the model, with the 3 Series being smaller than the 5 Series, and the 5 smaller than the 7. The next two numbers indicated the size of the engine, as measured by the cubic centimeters of displacement (space) in the cylinders for combustion. After that came assorted letters: "i" for fuel injection as opposed to conventional carburetors, "c" for cabriolet (convertible), "T" for touring (a fancy word for station wagon), "x" for all-wheel drive, and so on.

So the 318i meant a 3 Series car with a 1.8-liter fuel-injected engine. And the 325xiT meant a 3 Series all-wheel-drive station wagon with a 2.5-liter fuel-injected engine. Key components came from companies admired by the cognoscenti: transmissions from Getrag, steering systems from ZF Friedrichshafen ("Zed F," among the techno-knowing), and fuel injection from Bosch. A 1982 sales brochure for the 320i described the "Bosch K-Jetronic fuel-injection . . . hemispheric swirl-action combustion chambers . . . transistorized breakerless ignition-system . . . and four crankshaft counterbalances."[18]

Most Bobo-yup Bimmer buyers didn't have a clue about what all that stuff actually did. But "Getrag" and "Bosch" sounded great to those who appreciated the vibration-absorbing system on Rossignol skis. As Brooks observed of the inquiring—well, the acquiring—Bobo mind: "While cultivated people would never judge each other on the costliness of their jewelry, they do judge each other on the costliness of their gear . . . You have to prove you are serious enough to appreciate durability and craftsmanship. You have to show you are smart enough to spend the very most."[19]

BMW's marketing department understood that sentiment just as well as its engineers understood naturally aspirated multivalve engines. So the marketers raised prices. In 1968 you could buy a new BMW 2002 for $2,850. By 1982 you had to pay almost that amount ($2,620) just for the optional sport package (special suspension, upgraded sound system, and so on) on the BMW 320i. If you wanted to buy the entire car, the base price was $13,290.[20]

The higher prices didn't damp demand. They increased it. In 1970 BMW sold fewer than 10,000 cars in America, but its U.S. sales surged to more than 52,000 cars in 1982.

That was the year, as it happened, that the ailing Herbert Quandt died, not long before his seventy-second birthday. He had inherited a family industrial empire with an unsavory past, and restored it to prosperity after World War II. In 1959 he rescued an obscure, ailing little car company, and then oversaw its transformation, under Eberhard von Kuenheim, into one of the most prestigious automotive marques on earth. But even Herbert would have been amazed, it's safe to assume, when BMW's U.S. sales jumped 24 percent in 1985, to 88,000 cars, and raced past Mercedes-Benz for the first time.[21]

The rapid growth occurred despite an anti-yuppie backlash. In 1985 a woman in Seattle told the local newspaper that she was "just tired of their BMW's and Volvos, their Perrier water, their white wine, their self-conscious mellowness, their clothes from Abercrombie & Fitch and L.L. Bean and all the rest of it."[22] But there was no stopping the ascendancy of the yuppies, or of BMW.

A couple years later Henry Ford II visited Johanna Quandt, Herbert's third wife and widow. The man who had declined to accept Volkswagen, even for free, forty years earlier offered to buy the Quandts' controlling stake in BMW. But Johanna Quandt, who had been Herbert's secretary before becoming his wife, had developed considerable business savvy over the years by reading the business news to her nearly blind husband. She and the other Quandts weren't about to sell the family's crown jewel. Johanna rebuffed Henry with words only a fellow industrial aristocrat would understand: "Henry, what would I do with the money?"[23]

In 1983, as the yuppie phenomenon entered full flower, a movie called *The Big Chill* appeared. It's about a group of thirtysomethings who gather for a weekend to mourn the suicide of one of their college friends. The movie's soundtrack features songs from the mid- and late-1960s seared in yuppie memories: "Joy to the World" by Three Dog Night, "A Whiter Shade of Pale" by Procol Harum, and "You Can't Always Get What You Want" by the Rolling Stones. The last might have been a hard message to absorb for many yuppies, who were used to getting their own way. But it fit the movie's theme about the transition from college craziness to the sensibilities (if not always sense) of young adulthood. Millions of real-life baby boomers were making the transition depicted in the movie. Some of them were people who had driven Volkswagens in college in the 1960s but were graduating, as it were, to BMWs in the 1980s. Their personal and automotive journeys were intertwined: from bohemians to Bavarians, from hippies to yuppies, and from Beetles to Bimmers. Larry Schultz was one of them.

Schultz had grown up in Florida before attending a small college in Ohio and forming a rock band for "all kinds of sixties action," as he recalled more than forty years on. For a while the band toured with the Lovin' Spoonful, riding in a Volkswagen Microbus that exuded a whiff of marijuana thanks to occasional acts of vehicular herbicide. After an army stint at Fort Bragg, North Carolina, Schultz became a mergers and acquisitions attorney.[24]

"I did all the sixties stuff, the draft lottery stuff, the school stuff, the lawyering stuff, and the business stuff," Schultz explains. He found himself working on some deals with Fabian von Kuenheim, Eberhard's son, and spending time in Munich. "It was a real 'trip' from our Volkswagen band bus to BMW headquarters and hanging out with the von Kuenheims," he recalls. Along the way he bought a BMW 318i, falling in love with its assertive engine and crisp, precise handling. Three decades and many cars later, in his early sixties and semiretired back in Florida, Schultz was driving a BMW again.[25]

The hippie-to-yuppie trip didn't stop at America's northern border. In Toronto in 1965 Joseph Katz started his hippie phase early in life, getting kicked out of school the first day of seventh grade for showing up wearing love beads, a tie-dyed shirt, and pointy-toed "Beatle boots." The principal sent him home, where his room was decorated with aluminum-foil walls, a black light, posters of Jimi Hendrix, and *Playboy* centerfolds on the ceiling. Young Katz formed a band with his junior-high friends, playing jam sessions in the basement. "I was your typical little hippie kid," he would recall.[26]

On his sixteenth birthday, Katz bought his first car, a used Volkswagen Beetle, for $400. But he was just learning to drive a stick shift, so it took him four hours to get the car out of the underground parking garage where he picked it up. He kept trying, in vain, to master the manual transmission while climbing up the garage's steep ramps. A couple years later he traded the Beetle for a Microbus. "It was just very cool to drive a Bus," he recalls. Indeed, his teachers all came out to ogle it the first day he drove into the school parking lot.

By the time Katz got married in 1987, at age thirty-four, he had outgrown his hippie phase and entered his yuppie period. He got a BMW 325i. "It was light and nimble," he explains, "with a small, square, and simple look, and a hot engine." Later Katz tried a Mercedes for a few years, but dropped it because, unlike his agile Bimmer, "It drove like a tank." After that the financial realities of parenthood kept him out of luxury cars, but Katz's fondest automotive memories remained of his Beetle, his Bus, and his BMW.[27]

Just like its owners, the 3 Series evolved. Every six or seven years BMW developed a new generation of the car, each with more size and horsepower. The third generation, launched in 1990, marked a major styling breakthrough. BMW abandoned its steadfast adherence to boxy, squarish lines and adopted styling with sleek angles and curves. The Bobo-yuppies followed. In 1999 the company stopped selling four-cylinder engines in the U.S. and offered only sixes. At the time, gas cost less than bottled water (another yuppie staple), and Americans craved high horsepower. Not until 2011, when gasoline topped $3.50 a gallon, would BMW of North America reintroduce a four-cylinder engine. It had, in the company's words, "twin-scroll turbocharging, high-pressure direct injection and 260 pound-feet of torque."

During this time the 3 Series solidified its standing as the world's standard-setter for upscale, sporty driving. The new luxury marques from Japan—Acura, Lexus, and Infiniti—introduced models aimed at dethroning the 3 Series, but automotive writers dismissed them as "Wannabe-Threes." Such was the stature of the 3 Series that BMW executives lapsed into describing the car's attributes as "three-ness," whatever that meant.

In 1992, near the end of his long and remarkable career, von Kuenheim made one last bold stroke. He announced that BMW would build its first assembly plant outside of Germany, in Spartanburg, South Carolina. It was the culmination of his steady, relentless expansion of the company. But BMW had made its marque, as it were, by touting German engineering. The CEO had to reassure customers that American-made Bimmers would be just as Teutonic as the ones from Munich. "A BMW is a BMW," von Kuenheim declared at the groundbreaking ceremony, "even if it is built in South Carolina."[28] The man who found BMW a provincial company left it as a global one.

During the 1980s and 1990s the word "BMW" began surfacing with curious regularity in articles written by restaurant reviewers and food critics. Usually it was in close proximity to "goat cheese" and "radicchio" and, especially, "arugula." The tangy Italian green was becoming a staple in Bobo-yup salads, along with heirloom tomatoes, hydroponic watercress, and organic anything. A food writer in Chicago described the trend toward "status-conscious salads" by observing: "Reeboks on your feet, radicchio and arugula on your plate, a BMW parked outside—they all [go] together."[29] Half a world away in Australia, the *Sydney Morning Herald* described arugula, with appreciative enthusiasm, as "the botanical equivalent of the BMW."[30] It was high praise from down under.

By the time the Aughts rolled around there were some mishaps in the metaphor, as sometimes happens. In 2009 an aging yuppie had an accident in her BMW while driving through the college town of Boulder, Colorado. The vehicle spun out of control and plunged through the front window of a restaurant named . . . Arugula.[31] But the connection between the yuppie car and the leafy vegetable survived that unfortunate incident.

Also that year, by which time the BMW 330i had evolved into a car with a 300-horsepower engine and a price upward of $40,000, a gardening guru from Delaware mounted a mission of mercy to the ex–Soviet Republic of Georgia, in the remote Caucasus Mountains.

He brought seeds for six different varieties of arugula, so the natives could lift themselves from poverty by growing designer salads for chichi European restaurants. The yuppie Samaritan explained to his local newspaper that before he had discovered the commercial potential of arugula, "I was a redneck forced into a three-piece suit and a BMW."[32] It was a career trajectory only a yuppie could love.

By this time it was clear that millions of once-young but still upscale Americans, now around sixty years of age, wouldn't be returning en masse to driving Cadillacs or eating iceberg lettuce. Had a long-dead Irish poet named Yeats remained alive to observe this transformation, he might have marveled at the meaning of it all, and written a different version of one of his best-loved poems:

Picture then two yuppies, and behind them a third,
Riding in a BMW painted brilliant blue, like lapis lazuli.
Their car climbs and climbs, ever upwards,
Towards fields of free-range organic arugula,
Tenderly tended with tools
From the Smith & Hawken catalogue.

They are climbing towards the confluence of cars and cuisine
At the crossroads of the effete and the elite.
You might delight to see them there, staring at the passing scene,
When one asks for mournful music, and accomplished fingers
Send the sound track from *The Big Chill*
Piping through the Harman Kardon upgrade speakers.

In days long past BMW was an automotive also-ran
And arugula just a foul-smelling Italian weed.
Their simultaneous ascent surely was karma,

Because another name for "arugula," wise foodies know, is
   "rocket."
The yuppies' eyes grow wrinkled now,
Surveying fields lit by high-beam halogen headlights,
And their hair, their ancient glittering hair, is gray.

11

THE JEEP:
FROM WAR TO SUBURBIA, OR HOW TO LOOK
LIKE YOU'RE GOING ROCK CLIMBING WHEN
YOU'RE REALLY GOING TO NORDSTROM

If you took someone in the right demographic group—young, educated, and rich—you could tell whether they would buy a minivan or a Jeep by looking at their watch. If they had a Timex they were practical, unpretentious people: minivan buyers. If they wore a Rolex they would be Jeep buyers. They were more concerned with appearances than with practicality.

—R. S. Miller Jr., former Chrysler executive[1]

In 1986, while Chrysler basked in the success of the minivans and its corporate comeback, a spirited debate raged among the occupants of its executive suite, well out of sight of the press and public. The company's coffers were bulging. Lee Iacocca wanted to spend some of the cash to buy tiny American Motors Corporation. It had long been the Dog of Detroit.

The late-1950s glory days of George Romney and the Rambler were long gone, and AMC was mired in losses. A movie about the company could have been called *A River of Red Ink Runs Through It*. In 1981, during one of its recurring corporate crises, AMC sold a controlling interest in its stock to the French automaker Renault. Renault's major motivation for the deal was to fuel its own automotive ambitions in America. It had little to do with American Motors' best-known marque, which was Jeep.

Jeep had an illustrious past. The vehicle helped the Allies win World War II and had symbolized life on the front lines for ordinary GIs. But that past had given way, like the lives of many individual war heroes, to a less than distinguished civilian existence. The marque had endured a succession of corporate owners, all of them earnest but impecunious. Jeeps had stiff suspensions and harsh rides that made them known as "kidney busters," an all too accurate anatomical image. Their tendency to roll over at highway speeds prompted hundreds of lawsuits.

For more than forty years after World War II, Jeeps were déclassé, as a 1984 incident comically illustrated. AMC was supposed to supply Renault Alliance sedans as the official car for that year's Miss America parade in Atlantic City, but a production glitch required the company to substitute some Jeeps instead. "If that's what she has to ride in, then that's what she has to ride in," said a resigned pageant official from Alaska.[2]

A year later movie director Ron Howard approached AMC about using the Jeep factory in Toledo, Ohio, to film *Gung Ho*. AMC officials found the theme of comic culture clashes at a foreign-owned factory uncomfortably close to home, and rebuffed the request. "We're just not interested in participating in a story where an auto plant is having problems and a Japanese company comes in and saves the day," an AMC executive huffed.[3]

At about the same time an outbreak of sabotage hit the Toledo factory. Workers angry over a contract dispute threw wrenches into the machinery and damaged some of the vehicles. A restraining order by a federal judge restored order in the plant.[4]

Comedic mishaps. Worker vandalism. Never-ending financial woes. It was understandable why most Chrysler executives—including Hal Sperlich and almost the entire engineering team—wanted no part of American Motors. They viewed it as a longtime loser that could derail Chrysler's miraculous recovery, or siphon off money that the company could use to improve its minivans and develop other vehicles.[5] But Iacocca saw something different in ailing AMC. The minivan's success showed that many Americans aspired to lifestyles that demanded more versatile vehicles. In one significant way Jeeps were more versatile than minivans. Their four-wheel drive allowed them to venture beyond the pavement for camping, hiking, or other outdoor adventures.

It wasn't clear how many Americans were true outdoorsmen who knew that "hip boots" meant fly-fishing gear instead of stylish footwear. But a curious trend was afoot. Lots of upscale yuppies, the same people driving BMWs, started wearing L.L.Bean shirts, Patagonia windbreakers, and Timberland boots, even in places like Midtown Manhattan, where wilderness meant Central Park.

Wresting AMC from the French would involve traveling a road with enough potholes to daunt the sturdiest Jeep, but Iacocca determined to do it anyway. His vision would prove prescient, but it wouldn't be the last stop on Jeep's corporate journey.

"While the Jeep was probably the most spectacular single accomplishment" of the army's procurement system during World War II, a wartime government report concluded, "its conception and birth were not achieved without travail."[6] That put it mildly. But the vehicle's birthing pains are all but forgotten today.

One place they're remembered is Butler, Pennsylvania, a steel town thirty-five miles north of Pittsburgh. A roadside marker on Hansen Avenue designates where the Jeep was born. The U.S. Army needed to replace another hardy, go-anywhere transportation device: the mule. Mules had been the workhorses, so to speak, of the First World War, hauling everything from artillery to supply wagons to letters-from-home to and from the front.

But the success of Germany's blitzkrieg in the spring of 1940 startled the American army, not to mention the hapless French and British. One joke of the day was that Hitler had ordered 10,000 tanks from General Motors but said don't bother shipping them to Germany; the Wehrmacht would pick them up while passing through Detroit.[7] Gallows humor aside, military strategists became convinced that the new war, which America would soon join, would require a lightweight, all-terrain, fast-moving scout vehicle. And it had to carry a mounted machine gun, the driver, and at least one passenger. Army procurement officials asked 135 different companies—automakers, machinery manufacturers, and others— to bid on the contract. Only two companies did.[8]

The specifications were ridiculous. The army wanted a four-wheel-drive vehicle, a rarity at the time, with a four-cylinder engine, which was becoming another rarity, at least in America. Worse yet, the vehicle's weight was limited to 1,275 pounds, but it was supposed to carry a 600-pound payload *in addition* to a driver and passenger. That meant the total load could be almost as heavy as the vehicle itself. And the winning bidder would get just forty-nine days to submit a prototype for rigorous testing.[9]

Such impossible specifications would attract bids only from companies that were supremely confident or truly desperate. The American Bantam Car Company of Butler was the latter. It had been founded as American Austin Car Company in 1929, the year of the Great Crash on Wall Street, to build a U.S. version of the English Austin. Five years later the company failed. One of its salesmen acquired its remnants and relaunched it as American Bantam.[10]

There was little demand for cars in the Depression, and by 1940 American Bantam was barely hanging on. The army contract offered potential salvation, except for one problem: American Bantam had laid off most of its workers, and was employing only a skeleton crew at the time. The company contacted an independent engineer from Detroit named Karl Probst, who hesitated to take the assignment because he would be paid only if American Bantam won the contract.

But he signed on anyway, drove from Detroit to Butler, worked around the clock, and turned out blueprints for the new vehicle in five days. His design was more than 700 pounds overweight. But Probst figured—correctly, as it turned out—that the army would have to raise its weight limit. No vehicle under 1,300 pounds could meet the army's payload specifications. The only other bidder, Willys-Overland of Toledo, couldn't meet either the weight limit or the forty-nine-day deadline for producing a prototype. In early August the army notified Bantam it was the winner.

As it happened, Bantam barely made the forty-nine-day deadline itself. Finding axles that could handle the strain of four-wheel drive was difficult. It wasn't until September 22, 1940, the day before the deadline, that Probst completed testing the prototype. The next day he and a colleague climbed behind the wheel and drove the vehicle themselves to the army's Camp Holabird in Baltimore. They arrived with just a half hour to spare.[11]

Army bureaucrats called their new prize the General Purpose vehicle, or GP, which sounded like "jeep" when pronounced quickly. By coincidence, Eugene the Jeep was also the name of a cartoonish character—with floppy ears, a bulbous nose, and a pot belly—in the *Popeye* comic books of the day. Whatever the origin, GP or Eugene, the name stuck.

After America entered the war, the jeep—at first it was a generic term, with a small "j"—became as much of a GI staple as Spam. "It's as faithful as a dog, as strong as a mule and as agile as a goat," wrote war correspondent Ernie Pyle.[12] Cartoonist Bill Mauldin's wartime work featured Willie and Joe, two stubble-faced, bedraggled GIs, plus many a jeep. Perhaps his most memorable cartoon showed a grizzled sergeant pointing his pistol at the hood of his broken-down jeep, with eyes shielded and head turned away, to put the beloved vehicle out of its misery.

With their 60-horsepower engines, rugged transmissions, and four-wheel drive, jeeps pulled guns, hauled ammunition and people, and sometimes served as attack vehicles, all far beyond their intended mission as scout cars. Their hoods provided improvised clinics and altars in the field. In all some 650,000 jeeps were built during the war, but only 2,675 of them were made by Bantam.

The army determined it needed far more jeeps than the little Pennsylvania company could produce. So it gave the bulk of the production contracts to Willys-Overland, which named the engine in its jeeps the "Go-Devil," and to Ford, which called its own jeep prototype the "Pygmy." Bantam officials, disappointed that they didn't get more production orders themselves, had little recourse. After they were paid for their prototype, the jeep's design became property of the army, which was more concerned with getting enough vehicles to fight Hitler and Hirohito than with saving a little company in Pennsylvania.

Bantam did get some military contracts for special trailers designed to be hauled by jeeps and for other unglamorous items. Willys-Overland, meanwhile, tried to stake its claim to the brand name with illustrated advertisements showing jeeps in action.

"Invasions are 'meat' for the mighty Jeeps today," one company ad declared in 1943. It showed the vehicle storming ashore with U.S. troops at Salerno. The ad pointedly, and prematurely, spelled "Jeep" with a capital "J," as if it were a brand.[13]

In 1945, with the war winding down, Willys-Overland tried to reposition jeeps for civilian use with advertisements headlined "From Fighter to Farm Hand."[14] The company also launched a peacetime version of the vehicle called the CJ- (for "Civilian Jeep") 2A. Unlike the military jeep, it had modest creature comforts, including a tailgate and even a choice of colors, as opposed to the army's uniform olive drab.

The future, though, didn't seem promising. Jeep "deserves a longer life than it probably will enjoy," *Fortune* wrote in 1946. "As a fad it is selling even in passenger-car markets now, though by such standards it is uncomfortable and expensive. Generally, it is a piece of machinery, and has to be sold as such."[15]

The next two years brought more new models intended to be everyday passenger cars instead of farm tools, including the Willys Jeep four-wheel-drive station wagon and the Willys Jeep Truck. They were plain and functional, but a more stylized creature called the Jeepster followed in 1948.

The Jeepster was a "phaeton," a two-door touring car without a fixed top or even side windows, which made it sort of a cross between a roadster and a boxy Jeep CJ. It didn't even have four-wheel drive, but offered rear-wheel drive instead. Advertisements showed it on the glamorous playgrounds of beaches and country estates, but well-heeled Americans weren't moved to ditch their Cadillacs. Willys dropped the Jeepster in 1950, when war began in Korea and spurred demand for military jeeps again.

That same year Willys-Overland managed to trademark the jeep name. The effort required protracted wrangling with the aggrieved Bantam and even some congressional hearings, but after the legal and legislative smoke had cleared, jeep became Jeep. A year later Bantam folded, passing into history and out of memory, with its signal contribution to America soon forgotten.[16]

By that time a Jeep had become a television star. The *Roy Rogers Show*, which debuted in 1951, featured its clean-cut cowboy namesake and his faithful companions: wife Dale Evans, horse Trigger, and a Jeep named Nellybelle. In one episode Nellybelle started careening down a hill, out of control, with no driver behind the wheel. Roy hopped on Trigger, chased down Nellybelle, and leaped from his horse into the runaway vehicle, saving it from certain doom. "Roy, you shouldn't have taken a chance like that for Nellybelle!" admonished Dale. The cowboy sheepishly replied: "Well, she's part of the family, Dale."[17]

Over the next half century Jeep itself, like Nellybelle, would have many more close calls in its struggle for corporate survival.

Industrialist Henry J. Kaiser, who made a fortune building ships during World War II, had a sprawling industrial empire that included steel, aluminum, cement, gravel, and more. In 1951 his little car company, Kaiser-Frazer, began buying Jeep engines from Willys-Overland to install in its new small car named, with unabashed immodesty, the Henry J.

It was a cheap and rickety econo-car with a cramped backseat that musician Frank Zappa would remember from his boyhood, as he wrote decades later, as "the ironing board from hell."[18] With cars like that, it was no wonder that Kaiser-Frazer was a money loser, too small to compete against the Big Three giants of General Motors, Ford, and Chrysler. Kaiser needed either to get a bigger car company or get out of the business entirely. In 1953 he chose the former. Instead of just buying Jeep engines, he bought the whole company. Kaiser-Frazer merged with Willys-Overland and renamed it Willys Motors, with headquarters near the Jeep factory in Toledo.

The deal sparked some industrial aftershocks. It created an urge to merge among America's remaining little car companies, the so-called "independents" that weren't part of the Big Three, including struggling Nash-Kelvinator and Hudson Motor Car. In 1954 they merged to form American Motors Corporation, a company that would later play a major role in Jeep's future.

During its first thirteen years of existence, Jeep had changed owners as often as a well-worn jalopy. Bantam, Willys-Overland, Kaiser-Frazer, and Willys Motors: the corporate shuffles set a pattern for Jeep that endured. Overlooked amid all the maneuvering was an event called the Jeepers Jamboree in California, which was held just one month after Kaiser bought Jeep. It was a motorized mountaineering expedition by Jeep enthusiasts, who trekked from the town of Georgetown, in the Sierra Nevada foothills, to the hamlet of Rubicon Springs, just west of Lake Tahoe. The forty-five-mile journey covered terrain so rugged that the average speed in some stretches was just 3 or 4 miles an hour.

Jamboree participants used their Jeeps for organized fun instead of for fighting or farming. But Americans weren't yet flocking en masse to the great outdoors or even pretending that they were. Jeeps continued to be sold more for work than for play. Some even came with dashboard-mounted instructions on how to use the vehicle's engine to power farm implements. The Jeepers Jamboree would become an annual event, but the recreational market would be slow to develop.

In 1954 Willys Motors also launched the CJ-5, the newest in the Civilian Jeep line. It retained Jeep's characteristic round headlights, but was bigger and more refined than its predecessors, with wider and softer seats, curved fenders, and rounded hoods. Willys would build the CJ-5 for three decades, longer than the Model T Ford, a production run that would span presidencies from Eisenhower's to Reagan's. Longevity, however, didn't equal prosperity.

Despite Henry Kaiser's dream, Willys Motors wasn't strong enough to compete with the Big Three, at least not with a broad product line. By 1956, with tail fins sprouting on America's automotive landscape, the company decided to stop making regular cars, including the woeful Henry J, and to focus exclusively on Jeep. Even that didn't help much. The Jeep Maverick station wagon, launched in 1959, aimed at the market for everyday family transportation, but sales sagged because Jeep still had a farmhand image. Jeep brought out a model called the Surrey the next year. It was a tarted-up CJ with bright colors and a pastel-striped canvas top that made it look like a fruit basket on wheels. Farmers didn't buy it, nor did much of anyone else. In 1963 the company changed its name yet again: to Kaiser Jeep instead of Willys Motors, though the corporate ownership remained the same.

That year Kaiser Jeep launched the Wagoneer, in many ways the first modern Jeep. The Wagoneer offered the option of four doors and an automatic transmission instead of a standard stick shift, a first for any four-wheel-drive vehicle.[19] Other options included floor carpeting, instead of the standard rubber floor covering, and softer seats. The Wagoneer was modestly successful, and a more basic, two-door version called the Jeep Cherokee followed a decade later.

The Wagoneer and the Cherokee had creature comforts that brought Jeep a long way from the mud and blood of World War II. Jeeps were becoming civilized. One upscale version of the CJ-5, called the Tuxedo Park, came with chrome bumpers, mirrors, and hood latches. Buyers could choose convertible tops in matching or contrasting colors. Froufrou ruggedness, however, was still years ahead of its time.

In the Sixties and Seventies, camping, hiking, and other woodland pursuits were for the Boy Scouts in their quest to be "physically strong, mentally awake and morally straight."[20] None of that, especially the morally straight part, had much to do with Dick and Liz in Puerto Vallarta or Frank Sinatra's Rat Pack in Las Vegas, which were America's prevailing view of the good life in the early 1960s. Wilderness chic was oxymoronic in a country that hadn't yet discovered helicopter skiing, pastel Polartec fleece, or five-star fly-in hunting lodges.

So Kaiser Jeep trudged along, just like its vehicles. In 1967, faced with growing competition from the International Harvester Scout, the Ford Bronco, and other Jeep knock-offs, the company launched a new Jeepster series aimed at the "fun and recreation market."[21] Unlike the original Jeepsters, the new vehicles had four-wheel drive. A few years later a British rock band named T. Rex sang, "Girl, I'm just a Jeepster for your love."[22] Which presumably conveyed an expression of affection.

The Wagoneer and Jeepster, along with royalties from companies that manufactured Jeeps overseas, produced modest profits for Kaiser Jeep. But by 1970 Henry J. Kaiser had died. His son, Edgar, was running the Kaiser conglomerate. Seeing little future as a second-tier car company, Edgar sold Kaiser Jeep to American Motors that year for $70 million, giving Jeep yet another corporate owner. "Jeep can and will be a major profit contributor to American Motors," AMC's 1970 annual report promised.[23]

That wasn't saying much, considering the profit contribution, or lack thereof, from the company's other products. One memorable AMC advertisement from 1970 asked: "If you had to compete with GM, Ford, and Chrysler, what would you do?"[24] The company's answers, alas, were the Hornet and Gremlin, cars that could have made an automotive all-ugly team.

As the Seventies unfolded, AMC's annual report to shareholders rarely mentioned Jeep, except in the back pages. The company periodically tried gimmicks like outfitting Jeeps with blue-denim seat covers, which maybe made sense in a decade of Nehru jackets and double-knit polyester leisure suits. But the efforts didn't do much for sales. Nor did the publicity from the 1978 Expedicion de las Americas, a privately organized trek by hard-core off-roaders who drove five Jeeps from Tierra del Fuego to Alaska's Prudhoe Bay. Over 21,000 miles and five months the group traversed terrain that ranged from jungles to glaciers, a journey worthy of two future television shows: *Planet Carnivore* and *Ice Road Truckers*. Miraculously, the Jeeps and their drivers returned intact.[25]

AMC itself, however, was mired in financial quicksand. Throughout the 1970s the company never escaped the assessment that an automotive analyst offered early in the decade: "It will probably take a miracle to keep American Motors going."[26] As events unfolded, it didn't take a miracle. It took the French.

For a half century after World War II, the French government controlled the country's largest car company, Régie Nationale des Usines Renault (the government stills owns 15 percent). Renault's main mission wasn't to make cars but to make jobs. Sons and nephews of government ministers filled the management ranks. The factory floors were bloated with workers doing featherbedded jobs.

For a while during the 1950s Renault had been the largest foreign nameplate in America, until Volkswagen blitzed right by it, as it were. Renault's executives, however, continued to harbor visions of global *la gloire*. During the 1960s and 1970s, Renault tried to sell its cars in the United States and Canada through a distribution deal with AMC dealers.

But in the late 1970s AMC plunged into the latest of its recurring financial crises. The company's collapse seemed likely, in which case *la gloire* would be *la gone*, Renault would lose its American sales channel, and lots of jobs back in France would be threatened.

Renault responded by buying 47 percent of American Motors, injecting fresh cash and installing its own executives at AMC's helm. Jeep got yet another new corporate owner and AMC got a new nickname: "Franco-American Motors." It was a none too flattering reference to a brand of precooked canned spaghetti.

Controlling AMC, however, let Renault try new tactics to expand in the American market. Besides just exporting French cars to the United States, Renault would use AMC's factories to build cars developed by French engineers. The cars would use Renault components, of course, to keep Renault's French factories humming. Thus was born the 1982 Renault Alliance, a compact car so devoid of personality and performance that it was soon termed the "Renault Appliance." Neither AMC nor its cars were faring well on the nickname front.

They were stumbling on the public-relations front, too, due to a decision about headlights, of all things. On January 28, 1986, the last Jeep CJ-7 rolled off the assembly line in Toledo, ending a three-decade run for the CJ models that began with the original World War II jeep. The CJ's replacement was the Jeep Wrangler YJ, which had square headlights instead of the traditional round ones that had lit the way to victory during the war.

Hard-core Jeep enthusiasts said, "*Sacré bleu!*" Or something like that. Some of them started sporting bumper stickers and T-shirts that said: "Real Jeeps Have Round Headlights."[27] Adding

PAUL INGRASSIA

to the sacrilege, the Wrangler YJ was the first Jeep to be built in Canada instead of the United States. Rumors spread that the letters "YJ" stood for "Yuppie Jeep," which might have provided comfort to BMW except that it wasn't true. The letters were just a model code with no particular meaning.

As for the CJ, *tout de même,* it had outlived its time. Its narrow wheelbase and high ground clearance made it prone to rolling over at highway speeds, especially around curves. The product-liability lawsuits against Jeep were mounting into the billions of dollars. It was like Ralph Nader and the Corvair all over again.

The Wrangler YJ was lower and wider than the CJ, which made it more stable and comfortable at highway speeds. And the highways were where more Jeeps were confining their journeys. By 1986, 95 percent of Jeep drivers used their vehicles for driving around town as opposed to clambering over mountains, AMC's researchers found. That was up from just 17 percent eight years earlier. As for "frequent off-roading," only 7 percent of Jeep owners said they were doing it, down from 37 perfect in 1978.[28] It was like people started wearing hiking boots to the grocery store, which, in fact, many people were. The down-and-dirty workhorse of World War II was becoming a fashion statement.

In 1986, a developer in Dallas lost his Jaguar in a divorce and bought a new Jeep. He figured he came out ahead on both fronts.[29] Around the same time an AMC dealer near Chicago introduced a stretched 16-foot Jeep limousine with a television set, intercom, and wet bar. A Houston dealer unveiled his own Jeep limo that was 19 feet long. Everything being bigger in Texas, of course.[30]

By this time many Jeeps actually were being sold with two-wheel drive instead of four-wheel drive, rendering them incapable of handling anything tougher than a speed bump. A two-wheel drive Jeep looked the part but made no more sense than fake UGG boots in colors called "tomato" or "sorbet." But stuff like that was popping up everywhere. Radical chic was long dead in America. Wilderness chic had come alive.

In the mid-1980s sales of L.L.Bean's canvas "field coats" started surging in New York City, which puzzled the people who made them. Bean's field coats, designed for hunters, had oversized pockets for holding dead birds. There weren't many fields in New York. Most of the birds were pigeons, but they were for feeding, not shooting.

Bean's market-researchers delved into the mystery. They found that even though most New Yorkers were antihunting types, they liked to look like hunters. It was fashionable to traipse down 5th Avenue looking like you had just walked out of the woods, as long as what you were wearing wasn't itchy or smelly. Bean began a process its executives called "mainstreaming," which meant making the company's field coats suitable for people who did their hunting in the sales racks at Saks instead of in the woods of Maine. The oversized pockets for holding dead birds, deemed unnecessary, disappeared.[31]

Bean soon decided to mainstream much of its product line. Strap-on headlamps suitable for nighttime hunting and fishing in the deep woods became smaller and more comfortable for jogging on suburban streets after dark. The company made its sleeping bags lighter, more suitable for indoor slumber parties than camping out in the cold. Backpacks, which had been teardrop-shaped for maximum ergonomic hiking comfort, got squared-off contours to hold schoolbooks.[32] Even the company's original product, the rubber-and-leather Bean Boots, got an array of new colors: coral, "green tea," and bright yellow. Company founder Leon Leonwood Bean, long since dead, probably rolled over in his tropic-weight cargo pants.

Mainstreaming wasn't confined to Bean. Orvis transformed itself from a specialized maker of fly-fishing gear into a "lifestyle company" with a broad product line. The journey would lead to Orvis catalogs offering bison-leather baseball caps, "hemp indigo Montana morning jeans," and specially designed "anti-tail-wagging furniture" with sunken tabletops, to prevent friendly fidos from whacking over glasses of Côtes du Rhône.

In 1987 John Peterman started a catalog company to sell full-length leather duster coats worn by cowboys in the Old West. "Protects you, your rump, your saddle and your legs down to your ankles," his early catalogs declared. Peterman ran advertisements in the *New Yorker* and the *Wall Street Journal*. The ads explained that the coats could withstand "the winds of Wyoming and the blizzards of Wall Street," where covering one's ass was important. As it happened, Peterman sold more duster coats in the concrete canyons of lower Manhattan than in Wyoming.

Another adventure-clothing company, the Territory Ahead, touted its own similar jacket as "a hard-riding, easy-wearing saddle coat that takes the bluster out of a blue norther faster than you can say 'Saskatchewan.'" Which was pretty hard to do, fast or otherwise, after your second martini at Grand Central Terminal following a hard day of trading credit default swaps. But that was the point. Customers of these catalogs didn't have to do any of that country squire or outdoors adventure stuff. They just had to want to *look like* they did it. Catalog executives discovered that outdoor gear would sell to indoor people, and there were many more of them than true outdoorsmen. John Peterman wasn't so much selling clothes as selling dreams, or "living life the way we wish it was," as he put it.[33] That's what "mainstreaming" was all about.

"People want to be in the costume," L.L.Bean executives told one another. "They want to be part of the tribe."[34] Which meant wearing Bean Boots and a field coat on rainy-day runs to Target. Or buying the photographer's vest from Banana Republic and Zambezi twill cargo shorts from Orvis for crack-of-dawn treks to Starbucks for a Venti Java Chip Frappuccino.

America's emerging wilderness chic spelled opportunity for a young Californian named Yvon Chouinard, who was genuinely passionate about mountain climbing. In the 1950s he had started a little company to make carabiners and pitons. But the market for carabiners and pitons was limited.

In the 1970s, after naming his company Patagonia, Chouinard figured more people would buy outdoor clothing than mountain-climbing tools. It had to be "multifunctional technical clothing," as he put it, that kept you warm and dry even when the weather around you was cold and wet.[35] And it had to be fashionable. "We drenched the Patagonia line in vivid color," Chouinard would write. "Cobalt, teal, French red, mango, seafoam, and iced mocha. Patagonia clothing, still rugged, moved beyond bland-looking to blasphemous. And it worked." Between the mid-1980s and 1990 Patagonia's annual sales surged fivefold, from $20 million to $100 million.[36]

If pastel-colored Patagonia parkas could keep people alive on the slopes of Denali, they also could keep people comfortable in new indoor rock-climbing gyms. They started springing up around the country, complete with thirty-five-foot fake-mountain walls and readymade slots for inserting carabiners, pitons, and crampons.

The boom in outdoor recreation, and in simulated outdoor recreation, should have been a bonanza for AMC and Renault. Americans bought nearly 154,000 Jeeps in 1984, nearly double the year before.[37] But by the mid-1980s the French had soured on their American adventure.

Their only real goal was to sell Renault cars in America, which was like trying to sell Hamburger Helper in Paris. U.S. sales of Renaults plunged 35 percent in 1985. The AMC plant in Kenosha, Wisconsin, that built the Renault Alliance was a converted turn-of-the-century mattress factory that needed to be modernized. But by 1986 Renault had invested $645 million in AMC, more than twice as much as it had planned.[38]

Renault's despair spelled opportunity to Lee Iacocca. His own idea of outdoor recreation was lounging by the pool in Palm Springs, or maybe walking a few dozen steps from his limo to the corporate Gulfstream V. But Chrysler's CEO had a knack for sensing the pulse of America, as he had done with the Mustang and the minivan. Jeep, he figured, perfectly suited Americans' active, or pseudoactive, lifestyles. Besides, he couldn't resist the idea of rescuing an American icon from the French. Every other top executive at Chrysler thought buying AMC would be crazy. But Iacocca had the only vote that counted.

Negotiating the deal took a year. Renault wanted to stick Chrysler with $2 billion in liabilities from Jeep CJ rollover lawsuits. And the French wanted to use Chrysler, like AMC, as a channel for selling their cars in America. In the end Chrysler agreed to use an AMC factory in Canada to build a car engineered by Renault, and to split the rollover liabilities. In March 1987 Chrysler announced it would buy Renault's stake in AMC along with the rest of the company. Jeep got the sixth corporate owner of its forty-seven-year life span. America would soon embrace a new cultural craze.

When Chrysler bought American Motors in 1987, the best-selling model in the Jeep lineup was the four-door Cherokee. It left a lot to be desired. The interior was cramped and tacky. The most powerful engine was an anemic 135-horsepower V6 that was fine for crawling over rocks but was slower than a mule when traveling on the highway. Chrysler soon substituted a bigger engine, with 177 horsepower, and created an upscale version called the Cherokee Limited with leather seats, gold-paint body stripes, and gold-tinted aluminum wheel covers.

The trimmings made the Cherokee Limited look like a tarted-up tin can, but the extra horsepower and flashy trim proved surprisingly popular. Chrysler dealers couldn't keep them in stock. Some company executives got calls at home from their friends, wondering how they could get their hands on one. Chrysler was mainstreaming Jeep, and Americans were snapping up Cherokee Limiteds.

The rest of Chrysler's product line wasn't faring so well, just as Iacocca's colleagues had feared. The money spent to buy AMC had diverted dollars from product development, leaving many of Chrysler's cars outmoded. The Renault car that Chrysler had agreed to sell was a costly flop. But Cherokee sales jumped 33 percent in 1988 and another 5 percent the next year. The word "SUV" entered the Merriam-Webster dictionary as an acronym for "sport-utility vehicle."[39]

Chrysler's competitors, surprised by Jeep's success, rushed to respond. In 1990 Ford replaced its aging and clunky Bronco II with a new four-door SUV called the Explorer. It offered power locks, power windows, leather seats (heated, of course), and a sunroof, plus more interior space than the cramped Jeep Cherokee. The Explorer rushed past the Cherokee in sales.

Detroit's great SUV war began. In 1992 Chrysler countered the Explorer with the Jeep Grand Cherokee. It had sleek styling, an optional V8 engine, and a carlike "unibody" construction, instead of the Explorer's V6 and heavier body-on-frame truck architecture. The Grand Cherokee literally made a smashing debut at the Detroit Auto Show when vice chairman Bob Lutz drove one up a flight of steps, through a plate-glass window, and onto the show floor. Unbeknownst to onlookers, the stunt had been faked with tiny explosives rigged to shatter the glass at just the right moment.

Next came the merging of cars and couture, with a heavy dose of wilderness chic. Ford introduced the Explorer Eddie Bauer edition, with enough leather inside to make two or three duster coats. Jeep countered in 1995 with the Grand Cherokee Orvis Edition: moss green on the outside, "green and champagne" on the inside, and Orvis leaping-trout logos on the floor mats. It was the right equipment, one writer opined, for "climbing Mount Upscale."[40] Two years later Chrysler restored round headlights to the Wrangler, pacifying Jeep purists at last.

By the mid-1990s SUVs were proliferating into subspecies. They ranged from little "cute utes," as the car magazines called them, to extralarge "brute utes," such as the Ford Expedition and the Chevrolet Suburban. Japanese car companies were slow to respond, but they belatedly realized that SUVs weren't just a passing fad. Suzuki introduced an SUV called the Grand Vitara, which sounded like something from an old *Flash Gordon* episode, or maybe a minor potentate in the Ku Klux Klan.

Within a few years Toyota developed a lineup of five SUVs, ranging from the cute-ute RAV4 (which meant Recreational Activity Vehicle, four-wheel drive) to the bulging Land Cruiser that weighed nearly three tons. Many middle-aged women, tired of minivans and their attendant soccer-mom image, bought them. One New Jersey homemaker who had driven four different minivans over a dozen years bought a Toyota 4Runner that had front-end brush bars, a tubular-steel attachment designed for traversing the African bush country instead of the parking lots of Paramus. Another woman, in Sarasota, bought her $40,000 Range Rover to "go antiquing," and told the *Wall Street Journal* that she'd never even consider driving it off-road. "Ugh. Imagine the dirt," she explained.[41] Willie and Joe, Bill Mauldin's cartoon GIs, would have rolled their eyes.

Upscale SUVs started to supplant sports cars as fashion statements. In Mexico, Chrysler produced a video that showed a Porsche, bogged down in flooded streets, being passed by a Jeep Grand Cherokee pulling a water-skier, who "saluted" the hapless roadster with his middle finger.[42] In the United States, the video was shown often at Chrysler's U.S. sales meetings.

In 2002 Porsche launched its own SUV, the Cayenne, with a 405-horsepower GTS version that could zoom from 0 to 60 miles an hour in a neck-snapping 5.7 seconds. A hot-rod SUV was like a cross between a hippopotamus and a cheetah. But SUVs produced profits of $10,000 to $15,000 per vehicle, ten times the amount that car companies made on ordinary sedans. Big SUVs led to still bigger SUVs in the quest for still bigger profits. Less wasn't more anymore.

And more was still to come. In 1990, actor Arnold Schwarzenegger was in Oregon filming *Kindergarten Cop*, a role that perfectly prepared him for his next act as governor of California. He spotted a convoy of Humvee military vehicles, whose heavily muscled bodies were not unlike his own. When a civilian version debuted a couple years later, under the brand name Hummer, he decided to buy several.

Humvees and Jeeps had a lot in common, including corporate ancestry. AM General, which made the Humvee, had been a subsidiary of American Motors until 1983, when it was sold. Jeeps and Humvees both began as military machines with acronyms: "GP vehicle" for Jeep, and High Mobility Multi-purpose Wheeled Vehicle (HMMWV) for Humvee. Acronyms, like everything else in life, had gotten more complicated since World War II.

In 2003 AM General, in a partnership with General Motors, introduced a modified Humvee called the Hummer H2. Like L.L.Bean field coats, the Humvee was being mainstreamed, if you could use that term for a vehicle that weighed more than three tons and was wide enough to fill two parking spots in suburban shopping malls. It got about 12 miles a gallon. Environmental groups hated the Hummer H2, as did Chrysler, though for a different reason. Shortly after the H2 debuted Chrysler sued GM, claiming that the H2 copied the decades-old design of Jeep's front grilles, which featured seven vertical slots. A federal court ruled that a face couldn't be patented, even if it was the face of an SUV.

The biggest SUV ever wasn't the Hummer H2 but the Ford Excursion, which was launched in 2000. Nearly nineteen feet long and almost four tons in weight, the Excursion's 310-horsepower, 6.8-liter V10 engine got about 10 miles a gallon. But only if it wasn't carrying any cargo or passengers, and the driver was the size of a jockey.

The early years of the twenty-first century were the glory years for SUVs, the tail fins of the New Millennium. Both tail fins and SUVs had started out small before growing bigger and bigger, like hormone-fed sheet metal. Cheap gasoline made them both possible. And both would be a source of bemusement to future generations as potent symbols of mindless excess.

The SUV boom spawned backlashes from across America's ideological spectrum. The Hummer inspired a website called FUH2.com, which meant: "Fuck You and Your H2." Others protested with more reverent language. In 2001 and 2002, a group called the Evangelical Environmental Network launched an anti-SUV advertising campaign that asked, "What Would Jesus Drive?" The Bible didn't provide specific guidance, which was just as well. Any reference to "Jeep" in the ancient texts would have raised more fundamental questions.

As the SUV era approached its apogee, Chrysler was caught in a disastrous corporate drama. In 1998, nearly six years after Iacocca had retired, Germany's Daimler-Benz acquired the company for $38 billion. It was a huge price premium, thanks to the gusher of SUV profits being produced by Jeep. But the deal produced a clash of corporate cultures that was the biggest German-American conflict since the Battle of the Bulge.

The Germans and Americans at DaimlerChrysler, as the combined company was called, fought over issues ranging from the stupid (the size of their business cards) to the substantive (how many Mercedes parts could be used in Chrysler cars without undermining the status of Mercedes-Benz). The Germans ousted and replaced a succession of American executives. One U.S. journalist dubbed the company "Occupied Chrysler."[43]

Billions of dollars of losses and thousands of layoffs followed. Things got worse in 2005 after Hurricane Katrina struck New Orleans, sending gasoline prices soaring and demand for Jeeps plunging. Ford killed the Excursion, and the demand for gas-guzzling SUVs dropped, at least for a while.

In 2007 Daimler gave up on its American adventure and sold Chrysler, for a fraction of the price it had paid for the company, to Cerberus, a private-equity firm based in New York. Cerberus made the Chrysler mess even worse, losing billions more and earning itself a new nickname on Wall Street: "Clueless." On April 30, 2009, with America mired in its deepest recession in sixty-five years, Chrysler filed for bankruptcy under the auspices of the U.S.

government. The Treasury Department brokered a deal that transferred control of the company to the Italian automaker Fiat, the only car company in the world willing to wade into the Chrysler quagmire. Thousands more Chrysler employees lost their jobs, and Jeep got its ninth corporate owner.

It could have been a practical joke by Clio, the Greek goddess of history: Daimler and Fiat were based in countries that the doughty little Jeep had helped to vanquish during World War II. Willie and Joe would have rolled their eyes at that, too.

At least Jeep survived, unlike Hummer. When General Motors entered bankruptcy, a month after Chrysler, it put Hummer up for sale. No buyers could be found. Hummer went the way of tail fins.

Some things, however, remain unchanged. The Jeepers Jamboree continues after more than fifty years as an annual trek through the Sierra Nevadas, celebrating the days before "off-road" meant having a gravel driveway in the Hamptons. And a 1952 Jeep CJ still sits on display at the Museum of Modern Art in New York, labeled an American "cultural icon."

12

THE FORD F-SERIES:
COWBOYS, COUNTRY MUSIC, AND RED-MEAT WHEELS
FOR RED-STATE AMERICANS

The foremost high-speed-handling characteristic of pickup trucks
is the remarkably high speed with which they head from wherever
you are directly into trouble. This has to do with beer.

—P. J. O'Rourke, *Republican Party Reptile*[1]

Pickup trucks, like SUVs, can go off-road, but one can haul hay
and the other can't. They have other defining features, too. Land
Rovers are driven by people who fish for trout with high-modulus
graphite fly rods with silk ferrule wrappings. Pickups are driven by
guys who fish for bass with mail-order poles from Cabela's called
Whuppin' Sticks.

Or take the guy in New Canaan who drives a Range Rover and
the guy in Texarkana who drives a Ford F-150. Maybe they both
have two first names, but guess which one is Jean Paul and which
one is Jim Bob.

There is little scientific evidence to support these stereotypes.
But there is plenty of anecdotal evidence. Anyone who has driven

through Darien or Sausalito and tried to spot a pickup that didn't belong to a plumber can testify to that. So can anyone who has hung around the Amarillo Country Club, where the pickup trucks are driven by the members.

Texas is the capital of Pickup Truck America. Ford dealers there call the F-150, their full-sized pickup, the "Texas Mustang." Ford also sells a small, compact truck called the Ranger, but real Texans would no more drive a pygmy pickup than they'd go to Gilley's, the bar in *Urban Cowboy*, and order a Courvoisier.

Pickup Truck America is shaped like a giant backward "J" that runs from Walla Walla down to Waco, over to Ocala and up to Savannah. That geography happens to encompass Red State America, the bastion of cowboys, country music, and conservatism. This territorial overlap is what makes pickup trucks such curious cultural symbols. But before they became symbols pickups were just down and dirty work tools. Farmers and contractors used them to haul ladders, lumber, tools, hay, and herbicides. Not to mention the occasional cases of Budweiser and Lone Star.

For decades pickup drivers have fallen into two distinct and often warring camps: Ford guys and Chevy guys. Their rivalry would be expressed in bumper stickers that showed a mischievous boy with his pants partly down, showing his crack in back and peeing on "Ford" or peeing on "Chevy," depending on the truck owner's vehicular affiliation.

In the early 1970s, however, the status of pickup trucks began to change. Kids in Southern California started hauling their surfboards to the beaches in little compact pickups, imported from Japan, that were reliable and cheap. For a while the compact trucks even threatened the burly, full-sized models for dominance in the pickup market.

Pickups also started taking the stage, sometimes literally, in Nashville, the capital of country music. For years country music had been just an Appalachian sort of thing that evoked the *Andy Griffith Show* for people who believed in the agrarian myth. Or it evoked images of the hillbilly homosexual rape in the 1972 film *Deliverance* for those less benevolently inclined. But then country music went mainstream, just like J. Peterman duster coats and Patagonia windbreakers. In the late 1960s and early 1970s, country music stars Glen Campbell and Johnny Cash got prime-time television shows. They and other country singers, such as Loretta Lynn and Roy Orbison, made it big on the pop music charts. A decade later Barbara Mandrell recorded "I Was Country When Country Wasn't Cool."

By then country music really was hot. Or at least accepted beyond the locales where "violins" were known as "fiddles." Country songs touted traditional values with certain staple themes: Cheatin'. Lyin'. Double-crossin' men. Long-sufferin' mamas. And tough-haulin' trucks.

As this cultural phenomenon unfolded, Ford and Chevrolet signed country music stars to big-dollar promotional contracts to tout their respective trucks, a higher form of combat than bare-butt-peeing-boy bumper stickers. People started driving pickups for everyday transportation, just like regular cars, instead of for hauling hogs or plywood. Some owners even eschewed Confederate-flag decals and gun racks on their trucks. Especially if they also eschewed tobacco.

Mainstreamed pickups also went upscale, getting backseats, four doors, and cup holders that were used for Venti Lattes instead of Budweiser, at least some of the time. Then came hot-rod pickups, Harley-Davidson editions, and high-priced "specialty" trucks, as the car companies called them. Sometimes $70,000 pickups with in-dash navigation screens hauled $50,000 bass boats with electronic fish-finders.

Such redneck affluence was beyond the comprehension of people in the Acela Corridor or Marin County. Residents there were mystified when Americans twice elected a pickup-driving Texan as president of the United States. Pickup trucks, like minivans, carried heavy political payloads, as pickup truck cargo is called.

The first pickup trucks lacked many of the creature comforts that pickups have today. For one thing, they didn't come fully assembled. People who wanted to haul things had to buy a car chassis and a truck bed separately, and then bolt the two pieces together. It was like assembling a LEGO toy, but on a larger scale. Not until 1925 did Ford introduce a fully assembled truck called the "Model T Runabout with Pickup Body." A couple years later, the new Ford Model A offered a fully assembled pickup truck version from the start, powered by Ford's 40-horsepower, four-cylinder engine. In 1929 Chevrolet trumped Ford with a six-cylinder pickup, an advantage it touted with ads that declared: "A six for the price of four." [2]

The bigger engine and the pointed advertising slogan set a pattern of tit-for-tat one-upsmanship between Ford and Chevy trucks that would last the rest of the century and beyond. It would be a uniquely American rivalry. Even after Japanese automakers took half of the American car market and Japanese compact pickups got popular, the market for brawny full-sized pickups would remain largely a Ford-versus-Chevy contest, with a dash of Dodge thrown in. Rodeo stickers didn't look quite right on Tie-otas.

In 1930 sales of the Ford Model A pickup plunged 40 percent, and sales of other models didn't fare any better. It was the onset of the Depression, which brought searing scenes of Okies headed west to California with their possessions piled high into ramshackle pickup trucks. Their migration inspired John Steinbeck to write *The Grapes of Wrath*, chronicling the trek of the fictional Joad family in a cobbled-together pickup. Pickup trucks were anything but glamorous.

By the time Steinbeck published the novel in 1939, however, pickups were starting to get new features. Chevrolet gave its trucks wider and lower seats, so people could fit three abreast and sit upright without bumping their heads against the ceiling. The seats were made softer, too. Occupants would no longer feel they were bouncing on a bench. Creature comforts crept into pickup trucks, and there would be many more to come.

Chevrolet unveiled its first postwar pickup in 1947. It was a dimensional leap, seven inches longer than its prewar predecessors, and that was just the start. Over the next six decades Chevy's pickups would grow by another two feet, as if by manifest destiny, just like America itself.

In 1948 Ford responded with a whole new line of pickups called the F-Series, with three different models. The F-1 was the smallest, for light-duty work. The F-3 was a big bruiser, fit for hauling heavy payloads. The F-2 fit in between. In 1953 the company changed the trucks' nomenclature to F-100, F-200, and F-300, names that have evolved to F-150, F-250, and F-350.

In 1954 the F-Series trucks got a "Driverized Cab," which included such modern conveniences as armrests, a dome light, and sun visors. But the big news that year, pickup-wise, was that Ford upgraded its V8 engine, when Chevy trucks didn't yet have a V8. Ford's new engine delivered just 130 horsepower, less than most four-cylinder engines would provide a half century later. But Ford had climbed ahead on the pickup truck scoreboard.

The company's lead didn't last long, however. A year later Chevrolet added its new "small block" V8 to its line of pickup-truck engines. It was basically the same engine that Chevy's boss, Ed Cole, was using to add pep to the 1955 Chevy cars. With 162 horsepower, Chevy's V8 truck packed more power than Ford's. The Ford-versus-Chevy pickup wars gained new momentum, providing another glimpse of the corporate combat ahead. In 1957 General Motors made four-wheel drive available on its pickups for the first time. Ford, reeling from the Edsel debacle, took two years to catch up. But Ford outmaneuvered Chevy that year with a stylish new vehicle called the Ranchero.

The Ranchero combined the front end of a car with the body of a pickup truck. The concept was not unlike what the ancient Egyptians had done when they combined the head of a human with the body of a lion and called it the Sphinx. GM, caught unawares, took two years to introduce its own car-truck combo, the Chevy El Camino.

Just after Labor Day in 1960, another odd-looking vehicle took to America's highways, with John Steinbeck behind the wheel. The writer left his home in Shelter Island, New York, to take a lap around the circumference of the United States in a pickup that would have left the Joads slack-jawed.

It was a General Motors GMC with a V6 engine and an automatic transmission. The dark-green truck was outfitted with an oversized generator to power the slide-on camper where the aging author could cook, eat, and sleep. Fifty years later scholars questioned how much of that he really did in the truck, as opposed to "camping out" in roadside motels, but the fact that Steinbeck made the truck trek wasn't in doubt. His neighbors snickered that his journey was quixotic, so Steinbeck called his truck Rocinante, the name of Don Quixote's horse. If Roy Rogers could name his Jeep

Nellybelle, why shouldn't a literary legend name his truck after the steed in a seventeenth-century Spanish novel?

The pickup was "a beautiful thing, powerful and yet lithe," Steinbeck wrote. "It was almost as easy to handle as a passenger car."[3] He drove Rocinante across America's northern tier to Seattle and then down the coast to Salinas, California, his boyhood home. Along the way he steered clear of the new superhighways that had begun to lace the American landscape. "When we get these thru-ways across the whole country," the author observed, "it will be possible to drive from New York to California without seeing a single thing."[4] Except, as time would prove, McDonald's and Wal-marts and Chevron stations.

To return home Steinbeck took a southerly route through Texas, of which he observed: "By its nature and its size Texas invites generalities, and the generalities usually end up as para-dox—the 'little ol' country boy' at a symphony, the booted and blue-jeaned ranchman in Neiman Marcus, buying Chinese jades."[5] Sixty years later that observation, at least, was still valid.

Steinbeck's journey became a best seller, published in 1962, titled *Travels with Charley*, the name of his pet poodle and companion on the trip. Steinbeck had gone searching for America, just as Kerouac had gone searching for himself, and as P. J. O'Rourke would go searching for beer. The great American road trip was always about searching for something meaningful, more or less.

A year after Steinbeck published *Travels with Charley*, pickup trucks got tangled, bizarrely, in a trade dispute between the United States and Europe. Besides reshaping the pickup truck market, the spat would create some of the most comic episodes in the history of government bureaucratic futility, a rich history indeed. It all began when several European nations slapped a tariff on chickens imported from the United States, a brazen and unilateral act of pullet protectionism. In December 1963, just weeks after he became president, Lyndon Johnson retaliated with something that became known as the "chicken tariff."

Despite its name, LBJ's new tariff didn't apply to imported chickens. There weren't any at the time. The Europeans were keeping their chickens for domestic consumption instead of, um, fowling America's nest. So LBJ's new tariff applied to a weird assortment of goods: brandy, industrial starches, dextrin (used for envelope glue), and pickup trucks.

Pickups had nothing to do with chickens. But they had everything to do with politics, specifically with shoring up support from Detroit's car companies and the United Auto Workers union for Johnson's 1964 election effort. The chicken tariff slapped a 25 percent duty on imported trucks. The levy made compact pickups, almost all of which were imported, nearly as expensive as full-sized trucks, even though the compacts were only two-thirds the size. The combination of the tariff and the low gas prices of the 1960s meant there was little reason to buy little trucks. During the Sixties and early Seventies, compact pickups were about as common as cornfields in Manhattan (New York, of course, not Kansas).

But after the 1973 Arab oil embargo the price of gasoline doubled, and Americans started looking for more fuel-efficient vehicles. Car companies, sensing potential profits, started importing small pickup trucks. As companies are wont to do, they looked for technicalities to help them evade the tariff. One gambit was importing pickup truck cabs and pickup truck beds, separately, as two distinct pieces, which technically made them components instead of trucks. After both "parts" had cleared customs, they were hauled to nearby warehouses and snapped together, becoming—voilà!—a duty-free pickup truck.

The companies that resorted to such tactics weren't just Toyota and Nissan. Ford, General Motors, and Chrysler did it, too, even though they were prime proponents of the chicken tariff. The Big Three had deals with some of the Japanese car companies to make little pickups that could be imported as "components," and then reassembled and sold under the Ford, Dodge, and Chevy nameplates.

One such truck was the Chevy LUV (for "light utility vehicle"), which was built for GM in Japan by Isuzu and launched in the United States in 1972. Ford's counterpart was the Courier, made in Hiroshima by Mazda. The Dodge Ram 50 also was made in Japan, courtesy of Mitsubishi Motors. Detroit wasn't about to let a little hypocrisy stand in the way of profitability.

When the trade regulators closed that loophole, the car companies found another one. Subaru, for example, imported a little vehicle called the Brat. "With two seats welded to the bed," *Ward's Automotive Yearbook* reported in 1979, "it qualified as a multipurpose passenger vehicle, making it, technically, not a truck."[6] Once the "car" had cleared U.S. Customs, the second seat could be removed. Brat, indeed.

Thanks to such tactics, and to the rising price of gasoline in the Seventies, sales of compact pickups soared. In 1970 full-sized trucks outsold compacts by 20 to 1, but in 1980 the ratio shrank to just 2 to 1. Americans bought 630,000 compact pickups that year, a tenfold increase from a decade earlier.[7] Surprisingly, little trucks were becoming hip and fashionable. Not in Texas or Tennessee, where hip was where your blue jeans met your bull-hide belt, but out in California, on the beaches of Malibu and Orange County. Young people there could buy compact pickups for less than $7,000, thanks to the chicken-tariff evasions.

In the early 1980s, though, government regulators tightened

the trade regulations. With compact pickups so popular, Ford and General Motors decided to make them in America. Ford launched the American-made Ford Ranger to replace the Courier, and Chevy supplanted the LUV with a new truck called the S-10. They were taking clever advantage of the chicken tariff, or so it seemed until 1983, when Nissan started building compact pickups in a nonunion factory in Tennessee. Other Japanese car companies followed Nissan and Honda, which was making Accords in Ohio, and built their own U.S. factories. A measure intended to protect the Big Three and the UAW from competition, and enacted at their request, backfired. In 1986 and 1987 compact pickups outsold full-sized trucks, causing consternation at GM, Ford, and Chrysler. Big pickup trucks were their most profitable vehicles, and a wholesale shift to compacts threatened to undermine their corporate profit structures.

But in the late 1980s, the tide unexpectedly turned back to big trucks. Gas prices had dropped. Beyond that, compact pickups were getting bigger and more expensive, and their prices were climbing close to the price of big trucks. And the growing popularity of minivans and SUVs made Americans accustomed to driving big vehicles anyway. By 1995 full-sized pickups outsold compacts two-to-one again, and General Motors and Ford were transforming their big trucks into fashion statements.[8]

In television's Middle Ages of the late 1960s, forty years before *Real Housewives of Orange County* and *Jersey Shore*, prime-time programming consisted of variety shows instead of "reality" shows. Hosted by entertainers, comedians, or singers, they were pretty much alike. But in the summer of 1968 a fresh-faced country singer named Glen Campbell guest-hosted the *Smothers Brothers Comedy Hour*. He proved so popular that a few months later CBS gave him his own program, the *Glen Campbell Goodtime Hour*. It was the first prime-time network variety show built around country music. Campbell was well suited for bringing regional southern music into mainstream-American living rooms. An Arkansas native who started his career in the South, he moved to Los Angeles in the late 1950s and toured with the Beach Boys, allowing him to straddle the worlds of country and pop music. In 1967 he won Grammy Awards in both categories: for "Gentle on My Mind" in country and for "By the Time I Get to Phoenix" in pop. His television show lasted just three and a half years, but it started a trend.

In 1969 singer Johnny Cash also got a prime-time show, on ABC. For a couple years the two rival shows competed head to head. *The Johnny Cash Show* was canceled in 1971, but that same year John Denver, another rising country star, recorded "Take Me Home, Country Roads," his breakthrough hit, about a wistful return to West Virginia. Three years later Denver recorded "Thank God I'm a Country Boy," which described the joys, real or imagined, of a simple rural life. Meanwhile Loretta Lynn's signature song, "Coal Miner's Daughter," inspired a best-selling biography and, in 1980, an Oscar-winning movie by the same name.

A hardscrabble upbringing seemed essential for success in country music. Glen Campbell was the son of a sharecropper. Johnny Cash had picked cotton as a boy. Loretta Lynn was married at age thirteen. John Denver had it better as the son of a U.S. Air Force officer. But his real name was John Deutschendorf, which presumably was suffering enough.

The 1976 election of a peanut farmer and born-again Christian as president boosted country music's mainstreaming. Jimmy Carter hailed from the rural hamlet of Plains, Georgia, smack in the heart of the Bible Belt, the Redneck Belt, and the Pickup Belt. In 1977, Carter's first year in office, the University of Mississippi launched the Center for the Study of Southern Culture.

Just fifteen years earlier southern culture connoted Jim Crow laws and segregationist governors like Lester Maddox, one of Carter's predecessors in the Georgia statehouse. Ole Miss had been the scene of riots in 1962, when James Meredith became the first black student to enroll there. But with Jim Crow laws dying, a Georgian in the White House, and country music booming, southern culture was gaining respectability.

Journalists covering Carter soon discovered the teetotaling president's beer-drinking, bad-boy brother, Billy, who educated them on the difference between good ol' boys and rednecks. "A good ol' boy . . . is someone that rides around in a pickup truck—which I do—drinks beer and puts 'em in a litter bag," he explained, providing the scribes with welcome relief from the president's public piety. "A redneck's one that rides around in a truck and drinks beer and throws 'em out the window."[9]

Tossing beer bottles out of pickup trucks was getting easier because by the Carter years pickup trucks started getting extra doors and windows. In the early 1970s Dodge and then Ford introduced "Super Cab" pickups with two small, rear-hinged back doors that provided access to a small backseat used mainly for storage.

A step up from the Super Cab was the "Crew Cab," also known as the "Quad Cab," which offered four full-sized doors and a backseat suitable for adults. Early Crew Cab trucks had appeared in the late 1950s, but they didn't gain wide popularity until the mid-1970s. Giving pickups a backseat, more doors, and bigger interiors made them more versatile and broadened their appeal. In 1978 the Ford F-Series became the best-selling vehicle in America, supplanting GM's Oldsmobile Cutlass Supreme. It marked the first time a truck, as opposed to a car, topped the automotive-sales charts. No one then could have guessed how long the F-Series would stay there.

A prototype of the original Ford Mustang, which was launched on April 17, 1964. The backdrop is cheesy but the car's lines made it seem to be moving even when it was standing still. Beneath the catchy styling lay the architecture of the plain-Jane Ford Falcon, which was the key to the Mustang's low price and subsequent success. *(Ford Motor archives)*

Henry Ford II, grandson of the founder, who ran Ford Motor from the 1940s until the 1980s. The company's traumatic experience with the Edsel made Henry II skeptical of the Mustang initially, but later he basked in the car's success. *(Ford Motor archives)*

Lee Iacocca, the "father of the Mustang," on the cover of *Time* in April 1964, the week the car was launched. Iacocca made the cover of *Newsweek* that week as well. The Mustang was an instant sensation because it offered high style at a modest price. *(Courtesy of Time Inc.)*

Mustang meets mustang in this Ford publicity photo. In truth, the car's name was inspired by the P-51 Mustang fighter plane of World War II, not the horse. *(Ford Motor archives)*

Mr. Spock? No, it's John Z. DeLorean, GM's executive rebel and father of the Pontiac GTO, shown circa 1970. His sideburns, fashionable suits, and facelifts made DeLorean GM's in-house executive outlaw, although his success in reviving Pontiac propelled his career forward, for a while. *(GM Heritage Center)*

The 1964 Pontiac GTO convertible, America's original muscle car. Its enormous engine in a relatively small body violated GM's engineering rules, but DeLorean got around that by making the engine an optional item on the Pontiac Tempest LeMans. A couple years later the GTO became a model in its own right. *(GM Heritage Center)*

DeLorean in 1967 with the Pontiac Firebird, a spinoff of the Chevy Camaro, which in turn was GM's response to the Ford Mustang. *(GM Heritage Center)*

Soichiro Honda *(left)* and Takeo Fujisawa, the engineering and financial geniuses who built Honda Motor from a little motorcycle maker into one of the world's most successful car companies. Their contrasting personalities and occasional clashes belied a lasting partnership of mutual respect. *(Honda Motor archives)*

Honda Motor introduced the first Accord in 1976, when high gas prices sent many Americans searching for simple and reliable alternatives to Detroit's shoddy gas guzzlers. In November 1982 the Accord became the first Japanese car to be built in America, making it a visible symbol of a globalized economy. *(Honda Motor archives)*

Brad Alty in 2004, twenty-five years after the small-town Ohio boy started working for Honda in its factory near Columbus. He was among the first group of Honda's American employees to be sent to Japan for training. Today Alty is a senior manufacturing manager with Honda in Ohio. *(Honda Motor)*

The 1970 AMC Gremlin, which was designed on the back of a Northwest Airlines air-sickness bag and introduced on April Fool's Day. The glitch-prone Gremlin uniquely captured America's 1970s malaise. Brad Alty was driving one to his first day of work at Honda when it broke down, making him two hours late. *(Chrysler Historical Collection)*

Hiroyuki Yoshino, who was running Honda's American manufacturing operations when the Accord became America's best-selling car, the first for any non-Detroit vehicle. As a boy, Yoshino and his family endured a dangerous trek out of Manchuria at the end of World War II. Later he became president of Honda Motor. *(Honda Motor)*

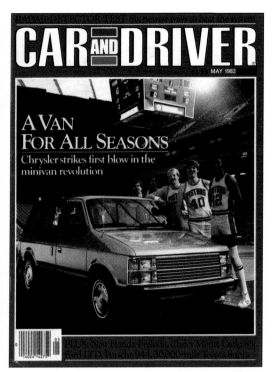

The May 1983 cover of *Car and Driver* magazine featured a revolutionary new vehicle, the minivan, which was small enough to fit in a garage but big enough to hold five Detroit Pistons basketball stars. Within a few years minivans became so ubiquitous that their "soccer mom" drivers became a powerful political force, avidly courted by candidates of both political parties. *(Car and Driver magazine)*

An ad for the 1984 Dodge Caravan, so named because it stood for "car and van." Minivans made station wagons virtually extinct, and cemented Chrysler's remarkable financial recovery from near-bankruptcy in 1980. *(Chrysler Historical Collection)*

Lee Iacocca, Chrysler's chairman and CEO, at a press conference in 1991, near the end of a remarkable career in which he championed the Ford Mustang, the Chrysler mini-vans, and Jeep. *(Reuters/John Hillery)*

The BMW 2002 from 1968, the year of its introduction. The car's peppy performance in a small package set the stage for the success of the BMW 3 Series as the quintessential "yuppie car" of the 1980s. *(BMW Archives)*

In 1973 BMW marked its new prosperity by opening a new corporate headquarters in Munich. The four cylindrical towers were designed to evoke BMW's powerful and efficient four-cylinder engines. *(BMW Archives)*

A BMW advertisement from 1976, the year that BMW launched the 3 Series, the successor to the BMW 2002. *(BMW Archives)*

Left: Herbert Quandt, the man most responsible for saving BMW from oblivion, in a photo from 1980, at age seventy. *(BMW Archives) Right:* Herbert (center) and Harald (right) Quandt, in 1966, celebrating the fiftieth anniversary of BMW. Seven years earlier the two half-brothers maneuvered to keep the struggling company from being absorbed by Mercedes-Benz. Herbert was virtually blind, and Harald was the stepson of Josef Goebbels, the Nazi propaganda chief. *(BMW Archives)*

The 1988 BMW 325i, when both the car and yuppie-dom were in full flower. *(BMW Archives)*

The 1941 Willys. Willys-Overland lost the Jeep design contest to American Bantam, but the Pentagon awarded the bulk of the production contracts to Willys anyway because it was a much bigger company. *(Chrysler Historical Collection)*

Patriotism was the theme, but commercial motives were evident in this 1944 Jeep ad, which ran in *The Saturday Evening Post* and other magazines. "Does the 'Jeep' figure in your post-war plans?" the text of the ad asks, suggesting "many personal business and pleasure uses" after the war. *(Chrysler Historical Collection)*

This ad for the twenty-fifth anniversary of the Jeep CJ-5 ("Civilian Jeep, 5") ran in 1979, when Jeep was a subsidiary of lowly American Motors, which in turn was controlled by French auto maker Renault. During the next thirty years, Jeep came to be controlled by companies from Germany and Italy, two of the one-time Axis powers that the Jeep helped to vanquish during World War II. *(Chrysler Historical Collection)*

The 1979 Jeep Wagoneer Limited, a vehicle that was ahead of its time. First introduced in 1963, the Wagoneer offered optional four-doors and automatic transmission. Such features would become standard when Jeeps came to symbolize outdoor lifestyles, even though the biggest bumps they usually hit were potholes on the way to the mall. *(Chrysler Historical Collection)*

The 1948 Ford F-1 pickup, the company's first post–World War II truck. For decades pickup trucks were work tools, but in the 1970s they started becoming fashion statements for "suburban cowboys." *(Ford Motor Archives)*

This 1953 Ford F-100 pickup was an ancestor of trucks that would become the best-selling vehicles in America (cars and trucks included) for more than thirty years, a record that continues still. *(Ford Motor Archives)*

Country music superstar Toby Keith with the Ford F-150 pickup. Ford sponsors Keith and scripts its top-selling truck into his concerts in ways that are "additive to the entertainment experience," the company says, not to mention being additive to Keith's bank account. Country music and pickup trucks began to enter the American cultural mainstream in the 1970s. *(Ford Motor Archives)*

The Ford F-150 Harley-Davidson edition comes in just one color—"dark amethyst," aka black, and features flame-like decals extending from the front wheel wells. It's the perfect truck for midlife-crisis escapes for middle-aged men, who can haul their Harley bikes to Harley conventions in Ford Harley trucks. *(Ford Motor Archives)*

Left: Toyota Motor patriarch Eiji Toyoda in the mid-1990s, near the end of his career. His vision to develop a car for the twenty-first century led to the Toyota Prius, the first commercially successful gasoline-electric hybrid car. *(Toyota Motor) Right:* Takeshi Uchiyamada, the "father" of the Prius. A Toyota lifer who grew up in postwar Japan, Uchiyamada was a surprise choice to lead the project that produced the Prius. *(Toyota Motor)*

Hiroshi Okuda, president of Toyota Motor, unveils the first Prius at the ANA Hotel in Tokyo on Oct. 14, 1997. Okuda's hard-nosed insistence on a car with revolutionary new technology spurred Toyota's drive to develop a gas-electric hybrid. *(Reuters/ Kimimasa Mayama)*

The second-generation Prius, launched in the U.S. in late 2003, propelled the car into the automotive mainstream. It was remarkably fuel-efficient yet roomy enough for a family of four. Its distinctive aerodynamic shape enhanced fuel economy, and also let drivers display their environmental sensitivity. *(Toyota Motor)*

Seeing double? No. This shows the gasoline engine *(left)* and the electric motor *(right)* in the second-generation Prius. The hybrid technology shifts the power load between the two engines, depending on the car's speed and driving conditions. *(Toyota Motor)*

Meanwhile, pickup trucks played supporting roles in two hit movies. In 1979's *The Electric Horseman*, Robert Redford played Sonny Steele, a washed-out rodeo cowboy who rescued a horse from a dreary existence. Redford saved the horse by setting him free in the wild, after spiriting him away from a corporate convention in a trailer hauled by a pickup truck. A year later, in *Urban Cowboy*, John Travolta rode the mechanical bull at Gilley's to prove his manhood. Between his bouts on the bull, Travolta got in fights and drove a pickup, proving his manhood further.

Pickups also appeared in the titles of country songs: "Ragged Old Truck," "Pickup Truck Song," "Big Ol' Truck," "Pickup Man," "That Old Truck," and "This Old Truck."[10] (The latter was by a recording artist from Australia, where parts of the outback are not unlike the Texas Panhandle.) If the songs' titles are distressingly similar, at least the themes are different. "Ragged Old Truck" is about a Texan's oppressive marriage to "that ol' heifer," which might have been a term of endearment for a cow but not for a wife. "Big Ol' Truck" is about falling in love "with a girl in a four-wheel drive."

The refrain on Joe Diffie's "Pickup Man" was: "There's just something women like about a pickup man." The double entendre was obvious. In 1995 Diffie recorded "Leroy the Redneck Reindeer," the tale of a heroic animal who saves Christmas by filling in for his ailing cousin Rudolph, after driving up to the North Pole in his pickup.

By 1990 Detroit's marketing managers decided to co-opt country music to sell trucks. Chevrolet licensed a heartland-rock song by Bob Seger, who happened to be a Detroit native. It went: "Like a rock . . . I was strong as I could be . . ." Comparing a motor vehicle to an immobile object, as opposed to an animal or an airplane, wasn't an intuitive advertising strategy. But guys who drank longneck Lone Stars didn't analyze it that way, or any other way. Seger's raspy-voiced wail was a hit. Chevy played it for more than a decade.

The song didn't vault sales of the Chevy Silverado ahead of the F-Series. But it did keep the Silverado hard on Ford's tailgate. Ford struck back in 1992 by signing a deal to become the official truck partner of the Dallas Cowboys. It worked so well that in 1999 Chevy stole the sponsorship from Ford for a few years.

In 1993 Dodge, long a distant also-ran in full-sized pickups, made a breakthrough. It launched a revamped full-sized pickup with a "drop-fendered" front that looked like muscular shoulders and evoked a menacing Mack truck. Dodge's designers had consulted with Clotaire Rapaille, a French-born medical anthropologist who believed burly trucks displayed man's repressed reptilian instincts.[11] Maybe it was psychological silliness, but Dodge's pickup sales soared.

Meanwhile, Ford started developing a new breed of specialty pickups that could do a lot more than just haul hay. In 1993 the company launched a truck called the F-150 SVT Lightning, a pickup truck with a hot-rod engine. The 1999 model got a 360-horsepower V8 that had a supercharger and an intercooler, two power-boosting devices usually reserved for sports cars. The SVT Lightning looked like a hay hauler, but it accelerated from 0 to 60 miles an hour in just 5.6 seconds. "This here is a truck that can run door handle to door handle with a BMW M3," *Motor Trend* wrote. "Can you say, 'Yuppie-scum beware'?"[12] Ford executives loved it.

A year later Ford launched another specialty truck, the F-Series Harley-Davidson edition, designed to haul hogs. Not hogs as in pigs, but the kind of hogs that had two wheels, black paint, and owners named Warlock or Mombo. But those were just noms de guerre. Many Harley-Davidson edition owners were accountants or urologists who rode motorcycles on weekends to indulge their inner outlaw. They could haul their Harley motorcycles to Harley conclaves in Harley trucks, in case anyone missed the point.

Like the Model T, the Harley-Davidson edition came in only one color. Officially it was "dark amethyst," though it was impossible to tell from black. The truck also had the Harley-Davidson logo stamped on the front fenders and embossed in the leather captain's chairs. In 2005 Ford added optional hot-red flame decals extending from the wheel wells. They were "the first factory-flamed trucks in the industry," the company boasted.[13]

Chevrolet counterpunched with the Avalanche, which weighed nearly three tons and was part truck and part SUV. Folding panels allowed the rear-seat area to open into the truck bed and become fully enclosed. The truck was like an adult Transformer toy. Chevy sold more than 93,000 Avalanches in 2003, the truck's second full year on the market.

Despite the Avalanche's success, however, Ford clearly held the upper hand in trucks in the early years of the twenty-first century. It became the official sponsor of Monster Jam events, in which pickup trucks jacked up on giant tires competed to see which could perform the most outrageous stunts. The company also regained the Dallas Cowboys sponsorship deal, winning it back in 2002 from the interlopers at Chevrolet. And Ford countered Chevy's Bob Seger on the country-music front by signing up two country artists of its own, first Alan Jackson and then Toby Keith. Keith's concerts were scripted with Ford's help "to integrate the brand into the concert in ways that are additive to the entertainment experience," a company spokesman explained.[14] It was also additive to Toby Keith's bank account.

In 2001 a Ford truck played a role in high-level international diplomacy. That June President George W. Bush met Russian president Vladimir Putin in Europe, declaring: "I looked the man in the eye . . . I was able to get a sense of his soul."[15] Five months later Bush drove Putin around his Texas ranch in his white Ford F-250 Quad Cab.

That year Ford also introduced another special version of the F-Series, the King Ranch edition, named for the biggest spread in Texas. The double-stitched saddle-leather seats, branded with the King Ranch longhorn logo, had more leather than most cows. The truck's center console was big enough for a poker game, and its interior had enough brushed steel to evoke a Sub-Zero refrigerator. It was the ultimate Texas truck with a Texas-sized price tag: $60,000 or more with such options as heated *and* cooled seats and a "man-step" in the tailgate that made it easy to climb into the truck bed, even for men wearing trim-fit designer jeans.

For city slickers, Ford introduced the F-Series Platinum edition, designed for "urban luxury," as the company put it. The truck featured shiny chrome trailer-hitch balls as opposed to ordinary steel balls, and a color-matched hard cover for the truck bed to keep urban thieves at bay.

Ford also signed on as the official truck sponsor of Professional Bull Riders, known as PBR, which has more followers in the Pickup Truck Belt than, say, squash or lacrosse. Such promotional ideas gestated in the Truck Room, a secret hideaway in a Ford office near Detroit that displays artifacts sacred to truckers. Those include lariats, cowboy boots, cowboy hats, Rapala bass lures, a fish-landing net, a canoe paddle, a duck decoy, a pair of dirty work gloves, and a T-shirt that says "Big Big Daddy." There's also a six-pack of the other PBR, Pabst Blue Ribbon beer.

Like many shrines, the Truck Room has rules for proper decorum: No froufrou coffee drinks or pink shirts can enter. No strawberry-kiwi anything is allowed. And wheeled luggage is forbidden, unless it's carrying golf clubs. The Truck Room is an exercise in "experiential marketing," as Ford describes it.[16] It allows the company's truck people to experience Texas and Alabama—say, over their lunch hour—without actually going there.

Experiential marketing helped inspire Ford's truck division to sponsor a TV reality show called *Dirty Jobs.* It shows working guys and gals using Ford trucks for tasks that require handling things that are icky and gooey, which many truck owners find entertaining. Far more F-Series drivers watched *Dirty Jobs* than watched, say, *Masterpiece Theatre.* It was natural, a Ford spokeswoman explained, to "include integrations of our vehicles into the show in a contextually relevant way."[17] Phrases such as "contextually relevant," it should be noted, are strictly forbidden inside the Truck Room. Offenders are fined $1, which is better than having to attend a Monster Jam.

The state of Texas is, in effect, a big Truck Room. One out of every eight vehicles sold in America is a pickup truck, but in Texas it's one out of four. Many Texans need pickup trucks for their work on ranches or oil rigs or construction sites. Others simply want to drive one because it's the local custom. Sometimes the difference between *needing* a pickup truck and *wanting* one can be blurry.

Bryan Jones, a professor of government at the University of Texas in Austin, drives a Ford F-150 on his five-minute commute to campus. Does he need a pickup truck? Well, he's a big man, he explains, and scrunching his six-foot-four frame into a regular sedan gave him back pains. He also has a "spread" of land in the country, out near Johnson City, and every so often he likes to drive out to make sure it's still there. A four-wheel-drive truck is just the ticket. "Being a professor I don't have this pickup identity thing," Jones avers.[18]

Many Texans do, however. Most people raised in the state raised a little hell in pickups while growing up. In the early 1960s John Williford was in high school in the tiny town of Shiner, the hometown of Shiner Beer. A favorite nighttime activity for him and his friends was getting a six-pack of Shiner, finding an empty pasture, turning on the high-beam headlights, climbing into the truck bed, and shooting jackrabbits. The shooting was easy but actually *hitting* the critters wasn't, beer or no beer. Jackrabbits dart to and fro like swallows, and pickup trucks don't. More than one boy fell out.

Decades later Williford was a lawyer, working in a downtown Houston office tower with a parking garage that was billed as pickup truck–friendly. Ceilings were high enough and the spaces wide enough to make it easy to park a truck. Williford figures half the vehicles in the garage are F-150s, Silverados, or Dodge Rams. Williford himself, however, drives a Mercedes-Benz sedan to work, preferring to reserve his extended-cab Chevy Silverado for fishing, hunting, hauling, and other weekend activities.

Like many Texans, Williford has definite views about what different trucks say about their owners. Guys with pickups decked out with running boards, decals, and hunting scenes etched in the rear window are "bubbas," as he puts it, who only drive their trucks to bars. "Especially if they're wearing three-hundred-dollar crocodile-skin boots," Williford adds, "which they probably bought with their last paycheck."[19] People of "means," which in Texas means landowners, know that running boards are good for nothing except collecting cow dung, Williford explains. Gun racks, he adds, are nothing but thief magnets. Williford keeps his guns in a toolbox stored discreetly in his Silverado's backseat.

Also, like many Texans, Williford has a ready supply of pickup truck tales. One involves the flood that hit the Hill Country north of San Antonio in 2004. "A couple of bubbas parked their F-150 on the low-water bridge where Hays County 305 crosses the Blanco River," he begins. "They set up lawn chairs in the truck bed, popped open a couple of beers, and waited to watch the water rise. Well it rose, all right, much faster than expected. The truck was swept downstream and quickly morphed into a marine amphib assault vehicle hurling down the Blanco. The bubbas bailed out over the truck-bed rails during one of its aquatic spins and were uninjured only because their nervous systems were nonreflexive due to copious quantities of Lone Star.

"The truck wasn't so lucky. It careened into the front of a new riverfront restaurant complex, owned by my friend Brooks, which had terraces, benches, a custom barbeque grill, and a Texas flag. The place was reduced to a prison-gang rubble pile with an upside-down truck on the lawn." [20]

It's a true Texas truck story, and it comes with a postscript as farcical as the incident itself. The car-insurance company refused to pay Brooks's property-damage claim by invoking the force majeure clause in the insurance contract. The incident was an "Act of God," the insurer claimed, as opposed to an act of two "bubbas."

Williford's other favorite pickup story involves his wife, who rarely rides in his Chevy truck. But one day, at his request, she drove it to a local truck-accessories store to get a new trailer-hitch ball installed on the rear bumper. She came home with a T-shirt that the store hands out to its customers. It read: "Hardest Balls in Texas." [21]

For the first half century of their existence, from the late 1920s to the late 1970s, pickup trucks were unglamorous work tools. But then country music got hot. Cowboys became cool. Marlboro Men stopped spending days in the saddle and started spending days in the saddle-leather seats of their F-150 King Ranch Crew Cab. The Platinum edition reflected the spirit of an age when borrowed money was called "leverage" instead of debt. Or where Microsoft millionaires tore down $5 million homes in Jackson Hole to build $10 million homes in their place.

The years 2004 and 2005 were the peak of the pickup boom. In both years Americans bought more than 3.2 million pickups, 80 percent of them full-sized trucks. More than 900,000 were various versions of the Ford F-Series. The profits were so lucrative that the Japanese automakers, whose cars were stealing customers from Detroit in droves, tried to crash the pickup truck party. In 2003 Nissan completed a new $1.4 billion plant in Mississippi to build the first full-sized pickup truck made by a Japanese company, the Titan.

The Titan targeted "the modern truck guy," as Nissan put it, meaning not farmers or contractors but suburban cowboys who drove their truck to work during the week and towed their boats or hauled dirt bikes on weekends.[22] The Titan, however, didn't have a choice of engines or a heavy-duty version with extra towing ability, like the Chevy and Ford trucks did. Besides, pickup buyers are loyal to their brands, as guys displaying peeing-boy bumper stickers can attest. Nissan never came close to its goal of selling 100,000 Titans a year.

In 2007 Toyota introduced its first full-sized pickup, the Tundra. Toyota produced it in a spanking-new American factory, just like Nissan, built in San Antonio, in the heart of Texas truck country. But before long the frames on the Tundra trucks began to rust, almost a parody of the quality lapses that plagued Detroit in the 1970s. Toyota had to recall the trucks and they flopped, just as the Titan had done.

A year later Dodge had its own pickup truck disaster when it launched an updated version of the Ram with a high-profile publicity stunt. In January 2008 the company shipped a herd of 130 longhorn cattle to Detroit and paraded them, with the new trucks, in front of the city's annual auto show. It was like a touch of Texas in Motown, except that some of the nervous steers started mounting each other, with local children watching and television cameras whirring. The new Dodge Ram got plenty of publicity, but it was mostly snickering references to horny longhorns and *Brokeback Pickup*.[23]

During the first decade of the twenty-first century, pickup trucks began to take on political as well as cultural significance. In the presidential election of 2000, NBC correspondent Tim Russert labeled states red or blue, depending on their political affiliation. The terms "red states" and "blue states" became instant political shorthand. Pickups became symbols of the proverbial little guys who drove their trucks to work, hung out at bars, and obeyed the rules (except maybe the laws against drinking and driving). Not many little guys drove SVT Raptors or Platinum editions, but who noticed?

In early 2010 the symbolic potency of pickups helped pull off one of America's biggest electoral upsets in decades. After the death of Massachusetts senator Ted Kennedy, the blue-blood Democrat from America's bluest state, the special election to fill his seat was won by an unknown Republican named Scott Brown. He campaigned in a dark-green 2005 GMC Canyon pickup. It was just a compact truck (although GM called it "midsize"), which would have been laughable in Texas, but in Massachusetts such distinctions didn't matter. "Mr. Brown presented himself as a Massachusetts Everyman," the *New York Times* wrote, "featuring the pickup truck he drives around the state in his speeches and [in] one of his television commercials . . ."[24]

On the night he was elected, Brown got a congratulatory phone call from President Barack Obama. "When I spoke to the president," Brown told his joyous supporters, "the first thing I said was, 'Would you like me to drive the truck down to Washington so you can see it?'"[25] The crowd roared. Ten months later, in the 2010 midterm congressional elections, a Tennessee candidate for the House of Representatives advertised himself as a "truck-driving, shotgun-shooting, Bible-reading, crime-fighting, family loving country boy."[26] The candidate happened to be a Democrat.

By that time America's pickup boom was fading, at least outside Texas and Tennessee. Sales started to soften in late 2005, after Hurricane Katrina hit New Orleans and gas prices soared. A year later a lobbyist for an automotive trade group was driving home on Interstate 495, the Capital Beltway that rings Washington, DC. He glanced to his right and saw, in the next lane, a Ford F-250 pickup being driven by a short, slender man wearing a blue blazer and a bow tie. The driver looked like a nerd, not the Marlboro Man. The lobbyist concluded the end of the pickup boom was nigh.[27]

He was right. In 2008 the price of gasoline passed $4 a gallon. That fall America plunged into its worst financial crisis since the westward journey of the Joads. The following spring brought the General Motors and Chrysler bankruptcies, alleviated by billions in bailout dollars provided by the U.S. and Canadian governments. Ford barely managed to escape the same fate.

In 2010 pickup trucks accounted for only 13 percent of the American car market, down from nearly 19 percent five years before. It was a tectonic shift in an industry where tenths of a percentage point mean billions of dollars. People who needed pickup trucks for work—farmers, ranchers, and contractors—still bought them. But those who had bought trucks to haul dirt bikes, or just to show off, returned to buying cars, if they were buying any vehicle at all.

The Ford-versus-Chevy rivalry remains dominant in the pickup market, as it has been since the 1950s. Both General Motors and Ford are making heavy-duty pickups that can tow nearly 25,000 pounds, which means a bass boat big enough to scare the fish out of the water. In 2011 the Ford F-Series remained America's best-selling vehicle, as it has been for thirty-three straight years. It's a record nobody anticipated when Ford's truck took the top spot in the late 1970s. Pickups aren't about to fade into oblivion like tail fins. But America's automotive sensibilities are adjusting to a new era.

13

AN INNOVATIVE CAR (THE PRIUS), ITS INSUFFERABLE DRIVERS (THE PIOUS), AND THE ADVENT OF A NEW ERA

It has become an automotive landmark: a car for the future, designed for a world of scarce oil and surplus greenhouse gases.

—*Fortune*, 2006[1]

You have a Prius. . . . You probably compost, sort all your recycling, and have a reusable shopping bag for your short drive to Whole Foods. You are the best! So, do we really need the Obama sticker?

—The *Portland Mercury*, 2008[2]

Cars had been cast as characters on TV shows almost since television began. Roy Rogers had his Jeep Nellybelle in the Fifties. Buzz and Tod drove their Corvette in and out of trouble on *Route 66* in the Sixties. Bo and Luke Duke barreled down country roads in their orange Dodge Charger, the "General Lee," in the *Dukes of Hazzard* during the early Eighties, always outwitting the comic-evil Boss Hogg.

All three cars served as faithful companions to their shows' heroic protagonists, sort of like horses but with more, well, horse-power. Cars weren't cast on TV as comic foils, at least not until the turn of the century, when HBO launched a sitcom called *Curb Your Enthusiasm.*

The show features a TV writer/producer named Larry, his environmental-activist wife, Cheryl, and, now and then, their Toyota Prius, the first successful gas-electric hybrid car. In one memorable episode Larry waves at another Prius driver, expecting a return salutation from a fellow member of the ecological elite. The other driver snubs him, and Larry gives chase. He gets his comeuppance, and the chase ends, after he accidentally, and stupidly, hits a dog.

TV heroes had changed a lot over fifty years. Then again, so had cars.

Toyota launched the Prius in Japan in 1997. The car ran on batteries at low speeds until a small gasoline engine kicked in above thirty miles an hour. The gas engine recharged the battery, which also captured the energy created by the car's braking action. So the battery never ran out of juice and the car never needed to be plugged in.

The Prius was successful in Japan, but only modestly so. Conventional small cars provided good fuel economy, too, but cost a lot less. And while the Prius's brakes captured energy, they also were "grabby," sometimes causing the car to lurch during routine stops. It took Toyota three years to address those and other issues. The Prius didn't debut in America until 2000. At the time, however, Americans were still enamored of gas-guzzling SUVs, and only 6,000 Priuses sold in the first year.

But then Toyota arranged for the Prius to play a starring role at the 2003 Oscars as limos for actors and actresses who wanted to display their commitment to the environment. The stars were less eager to display their commitment to private jets or their 30,000-square-foot mansions, but they were, after all, people who made their living pretending to be people they weren't.

The Prius's big breakthrough came in 2004, when Toyota launched the second-generation model. It had better tires, more room, more power, and better gas mileage, all in addition to sleek and instantly recognizable space-age styling. En route to success, however, the Prius got carjacked by the combatants in America's culture wars. Environmentalists embraced the car as a secular icon. NASCAR fans, in turn, smirked when young Al Gore III, son of the environmentalist ex–vice president, was arrested for hitting 100 miles an hour on an L.A. freeway in a Prius.[3] Which took more determination than going 100 in a GTO or a BMW.

America's culture wars notwithstanding, the Prius was a remarkable feat of engineering. Hundreds of Toyota engineers worked insane hours, sometimes nearly around the clock, for four years to develop the car. One man even died during the process. In the end Toyota achieved breakthroughs in technology deemed impossible for decades, even centuries. Unbeknownst to most Americans, even to Prius drivers, the quest for alternative-fuel vehicles had a long and circuitous history.

Historians trace the first plans for a steam-powered vehicle to the late seventeenth century, and a man named Ferdinand Verbiest. He was a Flemish Jesuit residing in the court of the Chinese emperor Kangxi.[4] It isn't clear whether the vehicle was ever produced.

Progress was slow. In the 1830s a Scottish inventor built a crude electric vehicle, but it was impractical and couldn't go far. At the beginning of the twentieth century, a young German developed a crude hybrid car that used a gasoline engine to spin an electric generator that in turn powered the wheels. The man was Ferdinand Porsche, later the inventor of the Beetle, which proved far more successful than his homemade hybrid.[5]

The Woods Interurban, a car built in Chicago around 1905, took a different approach to hybridization. It was designed with a removable electric motor that could be pulled out and replaced with a two-cylinder gas engine for longer drives between cities. Engine swapping, however, proved even more trouble than parallel parking. The car flopped.

Indeed, the early years of the auto industry, like the dawn of the Internet Age, were filled with lots of trial and mostly error. The first decade of the twentieth century brought the heyday of electric cars in America, but their batteries couldn't store enough energy to power cars over long distances, an issue that would persist for a century. Meanwhile, Henry Ford and others kept improving their gasoline-powered internal combustion engines, which proved to be the winning technology. One of the last alternative-fuel cars, the Stanley Steamer, died in 1924.

The 1950s brought the invention of electric golf carts, which conquered Florida the way Conestoga wagons had won the West. In 1966 General Motors scientists developed a couple of experimental electric cars, one of which had batteries powered by hydrogen-fuel cells, that could run on highways for limited distances. But this was America's Age of Innocence, before OPEC and Earth Day. Gasoline cost 25 cents a gallon, and an expensive electric car that couldn't go far was a nonstarter.

Then in the late 1960s three scientists at TRW, a car-components company in Cleveland, began working on a gas-electric hybrid system. "Even back then some of us were wondering about the long-term ramifications of using so much gas and oil," one of the scientists later explained.[6] To build the prototype the

TRW team cobbled together a Beetle engine, a Westinghouse generator, a direct-current motor from General Electric, and a Chrysler automatic transmission. (It really was a hybrid.) After it worked in the lab, they installed it in a beat-up 1962 Pontiac Tempest and got a patent.

They dubbed their invention the EMT, for "electro-mechanical transmission." It increased fuel economy by 30 percent and reduced tailpipe emissions by switching power back and forth between the gasoline engine and the electric motor, essentially what the Prius would do thirty years later. But the major car companies, both American and foreign, dismissed the EMT as too complicated and too expensive to build. The project was dropped.[7]

A few years later an inventor named Victor Wouk (the brother of novelist Herman Wouk) modified a 1972 Buick Skylark with his own gas-electric hybrid system. He sought government funding to develop the concept, but dubious federal officials stymied his effort.[8] In 1975, with America's new clean-air laws on the books, the U.S. Postal Service bought 350 electric Jeeps from American Motors to test in its delivery fleet. But it was basically a publicity stunt.

By 1990, despite all this experimentation, things stood about where they were back in 1890. Then General Motors unveiled an experimental electric car called the Impact, and announced that it would develop the vehicle for production. Environmental groups were positively exultant. So were California regulators, who decreed that 10 percent of the vehicles sold in the state would have to be emission-free early in the next century.

In 1994 GM offered two-week test-drives of prototype vehicles in Los Angeles. It expected fewer than 100 volunteers to sign up. As it happened, more than 10,000 people called in.[9] Environmental consciousness was increasing, and some Americans wanted an alternative to the internal combustion engine. The car, called the

General Motors EV1, launched on December 5, 1996, in Southern California and Arizona. But it had significant limitations, including a maximum range of 100 miles before it ran out of power. Another problem was refueling, which could only be done at special charging stations. Then there was size. The EV1 was just a two-seater, and unsuitable for hauling a family.

There were other issues, too. The EV1 could only be leased, not purchased, because GM wanted to retain ownership of the batteries. The price, around $500 a month, was enough to lease a couple of Chevy compact cars with seating for four. Also, the EV1's lead-acid battery, an industrial-strength version of ordinary flashlight batteries, prompted a recall for "thermal incidents," corporate-speak for engine fires. It was little wonder, then, that GM leased only 288 EV1s in the first year. But GM forged ahead anyway.

In 1999 it launched a second generation of the car with a more advanced nickel-metal hydride battery and a greater range before the power was spent. But the EV1 still required people to spend more money for a less capable car—in terms of size, range, and versatility—than a car with a gasoline engine. Not even many Sierra Club members wanted to pay more for less, no matter what they told public-opinion pollsters. In 2003, after a few more years of futility, GM decided to kill the EV1, only to learn that its good deeds wouldn't go unpunished.

GM's lawyers warned of potential legal liability if the used batteries went bad, so the company decided to repossess and crush all the EV1s on the road. Actress Alexandra Paul, a star on TV's *Baywatch* and an ardent fan of the EV1, helped lead a protest at the repossession center. She got arrested and got five days in jail. She also got plenty of headlines, of course.

Then a documentary film called *Who Killed the Electric Car?* posited an answer that was pure Hollywood: a conspiracy among

General Motors (which secretly wanted the car to fail), the oil companies (ditto), President George W. Bush (a tool of the oil companies), and California's environmental regulators (cowered by all of the above). The film stopped just short of fingering the second gunman on the grassy knoll.

The truth was more prosaic. The EV1 was a technological and commercial flop because few people wanted a car that might run out of juice. (Fifteen years later auto executives would label this concern "range anxiety.") In 2007, on the fiftieth anniversary of the Edsel, *Time* named the EV1 one of the fifty worst cars of all time. By trying to do well by doing good, General Motors had stumbled into one of its worst PR drubbings since the Corvair.

The real EV1 scandal was that GM wasted $1 billion to develop and produce an impractical car. Meanwhile, billions more had been blown by the U.S. government on a boondoggle called the Partnership for a New Generation of Vehicles. Launched in 1993, it was an unwieldy research consortium among eight federal agencies, several universities, and Detroit's Big Three car companies. The goal was to develop innovative new cars that would be more practical than the EV1 and reduce America's reliance on foreign oil. Excluded from the partnership were foreign car companies, including Toyota.

The Law of Unintended Consequences came into play. After being shunned by both Washington and Detroit, Toyota decided to launch an alternative-vehicle effort of its own. The unlikely leader would be a short, soft-spoken man whose career at the company had hit a dead end. Or so it had seemed.

Takeshi Uchiyamada, like most Toyota managers, was a company lifer. Born amid the drabness of defeated Japan in 1946 just outside Toyota City, he came of age during the country's postwar economic miracle. As a boy Uchiyamada loved mechanical things. He disassembled and reassembled transistor radios, and built model cars with wood and metal parts. He had to craft the parts by hand because the family couldn't afford prefabricated model kits like the boys in America had. One of his favorite homemade models was a 1959 Cadillac, which to him symbolized the wealth of the most powerful country on earth.[10] The future inventor of the world's first successful hybrid car was enamored of tail fins.

At nearby Nagoya University Uchiyamada got a degree in applied physics in 1969 and joined Toyota, where his father worked. Toyota City was just like Flint, Michigan, had been in the 1950s, a place where local boys attended the local university (the General Motors Institute, in Flint's case) and followed their dads into the company.

Uchiyamada's goal was to lead a vehicle-development team and take charge of developing new Toyota cars. His bosses had other ideas. They put him on a more specialized, and mundane, career path as an expert in reducing NVH, the automotive-engineering term for "noise, vibration, and harshness" in vehicles. He spent his days working with sound-dampening insulation, industrial screws and bolts, and similar stuff.

It was in this role that, in 1979, the thirty-two-year-old Uchiyamada made his first trip to America to present a technical paper to the Society of Automotive Engineers. In New York and then in Detroit, America's multilane freeways amazed him.[11] Though the young engineer didn't know it, the year 1979 would be an inflection point. Detroit's once dominant car companies began their long decline, and their Japanese competitors, including Toyota, launched their remarkable ascent.

During the next fifteen years Uchiyamada's career progressed predictably but gradually upward in Toyota's management bureaucracy. He gave up on his dream of leading a vehicle-development team. Entering his late forties, he had never even served as a subordinate to a team leader. His bosses shocked him, then, in the autumn of 1993, when they tapped him to run Vehicle Development Center 2, a banal name for a group with an ambitious assignment. The team's task was to develop Toyota's signature car for the approaching twenty-first century, a car big enough for a family but with a quantum leap in gas mileage and a dramatic reduction in emissions.

The early 1990s were the beginning of the SUV boom in America, then the world's largest car market. Toyota had missed the trend, and was scrambling to catch up. But even as SUVs gained popularity, Eiji Toyoda, Toyota's aging chairman and the patriarch of its founding family, feared a future in which oil would be expensive and scarce. On the cusp of his retirement, Toyoda still wanted his company to produce a breakthrough in technology.

Uchiyamada was an unconventional choice to lead this effort, but that was just the point. His bosses believed that his lack of experience on a vehicle-development team would make him less likely to be bound by conventional wisdom. Thus Project G21, a name that stood for "Global 21st Century," was born.

Initially Uchiyamada had just ten engineers on the team. The group held its first meeting on February 1, 1994, but didn't get off to a fast start. "Let's first set the position of the driver's buttocks," one team member suggested at an early meeting.[12] The smart-ass answer would have been to position them right behind the driver, but these guys were engineers, after all. With due deliberation, they decided to set the seat height 22.64 inches off the car's floor.

This was hardly the stuff of revolution. At the outset Uchiyamada approached his assignment conservatively. He envisioned a new car that used improved conventional technology: a highly efficient gasoline engine, an advanced transmission, and light-weight materials throughout. Combining all that had the potential to produce a small car with gas mileage 50 percent better than the Toyota Corolla, he figured. And by using existing technology, Toyota could bring the new car to market faster and more cheaply than would be possible with a new technology.

As a sop to the bosses, however, the G21 team decided to develop a "concept car" that used hybrid technology for display at the Tokyo Motor Show in October 1995. It would be a "static prototype." The engine could be started but the car couldn't move. It would be all show but no go.

In late 1994 Uchiyamada presented that plan to Toyota's top management. His bosses liked the concept car so much that they decided it should be developed for mass-market production. They wanted a futuristic car that would justify the name they had chosen: Prius, Latin for "to go before" or "lead the way." They insisted on a car that would get twice the gas mileage of the Corolla, not just 50 percent more, and would reduce vehicle emissions.

"That's outrageous," Uchiyamada blurted out, as the enormity of the challenge became clear.[13] Toyota's senior executives, however, wouldn't budge. They told Uchiyamada that if he didn't agree, they would sack him as the G21 team's chief engineer. Faced with that alternative, Uchiyamada relented.[14] With all-out effort and lots of luck, he figured, the new car might be ready by 1999.

His team, which had been expanding, pored over eighty different hybrid designs. They were divided into "series" and "parallel" hybrids, in the techno-talk of automotive engineers. Series hybrids operate sequentially (thus the name). The car is always powered by an electric motor, but when the batteries run low a gasoline engine delivers power to the motor. In parallel hybrids, the car can be powered directly either by the gas engine or the electric motor, or the two power sources can power the car together.

Series hybrids are simpler to develop and build but require more robust batteries, which presents a major challenge. The engineers who had developed GM's EV1 could testify to that. The Prius team opted for a parallel system, figuring that the complexity issue could be handled more easily than the challenge of developing more-powerful batteries. But a more difficult challenge would soon surface.

In August 1995 Toyota got a new president, Hiroshi Okuda, the first chief executive from outside the Toyoda clan in nearly three decades. He was an aggressive man, believing that, after three decades of rapid growth, Toyota had become plodding and predictable. Okuda wanted to make a dramatic statement by launching the Prius in 1997, the same year the United Nations conference on climate change would convene in Kyoto, just 75 miles from Toyota's headquarters. It would mean accelerating the Prius's launch by more than a year.

Once again Uchiyamada was aghast. Developing a normal new car takes three or four years. Applying the same timetable to untested technology seemed crazy. He said he couldn't guarantee the 1997 launch date, but Okuda insisted that it be the target anyway.[15] The one concession he made was to let Uchiyamada draft anyone in the company, including Toyota's best engineers, for the project. Well, that was big of him.

From mid-1995 onward Uchiyamada's growing group launched into marathon days that often stretched sixteen hours or more. One engineer assigned to the team went home to tell his wife the good news about his new job. Then he told her the bad news: he would move into a company dormitory to be on call at all hours.

His confidence in meeting Okuda's deadline, he confessed, was "less than five per cent."[16]

Privately, Uchiyamada thought even 5 percent was optimistic, but he couldn't say that. As he later put it, "I could not say to my team that our goal is impossible, but top management insists on it anyway."[17] Before he tried to convince his team that the deadline could be met, Uchiyamada decided, he had to convince himself. He found inspiration by reading the accounts of two quests to accomplish tasks that had seemed impossible.

One was an all-out effort by Japanese engineers to develop a jet fighter plane near the end of World War II. They completed the plane, the Kyushu J7W1 Shinden, in less than one year, though it came too late to affect the war's outcome. The second was America's Apollo space program. In 1961, when President Kennedy set the goal of putting a man on the moon within ten years, the technology for the mission didn't exist. That situation was no different, Uchiyamada realized, than the quest for a hybrid car.

His reading prompted him to adopt a method of project management used by NASA. After a target date was set for the final goal, the project leader worked backward to develop a timetable for interim milestones. Dividing the project into chunks would make it easier to manage, at least in theory. But in reality, interim milestones could slip when arrangements didn't go according to plan, as Uchiyamada soon learned. His group completed an early Prius prototype on schedule in late 1995, but when the team took the car to the test track, it wouldn't run.

For weeks the car stood still, confounding Uchiyamada and his engineers. By late December Uchiyamada figured that his team wouldn't get it to run until sometime in January, or even later. But on Christmas Eve, after fifty-nine days of frustration, one of his engineers climbed behind the wheel and, as if by a miracle, the car moved slowly down the track.[18] The success was short-lived. The car stalled out after just 500 yards, and wouldn't move again. But it was clear that the Prius prototype could work.

The battery was the team's biggest technical challenge. It had to be powerful, reliable, and durable, a difficult combination. The first battery had only half the power that the car's electric motor would require. To tackle the reliability issue, meanwhile, the team visited battery experts at Panasonic, the Japanese electronic-products giant. But there the Toyota people found that Panasonic's battery-failure rate was far higher than what Toyota deemed acceptable. Battery failure in a car, after all, could cause injury or death, while battery failure in an HDTV would only mean missing the latest episode of *The Simpsons*.

As for durability, Uchiyamada decided the battery had to last for the life of the car. Few people would buy a Prius if they faced the prospect of buying a new battery every few years. So his team developed techniques to cram ten years of durability testing into just a couple years. Still, the first batteries proved vulnerable to extremely hot and cold temperatures, which produced some comic episodes. One hot summer day, when Okuda was scheduled for a test drive in a prototype Prius, the engineers rigged the car by parking it in the shade beforehand and blowing a fan on the battery to make sure it would work.[19] Score one for the nerds.

In cold-weather testing in Hokkaido, the northernmost island in the Japanese archipelago, the early batteries died when the temperature fell below 14 degrees Fahrenheit, the equivalent of a balmy winter's day in Minneapolis. The engineers joked among themselves that no one should take a Prius on the test track without winter coats and emergency radios. Comic relief was a good thing for men working under stress.

But there was tragedy as well as comedy. In January 1997 an

engineer named Masahito Ninomiya boarded a company helicopter at Toyota's technical facility near Mount Fuji, where he
had been doing additional cold-weather testing on the Prius.
Ninomiya, thirty-seven years old, was heading back to headquarters to report his results. He never got there. The chopper crashed
en route, killing all eight people on board.[20]

By then, with the launch less than a year away, Prius prototypes were running on Toyota test tracks twenty-four hours a
day, seven days a week. Once Uchiyamada paid a late-night visit
to a track, only to have a junior engineer tell him to get lost. "You
are distracting us," the young man told his boss. "We will let you
know the results in the morning."[21]

Step by step the technical obstacles were surmounted. The
G21 engineers developed a system with a single gasoline engine
and two electric motors, one to power the wheels and the other
to control the transmission. The transmission governed the power
source, keeping the Prius running on electricity at low speeds, then
blending in power from the gasoline engine above thirty miles an
hour. The concept wasn't too different from what the American
scientists at TRW had dabbled with decades earlier, but this time
sophisticated electronics made it practical for production.

There were other engineering breakthroughs, too. The car's
system captured "regenerative" energy when the car was braking,
which boosted battery power. The gasoline engine, meanwhile,
was engineered to turn off when the car came to a stop, and restart
automatically when the driver accelerated again. A monitor on the
dashboard showed the driver whether the car was running on electricity or gasoline.

The Prius had two radiators: one cooled the gas engine, and the
other the electric motor. Fitting all that under the hood was like

cramming books into an overstuffed briefcase. But after lots of trial and error, the engineers found a way.

As the launch date neared, Uchiyamada learned that Toyota's sales department planned to sell just 300 cars a month. It seemed a minuscule target considering the herculean engineering effort, and hardly a statement of Toyota's faith in twenty-first-century technology. After some cajoling, he convinced the sales people to raise their target to 1,000 cars a month. It was less than one-tenth of one percent of Toyota's annual sales in Japan, but at least it was a four-digit number.[22]

Toyota unveiled the Prius in Tokyo on October 14, 1997. Uchiyamada drove the car into a ballroom at the ANA Hotel, with Okuda in the backseat. The hundreds of journalists on hand marveled that the car was quiet enough to drive indoors and, according to Toyota, would go 66 miles on a gallon of gas. That was double the gas mileage of the Corolla, just as Okuda wanted. The comparable mileage ratings, though accurate, were based on controlled lab tests. American regulators would later revise the measurement methods and rate the Prius just above 40 miles a gallon in city-highway driving, which was closer to what drivers got in everyday use. But using either method, the Prius's fuel economy still far exceeded that of any other car on the road.

Toyota priced the Prius at $18,000, only 15 percent more than the price of a Corolla. Auto analysts had expected a price premium of at least 30 percent. Toyota conceded it would lose money on the car at first, but explained that it was an investment in the future. To drive home that point, the Prius left the stage at the ANA to the tune of "When You Wish Upon a Star."[23]

The car went on sale in Japan in December, a few days after the Kyoto climate conference began. Soon afterward Toyota doubled the car's production, but it was still just 2,000 cars a month. Priuses were so scarce that it took a couple months before Uchiyamada spotted one driving on the streets of Nagoya, just like any other car. He was elated.

In February 1998 the G21 team celebrated with a party, and Okuda sent *sake* in salute. A couple months later a G21 engineer picked up Eiji Toyoda in a Prius and took him on a ride through the Mikawa Highland near Toyota City. The patriarch whose vision had launched the Prius project was pleased.

In America launching a new car that had no direct competition might seem like every automaker's dream. But Toyota realized it had to explain newfangled technology to a skeptical American public. Memories of the GM EV1's well-publicized limitations didn't help.

Toyota started early, displaying the Prius at an electric-vehicle show in Orlando the same month it went on sale in Japan. A few months later, in early 1998, the company sent two Priuses on a sixteen-city tour of the United States, inviting journalists, academics, and environmentalists to see the new car. "Where do you plug it in?" was the most frequent question. The answer, of course, was nowhere, but people couldn't believe that the brakes and the engine would keep the battery fully charged.[24] There were other issues, too.

One was the styling, which was developed at Toyota's design studio in California. The car's bullet-nosed front and chopped-off back end made the Prius look like a Gremlin from outer space. Toyota's American managers deemed the design "polarizing."[25] The interior design, meanwhile, was quirky. The gauges sat in the center of the dashboard instead of behind the steering wheel. The shift lever looked like a joystick. The overall effect evoked a video

game instead of a car. What's more, the car's acceleration was gla-
cial and its footing seemed uncertain. One automotive journalist
who took an early test drive deemed the Prius "a boring econo-car
with good gas mileage."[26]

But Toyota forged ahead. The company aimed its message at
"early adopters," marketing-speak for people eager to try new
technology. In 1999 the company let people volunteer to drive
the Prius for a month and provide detailed feedback about the car.
"The Prius feels like a hot-rod golf cart . . . pleasant enough, but
it doesn't provide me with the driving thrill I seek," reported a
woman in Seattle.[27] Indeed, the Prius needed nearly 13 seconds to
accelerate from 0 to 60 miles an hour, making it 30 percent slower
than a Corolla.

What's more, the Prius's U.S. price of $19,995 was some 30
percent more than the basic Corolla. The executives at Toyota's
American sales office in Torrance, California, worried that despite
the Prius's advantage in gas mileage, the price gap was too great.
The company's headquarters in Japan suggested adding a small
solar panel to the car's roof to give the Prius a touch of high-
tech pizzazz. But the Americans objected that it would add $100
to the price of the car. The idea was dropped. The company did,
however, include maintenance and towing plans in the price of
the Prius, at no extra cost.

The biggest crisis hit in December 1999, just seven months
before the Prius would go on sale in America. Honda upstaged
its larger rival in the United States by introducing a gas-electric
hybrid of its own called the Insight. It was more a novelty than a
practical car. A tiny two-seater, just like the EV1, it used a less sophisticated hybrid system than the Prius. Still, it gave Honda bragging rights. In Torrance, Toyota executives nicknamed the Insight
the "In Spite."[28]

Despite the setback, the plans for the Prius launch continued. Toyota set up a special Internet ordering system that required people to order their car online and pick it up from a nearby dealer. It was a departure from the usual system of simply shipping cars to dealers and letting them do the selling, but it seemed suitable for a new high-tech car that would be in limited supply at the outset. At the last minute, Toyota's sales executives worried that customers and dealers would object to the hassle and almost canceled the plan, but then they relented.

On June 30, 2000, Toyota's sales and marketing managers gathered around a computer console in Torrance to watch the first Prius orders come in. They had worked on the Prius for nearly three years, and had mounted an unprecedented effort for a car with an initial U.S. sales target of just 1,000 a month. So much about the Prius was so different that nobody knew what to expect. As the attendees held their collective breath, the screen started flashing with orders.

"We went from fear . . . then to joy . . . and then to elation," one Toyota manager later recalled. "The orders just lit up the website."[29] The first three months of Prius production were sold on the very first day. The car's American launch was a success to savor. The only drawback was that most Prius buyers would have to wait months to get their cars. To thank them for their patience, Toyota started sending the buyers periodic tokens of appreciation: key chains, mouse pads, and other trinkets. One item was a rear-window decal, the sort of thing that usually says "HARVARD" or "BROWN" or, for the less favored, "MANKATO STATE." The Prius decal, born in a burst of enthusiasm among Toyota's marketing managers, said: "EAT MY VOLTAGE."[30]

It could have been worse, something like "UP YOUR WATTAGE" or "KISS MY CURRENT." But it seemed bad enough to

Toyota's executives, who hadn't seen the decals before they were sent out. They ordered the first decal recall in automotive history.

That comic mini-debacle aside, Toyota had every reason to celebrate. It had launched the first practical mass-market hybrid car in the world's largest automotive market. And it had done so right under the nose of Detroit, then too busy counting its prof-its from selling SUVs. Toyota sold nearly 16,000 Priuses in the United States in 2001, the car's first full year on the market. One early buyer was Victor Wouk, the inventor whose hybrid had been rejected nearly forty years earlier.[31] Prius sales topped 20,000 in 2002. It wasn't a lot, but the car was making a statement. And the Prius was being discovered, like an aspiring starlet, in a place that's always searching for new hits: Hollywood.

In the spring of 2002 actor Tom Hanks hosted a fund-raiser for en-vironmental causes in Hollywood attended by an enviable A-list. Bill Clinton, Bobby Kennedy Jr., and Rob Reiner were there, among others. In years past everyone would have arrived in Fer-raris, Jaguars, BMWs, or stretch limos. But this time many of the guests arrived in Toyota Priuses. So many, in fact, that guests joked that they might accidentally drive off in somebody else's car.[32]

Almost overnight, the Prius was becoming a status symbol among entertainment's elite. "The list of Hollywood's hybrid-come-lately owners reads like the table of contents of *People* magazine," the *Washington Post* reported. "Cameron Diaz, Leon-ardo DiCaprio, Carole King, Billy Joel, David Duchovny and Bill Maher, to name-drop a few."[33] DiCaprio and his family actually owned four, a Titanic number, as it were. The Prius burst onto center stage at the 2003 Academy Awards. Toyota provided a small fleet of Priuses to ferry stars and starlets to the Kodak Theatre, where they disembarked with the TV cameras rolling. So many

Priuses were pulling up to the red carpet that it could have been named Best Supporting Car.

For Toyota this was marketing nirvana. "It gave Hollywood types the chance to show their commitment to the environment," a company executive explained. "And it gave us publicity and endorsements." [34] Better yet they were free endorsements, which was no small matter when Toyota was selling only 25,000 Priuses a year in the United States, nowhere near enough to justify an expensive advertising campaign.

By the time of its Academy Awards debut, the Prius was ensconced on HBO's *Curb Your Enthusiasm*, one of the hottest sitcoms on television. The brainchild of Larry David, the chief writer and cocreator of *Seinfeld*, *Curb* is about a curmudgeonly Hollywood writer whose foibles and schemes manage to annoy just about everyone. As the show's lead character, Larry David, plays himself. David's real-life wife, Laurie, was an ardent environmentalist. Big SUVs "should just have 'PIG' spray-painted on them," she once told an interviewer.[35] The Davids had three Priuses: one for Laurie, one for real-life Larry, and one for Larry to drive in the show. "The best gift he ever gave me was making his character drive a hybrid car," Mrs. David once said.[36] (The couple later divorced.)

Larry drives his Prius everywhere on *Curb*: to work, to Dodger Stadium, and to a klutzy encounter with a drug dealer to buy marijuana to ease his father's glaucoma. The car mostly plays a background role, but on the episode of February 29, 2004, it stepped into the spotlight.

Larry and his manager, Jeff, are driving the car to search for Jeff's German shepherd, Oscar, who had run off. Larry spots another Prius driver and waves, only to be disappointed. Then the scene unfolds:

LARRY: See that? I waved to a guy in a Prius and he didn't wave
        back!

JEFF: I don't wave to people in the same car as me.

LARRY: We're Prius drivers; we're a special breed. . . . I want to
        see what's up with that guy.[37]

So the chase begins, only to end when Larry bumps into a dog that happens to be Oscar. The dog survives the mishap. His bark is temporarily impaired but, unfortunately for Larry, not his snarl.

While Toyota reveled in the free Prius publicity, not every Prius owner was pleased with Hollywood's embrace. "I didn't want people thinking I had become a mindless follower of some movie star," fretted a middle-aged mathematician in Virginia who bought his Prius in 2003. "That whole thing was a turnoff."[38]

Prius preening was a turnoff to some people in Hollywood, too. Some movie stars were more enamored of GM's Hummer H2 than the Prius. The Prius-versus-Hummer conflict burst into view at the 2004 Oscars. The two vehicles vied in the valet lines at parties after the ceremonies, with their respective occupants eyeing each other warily. "It's Hummer versus hybrid, Hollywood hedonism versus holier-than-thou Hollywood political correctness," the *New York Times* observed.[39] The fault line wasn't just liberal versus conservative. It was established star (Prius) versus the *nouveau* famous (Hummer). And it was old money versus new, which would be a couple of decades versus a couple of years in Hollywood terms.

Hollywood's most famous Hummerite was Arnold Schwarzenegger, who had seven Hummers, at least until he became governor in 2003. Then he bid some of them "*Hasta la vista*, baby," and dutifully set an environmental example for Californians by cutting his personal flotilla from seven Hummers down to three.[40] The Sierra Club still wasn't mollified.

A couple years later, California's legislature passed a law allowing Prius owners to drive without passengers in the carpool lanes on California's freeways. This was no small matter. During rush hour it was hard to tell the difference between the freeways and the parking lots in California, which made carpool lane privileges just as coveted as tickets to the Oscars. Not surprisingly, the proposed law stirred controversy. The *Sacramento Bee* labeled it "hybrid hypocrisy," arguing that California's carpool lanes "were not created primarily to save fuel, but to reduce congestion."[41] The "Gubernator" signed the law anyway.

By the summer of 2005 the carpool lane law and the Hollywood hype were creating an anti-Prius backlash well beyond California. Some people started calling the Prius and its drivers the "Pious." The derision wasn't exactly what Toyota wanted from its Hollywood marketing strategy, but there was more to come. The zenith, or perhaps the nadir, of the anti-Prius backlash came on a cable TV show called *South Park*, an absurdist animated sitcom about a mythical Colorado town with mixed-up kids, screwed-up adults, and lots of scatological humor.

In the episode of March 29, 2006, Gerald Broflovski, one of the grown-ups in South Park, buys a Prius. Before long he starts handing out fake parking tickets to the town's SUV owners for "failure to care about the environment." This stunt doesn't make him popular. One of Gerald's neighbors tells him he's become so smug that he must love smelling his own flatulence. Gerald takes offense and moves his family to San Francisco. Sure enough, his new neighbors there all drive Priuses. They also have the habit of passing gas and bending over to inhale.

Meanwhile, back in South Park, Priuses have become so popular that the town develops a huge cloud of "smug," which then merges with two other smug clouds. One is moving in from San

Francisco and the other from Los Angeles, where it was created by George Clooney's self-righteous acceptance speech at that year's Oscars. The Broflovskis manage to escape San Francisco just before the city disappears, as the TV weatherman reports, "into its own asshole."[42]

The episode was hilarious to people who liked jokes about butts and farts, and to people getting tired of granola-fed dilettantes flaunting their Priuses as signs of moral superiority. Presumably there were more of the latter, at least among people over twelve. But by the time the *South Park* Prius episode aired, the Prius was turning the cultural corner. Toyota's car of the future was starting to attract buyers whose main motive was saving money, not saving the planet.

In late 2003, the same year that General Motors killed the EV1, Toyota started previewing the second-generation Prius in selected venues. Among them were Whole Foods supermarkets and the International Yoga Convention. The Daytona 500 and the World Hunting Expo didn't make the list.

The company also displayed the new Prius, which was the 2004 model, at the U.S. Electric Vehicle Association convention in Long Beach, California. With its gas-hybrid engine, the Prius wasn't truly an electric car, but that didn't matter. A couple of Hollywood actor-activists, Ed Begley Jr. and Rob Reiner, sang the Prius's praises. When Toyota offered test drives to the public, the lines snaked around the building as people waited two or three hours for their turn. "Maybe we've got a mainstream hit here," a Toyota marketing executive said to his colleagues.[43] He was right.

The new Prius was six inches longer than the original and could go from 0 to 60 miles an hour in only 10 seconds, an improvement of 25 percent. The new car got 46 miles a gallon compared to 41 miles per gallon for the first-generation Prius, according to the Environmental Protection Agency's revised testing methods. The mileage improvement came in part from the new, sleek styling, which reduced wind resistance. The car also looked different from anything else on the road, which let Prius drivers flaunt their automotive greenness. Toyota left the base price unchanged at $19,995.

In 2004 Toyota sold 54,000 Priuses in the United States, more than twice the sales of the previous year. A year later Hurricane Katrina slammed into New Orleans, and into the American psyche as well. Gasoline finally became more expensive than Evian and Pellegrino. In 2005 Prius sales doubled again to 108,000 cars in the United States, cracking six figures for the first time. And the Prius set another new sales record, more than 181,000 cars, in 2007. That was the year that young Al Gore III got arrested for topping 100 miles per hour in a Prius, but he wasn't the only Prius driver pinched that year for triple-digit speeding.

On the night of March 28, a California highway patrolman flagged down a Prius going more than 100 miles an hour on a freeway. Word of the infraction reached Gary Richards, writer of the Mr. Roadshow column in the *Mercury News* of San Jose, California, the hometown of the offending driver. His name was Steve Wozniak, aka "The Woz," a cofounder of Apple Computer.

Mr. Roadshow hopped on the story. He shot an e-mail to Wozniak, asking whether he had really hit 105 in his hybrid. "Not true," replied the Woz. "104 mph." He told the judge he had gotten mixed up between miles and kilometers on the speedometer. The judge smiled and fined him $700 anyway.[44]

Mr. Roadshow wanted to know how the Prius had handled at 104. "I was surprised to discover that the Prius was very stable, even with major gusting winds," Wozniak told him. "Being used to a Hummer, I expected the opposite."[45]

Now *this* was the real news. Steve Wozniak, beloved techno legend and guru geek, owned and drove both Priuses and Hummers. It was like cheering for both the Red Sox and the Yankees, or even like owning both Macs and PCs. If this had been half a century earlier, the Woz might have owned both a Volkswagen Beetle and a tail-finned Cadillac.

To anyone tired of America's culture wars, here was reason for hope. Maybe the lion could lie down with the lamb, and maybe an obnoxious redneck who drove a Ford F-150 King Ranch edition could have a beer with an insufferable Hollywood liberal who drove a Toyota Prius, though they might argue whether the beer should be Stella or Lone Star.

By 2008, more and more Prius buyers weren't people like Larry and Laurie David. Instead they were people like Scott and Katie Jackson of Edwardsville, Illinois, a tiny town near St. Louis. In 2008 the young salesman was putting a lot of miles on his Ford Explorer Sport Trac, an SUV with a short pickup truck bed. In other words, it was a lot closer to a Hummer than a Prius. Young Jackson loved the Sport Trac. It made him feel like he owned the road, though he couldn't afford to own much else. It was hard to buy groceries when he was shelling out $450 a month for gas.

His wife, Katie, started crunching numbers and concluded that buying a new Prius would cut their gasoline bills so much that the car would almost pay for itself. The Jacksons decided to dump the Sport Trac and order a Prius. The car was in such high demand that it took two months to arrive at the dealer, who then gave them just twenty-four hours to come down and sign the papers. Otherwise, he warned, he would sell their Prius to somebody else. It was a far cry from 2000, when Toyota managers worried that nobody would want the Prius.

The Jacksons paid $30,000 for their Prius, not exactly a bargain.

But it came with a rearview camera, keyless ignition, auxiliary audio jack, satellite radio, fifteen-inch alloy wheels, cruise control, and more. Guy stuff and tech stuff, basically. Still, Scott Jackson had some adjustment problems. He missed looking down on other drivers from the lofty perch of his Sport Trac. And he grimaced when friends asked whether he was carrying a purse. "Let's face it," he sighed, "the Prius isn't the most masculine vehicle on the planet."[46]

But the bottom line was, well, the bottom line. The Jacksons found themselves saving about $350 a month on gasoline, which pretty much paid for the Prius. For most people outside Hollywood and Sausalito, that was the essence of the car's appeal. They appreciated the Prius's environmental benefits, but they liked the benefits to the family budget even more. If they saved gasoline and reduced CO_2 emissions in the process, however, did it matter that there were more Scott Jacksons than Larry Davids driving Priuses? At least America's dogs would be safer.

In January 2011, Takeshi Uchiyamada arrived in Detroit for the city's annual auto show. Like his car, Uchiyamada had come a long way since the first Prius appeared in 1997. At age sixty-four, he was an executive vice president of Toyota and a member of its board of directors.

The Prius, now in its third generation and rated at 51 miles per gallon, was on display at the show, along with competing cars that testified to the Prius's pervasive influence. Honda had a new version of the Insight, with a rear seat and four doors, and the CR-Z "sport hybrid," a small SUV. Ford was displaying hybrid versions of its Fusion sedan and its Escape SUV.

Motor Trend's Car of the Year was the Chevrolet Volt, a hybrid with a plug-in feature that allowed it to run longer on electricity, and thus go more than 90 miles on a gallon of gas. A Silicon Valley car company called Tesla showed its all-electric roadster, priced upward of $100,000. Its chief competitor was Fisker, a high-tech hybrid that cost nearly as much. Nissan, meanwhile, was preparing to launch the LEAF, the first mass-market all-electric car. All of these cars, and others like them, showed that the automotive-propulsion revolution begun by the Prius, and with roots that went back centuries, was evolving from novelty to reality.

At the auto show Toyota announced it would expand the Prius from a single model to an entire family of cars. Among them would be a Prius station wagon and a Prius plug-in hybrid, similar to the Volt. The new models would mark the next phase of the Prius's improbable journey.

The day after the announcement, Uchiyamada reflected on how the Prius might be viewed some fifty years hence, when cars would be running on who-knows-what. "It is hard to forecast," he said through an interpreter. "My hope is that when people look back, they will say that because of the Prius, automakers all over the world took environmental issues seriously," he said. "And because of the Prius, things began to change."[47]

AFTERWORD

The hardest part about writing this book wasn't deciding what cars to include. It was deciding what cars to leave out. My selections will disappoint some people, especially fans of iconic automobiles not included.

But this book isn't intended to be about great cars, fast cars, or famous cars, although it contains some of each. Instead it's about the automobiles that have influenced how we live and think as Americans. The cars in this book either changed American society or uniquely captured the spirit of their time. By those criteria most cars, even those regarded as automotive icons, fall short.

Among them is the Chrysler Airflow, built from 1934 to 1937. It pioneered the use of streamline design in automobiles. But the Airflow was a commercial flop, and it is impossible to cite any broad impact that the car had on American life. The only two pre–World War II cars in this book are the Ford Model T and the La-Salle. Other important prewar cars existed, of course, but the pace of cultural and societal change in America accelerated only after the war.

Ed Cole's 1957 Chevrolet Bel Air remains a classic car to this day, thanks to its beautiful and proportioned lines, art deco colors, and peppy "small block" V8 engine. It was an important automobile, but that isn't enough. The '57 Chevy didn't have the cultural impact of the tail-finned Cadillacs.

As for the Ford Thunderbird, Robert McNamara's decision to add a backseat enhanced the car's commercial success. In 1964 the 'Bird inspired a hit song, "Fun, Fun, Fun" by the Beach Boys. But the Thunderbird abandoned the American sports car arena to the Corvette, which uniquely reflected America's renewed zest for life after the twenty-five-year trauma of Depression and war.

The Chevy Camaro is a favorite of boy racers and grown-up racers, but it began life as a belated, and slightly desperate, response to Ford's success with the Mustang. The AMC Gremlin was a closer call, to my mind, because its ungainly ineptitude captured the feelings, frustrations, and failures of America in the 1970s.

Besides that, during the course of my research I ran across a couple of Gremlin stories I loved. One came from comedian Jon Stewart. He told me his first car was a 1975 Gremlin. He put a bag of limestone in the backseat to help provide traction on wet roads, and some of the limestone started spilling out of the bag. On the day of his high-school graduation, Stewart put his cat in the car to take it to the vet. The cat mistook the limestone for kitty litter. The smell probably outlived the poor cat.

The other story occurred at a display of classic cars at Greenfield Village near Detroit, where a man from Ohio displayed his Gremlin alongside beautifully restored Corvettes, Edsels, and bat-wing 1959 Chevy Impalas. But the lowly Gremlin drew a steady stream of admirers, including one middle-aged woman who revealed a piece of her family lore: that she had been conceived in the backseat of a Gremlin. Her parents must have been particularly lithe.

So why not the Gremlin? The lasting legacy of the 1970s was the beginning of the American revival that occurred in the 1980s, a story best captured by Honda's success in Ohio. Honda didn't start making cars there until 1982. But the planning began in the mid-1970s, and the motorcycle factory that was the precursor to the auto-assembly plant in Marysville launched production in 1979.

The 1986 Ford Taurus revolutionized American automotive styling with its sleek, curvy lines. But it's hard to point to an impact the car had beyond that. The Toyota Camry has been a best-selling car for a couple of decades. But commercial success doesn't equate to cultural impact. One of my friends suggested including his high-school car, the Dodge Dart, which was ubiquitous in the 1970s. Obviously he had a deprived childhood.

I selected the fifteen cars here because each one left an enormous imprint on American life. In some cases their impact was palpable. The Model T Ford made America a mobile society and helped develop the American middle class. Other cars—the Volkswagen Microbus and the Chrysler minivans, among others—had a mostly symbolic impact. But symbols are part of the social glue that keeps a large and diverse nation together.

During the course of my research and writing, several friends have asked which car in this book is my personal favorite. It's a tough choice, and the best I can do is narrow it down to two.

One is the Ford Mustang. I'm a baby boomer who loved the car from the minute it appeared, and it symbolizes America in so many ways. Mustang clubs exist in most countries around the world. The car's champion was Lee Iacocca, the son of immigrants, and a man whose success embodies the American dream. And forty years after the first Mustang appeared, the latest version of the car was developed by another son of immigrants, Vietnamese-American Hau Thai-Tang. His personal story is as compelling as the car itself.

Thai-Tang saw his first Mustang at a USO show for the American army in Saigon in 1971, when he was five. Four years later Saigon was about to fall to the Vietcong. Young Hau's father had fought in the South Vietnamese army and his mother had worked for an American bank, which made the family a likely target for "reeducation."

The Thai-Tang family was fortunate to be put on the American evacuation list. In July 1975 American officials told them to listen for Bing Crosby's "White Christmas" being played on Armed Forces Radio. That would be their signal to head to the pickup point.[1] The family left Vietnam by helicopter, spent several weeks in refugee camps, and then settled in Brooklyn, where nine-year-old Hau started public school knowing just a few words of English.

He learned quickly, later graduated from Carnegie Mellon University, and joined Ford as an engineer. There his story came full circle when he led the team that developed the 2005 Mustang, the first new Mustang in more than a decade. The job "exceeded my wildest dreams," Thai-Tang later said. "This car embodies freedom."[2] Hau Thai-Tang is a senior Ford executive today.

My other favorite car is the most flawed automobile in this book, the Chevrolet Corvair. Its story is both remarkable and remarkably sad. The Corvair represented a genuine effort to develop a fuel-efficient car of the future, with all the elegant simplicity of the Volkswagen Beetle but with more size and greater functionality. The man behind it, General Motors' Ed Cole, was an engineering genius with the courage to challenge his company's formidable bureaucracy. It's impossible not to like someone whose motto was "Kick the hell out of the status quo."

The man who brought the Corvair down, Ralph Nader, had equal vision and determination. He believed corporate America was callous, which was and is sometimes true, and that product-liability lawsuits provided a powerful tool to prevent companies from putting profits ahead of people.

Cole and Nader were brilliant and worthy adversaries, a pro-tagonist and antagonist who could have been characters in classic Greek drama. Cole's hubris doomed the Corvair. But his leading role in developing catalytic converters and getting the lead out of gasoline was a story of redemption that has gone largely unrecog-nized. Nader fell victim to his own brand of hubris, seeing malfea-sance of some sort behind every corner, and has lived to become the cranky old man of American public life.

Cole-versus-Nader was a clash of titans that launched Amer-ica's litigation society, the vast expansion of government's regula-tory apparatus, the discrediting of corporate America in the minds of many people, the waning of engineering innovation at General Motors, and finally Nader's role in tilting the razor-thin 2000 pres-idential election to George W. Bush.

Historians rightly agree that the Model T Ford is the most important car in American history. The more elegant question is which car ranks second in importance. Because its story is so dra-matic, and because its influence has been so lasting and pervasive, the fatally flawed Corvair gets my vote.

Now let the arguments begin.

ACKNOWLEDGMENTS

There are many people who played critical roles in bringing *Engines of Change* to life, more than I can thank by name in the space allotted here. I'm especially indebted to my very capable editor at Simon & Schuster, Ben Loehnen, and to his assistant, Sammy Perlmutter, for their interest in this book, as well as their guidance, help, and friendship. My agents Andrew Wylie and Scott Moyers (who has since left the Wylie Agency for Penguin Press) were incredibly supportive and resourceful. I have the highest regard for them both.

I gratefully acknowledge the help of many people who know and love automobiles. Among them are Hal Sperlich, Paul Lienert, Csaba Cera, Joe White, Jean Jennings, Bob Casey, Bob Lutz, Jerry Burton, Jerry Palmer, Jim Fitzpatrick, Carsten Jacobsen, Jack Harned, Larry Kinsel, Bud Liebler, Soichiro Irimajiri, Shin Tanaka, Bill Hoglund, Arv Mueller, Doug Scott, Steve Harris, and Steve Miller. Miles Collier was generous with access to the marvelous Collier Museum and Library. Mark Patrick, the Collier librarian, provided invaluable support. The wonderful people at the Detroit Public Library's National Automotive History Collection—Gina Tecos, Barbara Thompson, Patrice Merritt, Paige Plant, and others—went beyond the call of duty to help me. I received generous help from knowledgeable public relations staffers at several car companies and advertising agencies: Tom Wilkinson at General Motors, Joe Tetherow at Toyota, Gualberto Ranieri at Chrysler, Ed Miller at Honda, Bill Collins at Ford, and Pat Sloan at DDB Worldwide. The Benson Ford Research Center at the Henry Ford Museum in Detroit provided access to its marvelous archives. Three of my close colleagues at Reuters who travel mostly by subway, Steve Adler, Stuart Karle, and Jean Tait, at least pretended to enjoy my favorite automotive anecdotes.

Several close friends provided ideas, encouragement, and hospitality on my research travels: Logan and Edrie Robinson, Keld and Jytte Scharling, Claus and Helga Hansen, Joe McMillan, and Gary and Leslie Miller. I was very fortunate to have had wonderful interviews with two key figures who, sadly, passed away during the course of my research. One was Don Frey, one of the fathers of the Ford Mustang, and the other was Chuck Jordan, who designed the 1959 Cadillacs with the tallest tail fins ever. I'm sure they are cruising in their favorite cars somewhere above the clouds.

Finally there is my family: wife Susie; sons Adam, Charlie, and Daniel; grandson Jasper; brother Larry and sister-in-law Vicki; and Uncle Tony Ingrassia, the first journalist in the family.

NOTES

ADOH   Automotive Design Oral History, Benson Ford Research Center, HFM

HFM   Henry Ford Museum, Dearborn, MI

NAHC   National Automotive History Collection, Detroit Public Library

CHAPTER ONE

1. John Steinbeck, *Cannery Row* (New York: Viking Press, 1947), 41.
2. Henry Ford with Samuel Crowther, *My Life and Work* (Garden City, NY: Garden City Publishing Co., 1922), 73.
3. Steinbeck, *Cannery Row*, 41.
4. Henry Ford III (speech, July 21, 2008). Text provided by Ford Motor Co.
5. *The New LaSalle*, General Motors sales brochure, 1929, NAHC.
6. General Motors Corp., "Design History of General Motors," May 12, 2006, http://www.worldcarfans.com/10605127100/design-history-of-general-motors.
7. Peter C. T. Elsworth, "Ford's Model T: It All Began 100 Years Ago," *Providence Journal*, April 19, 2008, http://www.projo

.com/projocars/content/CA-MODELT_04-19-08_CE9Q41L_
v24.2385320.html.

8. HFM, http://www.hfmgv.org/exhibits/showroom/1896/quad.html.

9. Ford Motor Co., http://corporate.ford.com/about-ford/heritage
/vehicles/quadricycle/675-quadricycle.

10. Ford, *My Life and Work*, 42.

11. Robert Casey, *The Model T: A Centennial History* (Baltimore: Johns
Hopkins University Press, 2008), 14–16.

12. Ibid., 17.

13. "Brush Cars," http://remarkablecars.com/main/brush/brush.html.

14. Ford Motor Co. sales brochure, 1909, HFM, http://www.hfmgv.org
/exhibits/showroom/1908/lit.html.

15. *The Original Ford Joke Book* (Binghamton, NY: Woodward Publishing Co., 1915; Vintage Antique Classics, 2006), 23, http://www
.vintageantiqueclassics.com/fordjokebook/.

16. Casey, *Model T*, 53.

17. Harry Barnard, *Independent Man: The Life of Senator James
Couzens* (New York: Scribner, 1958; Detroit: Wayne State University Press, 2002), 91.

18. Ibid.

19. Editorial, *Wall Street Journal*, January 7, 1914.

20. *Ford Joke Book*, 4.

21. "Sentiment In Business," *Bismarck Tribune*, November 8, 1927, 4.

22. "Seven Youths in Gang Steal 25 Machines," *Richwood Gazette*
(Richwood, OH), September 22, 1927, 4.

23. Michael Lamm and Dave Holls, *A Century of Automotive Style: 100
Years of American Car Design* (Stockton, CA: Lamm-Morada Pub.
Co., 1996), 89.

24. Ron Van Gelderen and Matt Larson, *LaSalle: Cadillac's Companion
Car* (Columbus, OH: Cadillac & LaSalle Club, 1999), 15.

25. Ibid. (Originally from *New Yorker*, March 1927).

26. "The Woman's Reagan 'Par le Sport'" *Vogue*, June 1929, 56.

27. Van Gelderen and Larson, *LaSalle*, 45.

28. William L. Mitchell, April 8, 1987, ADOH, vol. 1, 3.

29. Harley J. Earl, "I Dream Automobiles," in *The Saturday Evening
Post Automobile Book*, ed. Jean White (Indianapolis: Curtis Pub.
Co., 1977), 46.

30. Mitchell, April 8, 1987, ADOH, 9.

31. David R. Holls, April 2, 1987, ADOH, vol. 1, 8.
32. Richard A. Teague, January 23, 1985, ADOH, vol. 2, 58.
33. Mitchell, April 8, 1987, ADOH, 19.
34. Van Gelderen and Larson, *LaSalle*, p. 177.
35. Author interview with Raymond Paske, August 2008.
36. James D. Bell, "Companion Car to Cadillac," *Automobile Quarterly* 5, no. 3 (Winter 1967), 311.
37. Earl, "I Dream Automobiles," 46.

CHAPTER TWO

1. The definitive book about Zora Arkus-Duntov is *Zora Arkus-Duntov: The Legend Behind Corvette*, published by Jerry Burton in 2002. Burton had extensive access to Duntov and his wife, Elfi, when both were still alive. He's also, fittingly, a member of the Corvette Hall of Fame. Besides relying extensively on Burton's book as source material for this chapter, I was fortunate to have interviewed Burton himself, who generously provided additional perspective and insight.
2. Philip Booth, "Route 66—Television on the Road toward People," *Television Quarterly* 2 (Winter 1963), 9.
3. Dan Jenkins, "Talk About Putting the Show on the Road," *TV Guide*, July 22, 1961, 14.
4. Television Reviews, *Variety*, October 12, 1960.
5. Tom McCahill, "MI Tests the Chevrolet Corvette," *Mechanix Illustrated*, May 1954, 202.
6. "Jack Kennedy—The Senate's Gay Young Bachelor," *Saturday Evening Post*, June 13, 1953, headline on cover.
7. Zora Arkus-Duntov, untitled confidential memo to Ed Cole and Maurice Olley, October 15, 1954, National Corvette Museum, http://www.corvettefever.com/featuredvehicles/corp_0909_1956 _chevrolet_corvette/index.html.
8. General Motors Corp., First Quarter Report to Shareholders, May 1953, NAHC.
9. Jerry Burton, *Zora Arkus-Duntov: The Legend Behind Corvette* (Cambridge, MA: Bentley Publishers, 2002), 14–16.
10. Coles Phinizy, "The Marque of Zora," *Sports Illustrated*, December 4, 1972, http://sportsillustrated.cnn.com/vault/article/magazine /MAG1086825/index.htm.

11. Ibid.

12. Ibid.

13. Burton, *Zora*, 134.

14. William L. Mitchell, August 1984, ADOH, http://www.autolife
.umd.umich.edu/Design/Mitchell/mitchellinterview.htm.

15. David Halberstam, *The Fifties* (New York: Villard Books, 1993),
488–89.

16. Burton, *Zora*, 160.

17. Zora Arkus-Duntov, "Thoughts Pertaining to Youth, Hot-Rodders
and Chevrolet," internal General Motors Corp. memo on display at
the National Corvette Museum, Bowling Green, KY.

18. Ibid.

19. "GM Workers Beat, Expel Red Suspect," *Detroit Times*, June 17,
1954.

20. "Millionaire at High Speed," *Time*, April 26, 1954, http://www.time
.com/time/magazine/article/0,9171,860655,00.html.

21. Zora Arkus-Duntov, memo to Cole and Olley.

22. Ibid.

23. Kenneth Rudeen, "Fantastico is for Fangio," *Sports Illustrated*, April
1, 1957, http://sportsillustrated.cnn.com/vault/article/magazine
/MAG1132468/index.htm.

24. Author interview with Jerry Palmer, former General Motors design
executive and colleague of Duntov, July 2008.

25. Lamm and Holls, *Century of Automotive Style*, 173.

26. C. Edson Armi, *The Art of American Car Design: The Profession
and Personalities* (Pennsylvania State University Press, 1988), 37.

27. Lamm and Holls, *Century of Automotive Style*, 173.

28. Jan P. Norbye, "Mr. Duntov and His Cars," *Car and Driver*, November 1962, 41.

29. Burton, *Zora*, 271.

30. Booth, "Route 66," 6.

31. Jenkins, "Talk About Putting the Show on the Road," 14.

32. Palmer, interview.

33. Ibid.

34. Phinizy, "Marque of Zora."

35. Jerry Flint, *The Dream Machine: The Golden Age of American
Automobiles: 1946–1965* (New York: Quadrangle/New York Times
Book Co., 1976), 132.

36. Rick Ratliff, "Mr. 'Vette's Back on the Fast Track," *Detroit Free Press*, June 18, 1980.
37. Author interview with Jim Perkins, August 2008.
38. George F. Will, "Corvette King Revved Up America," *Washington Post*, April 28, 1996.
39. Andrew Peyton Thomas, *Clarence Thomas: A Biography* (San Francisco, Encounter Books: 2002), 557.
40. Author interview with Shelby Coffey III, July 2008.

CHAPTER THREE

1. Chrysler Corp. sales brochures, 1957, NAHC.
2. William H. Whyte Jr., "The Cadillac Phenomenon," *Fortune*, February 1955, 181.
3. "The Cellini of Chrome," *Time*, November 4, 1957, http://www.time.com/time/magazine/article/0,9171,867903,00.html.
4. Teague, ADOH, 46.
5. Peter Grist, *Virgil Exner: Visioneer; The Official Biography of Virgil M. Exner, Designer Extraordinaire* (Dorchester, UK: Veloce Publishing, 2007), 39.
6. "Up from the Egg," *Time*, October 31, 1949, http://www.time.com/time/magazine/article/0,9171,801030,00.html.
7. Virgil Exner Jr., August 3, 1989, ADOH, 75.
8. David Riesman and Eric Larrabee, "The Executive as Hero," *Fortune*, January 1955, 108.
9. David Sarnoff, "The Fabulous Future," *Fortune*, January 1955, 83.
10. "First Among Equals," *Time*, January 2, 1956, http://www.time.com/time/magazine/article/0,9171,808128,00.html.
11. Lawrence R. Hofstad, "The Future Is Our Assignment," *The Greatest Frontier: Remarks at the Dedication Program, General Motors Technical Center; Detroit, Michigan*, May 16, 1956, http://history.gmheritagecenter.com/wiki/uploads/c/c9/The_Greatest_Frontier_LoRes.pdf.
12. "Stylist on the Spot," *Look*, September 1954, 122.
13. "Cellini of Chrome."
14. Raymond Loewy, "Jukebox on Wheels," *Atlantic*, April 1955, http://www.theatlantic.com/magazine/archive/1955/04/jukebox-on-wheels/3944/.

15. Chrysler Corp. sales brochure, 1956, NAHC.
16. "Road Test: Plymouth Four-door Hard Top," *Motor Life*, April 1956, 48.
17. Chrysler Corp. sales brochure, 1956.
18. Author interview with Chuck Jordan, June 2008.
19. Lamm and Holls, *Century of Automotive Style*, 38.
20. Teague, ADOH, 45.
21. Chrysler Corp. sales brochure, 1957.
22. Ibid.
23. Virgil M. Exner, "Styling and Aerodynamics" (speech, Society of Automotive Engineers, September 14, 1957), box 4, Exner's personal papers, Benson Ford Research Center, HFM.
24. Ibid.
25. Virgil M. Exner, "Style Sets a Winning Pace," Tobe Lecture in Retail Distribution, Harvard Business School, December 12, 1957, HFM archives.
26. Ibid.
27. Vance Packard, *The Hidden Persuaders* (New York: Pocket Books, 1957), 97.
28. John Keats, *The Insolent Chariots* (Philadelphia: Lippincott, 1958), 53.
29. "Cellini of Chrome."
30. "The Shape of Things to Come," *Motor Life*, August 1959.
31. Chrysler Corp., *The 1959 Cadillacs*, Competitive Assessment Report, NAHC.
32. Jordan, interview.
33. General Motors Corp., Cadillac press release, 1959, NAHC.
34. Mitchell, April 8, 1987, ADOH.
35. Author interview with Leif Kongso, May 2008.

CHAPTER FOUR

1. "Volkswagen Microbus," *Car and Driver*, June 1970, 76–77, 92.
2. "The German People's Car," *Autocar*, February 10, 1939, 12.
3. Volkswagen advertisement, *Rolling Stone*, September 21, 1995.
4. *Small World* (Volkswagen of America magazine, introductory issue), 1962, NAHC, 9.

5. Josef Ganz's contributions to a German people's car are described by Paul Schilperoord in *The Extraordinary Life of Josef Ganz, the Jewish Engineer Behind Hitler's Volkswagen*, RVPP Publishers, December 2011.
6. Phil Patton, *Bug: The Strange Mutations of the World's Most Famous Automobile* (Cambridge, MA: Da Capo Press, 2004), 26.
7. Arthur Railton, *The Beetle: A Most Unlikely Story*, Verlagsgesellschaft Eurotax AG, 1985), 18.
8. Walter Henry Nelson, *Small Wonder: The Amazing Story of the Volkswagen* (Boston: Little Brown, 1965), 45.
9. Ibid., 74.
10. Frank Rowsome Jr., *Think Small: The Story of Those Volkswagen Ads* (Brattleboro, VT: S. Greene Press, 1970), 32.
11. "German Car For Masses," *New York Times*, July 3, 1938.
12. "A German War Vehicle," *Automobile Engineer* 34, no. 451 (July 1944), 259.
13. Patton, *Bug*, 82.
14. "Builder of the Bug," obituary of Heinz Nordhoff, *Time,* April 19, 1968, http://www.time.com/time/magazine/article/0,9171,838254-1,00.html.
15. Railton, *Beetle*, 109.
16. Obituary of Ivan Hirst, *Guardian*, March 18, 2000, http://www.guardian.co.uk/news/2000/mar/18/guardianobituaries.
17. Patton, *Bug*, 84.
18. Volkswagen files, 1950, NAHC.
19. "Hitler's Flivver Now Sold in the U.S.," *Popular Science*, October 1950, 162.
20. Author interview with Holman Jenkins Sr., February 2008. Mr. Jenkins is the father of Holman Jr., a *Wall Street Journal* columnist and former colleague of the author.
21. Mark Tungate, *Adland: A Global History of Advertising* (London: Kogan Page, 2007), 54.
22. Nelson, *Small Wonder*, 232.
23. Volkswagen Beetle ad, 1959, courtesy of DDB Worldwide.
24. Ibid.
25. Clive Challis, *Helmut Krone. The Book.: Graphic Design and Art Direction (Concept, Form and Meaning) After Advertising's Creative*

Revolution (Cambridge, UK: Cambridge Enchorial Press, 2005), 65.

26. Ibid., 1.

27. DDB ad.

28. Railton, *Beetle*, 162–63.

29. Author interview with Bob Kuperman, June 2008.

30. DDB ad.

31. Volkswagen of America, press release, 1962, NAHC.

32. Challis, *Helmut Krone*, 61.

33. *Small World*, 1969, NAHC, 7.

34. Author interview with former Volkswagen of America employee, who wished to remain anonymous, June 2008.

35. DDB ad.

36. Volkswagen sales brochure, 1969, NAHC.

37. Dan Greenburg, "Snobs' Guide to Status Cars," *Playboy*, July 1964, 66.

38. Ibid.

39. "Volkswagen Microbus," 76–77.

40. Bob Weber cartoon from *The New Yorker*, in *Think Small*, Volkswagen of America booklet, 1967 (no page numbers).

41. DDB ad.

42. Jim Jones, "The 'All-New' VW," *Newsweek*, October 20, 1969, 98B.

43. Heinz Nordhoff (speech to Economic Club of Detroit), in *VW Weathervane*, Volkswagen employee magazine, 1962 Commemorative Issue, 24.

44. John Muir and Tosh Gregg, *How to Keep Your Volkswagen Alive: A Manual of Step by Step Procedures for the Compleat Idiot*, 1981 ed. (Santa Fe, N.M.: John Muir Publications Inc.), 3.

45. Railton, *Beetle*, 142.

CHAPTER FIVE

1. "Executives: G.M.'s New Line-Up," *Time*, November 10, 1967, http://www.time.com/time/magazine/article/0,9171,837554,00.html.

2. "The New Generation," *Time*, October 5, 1959, http://www.time.com/time/magazine/article/0,9171,894298,00.html.

3. Karl Ludvigsen, "SCI Analyzes Ed Cole's Corvair," *Sports Car Illustrated*, November 1959, 23.

4. Donald MacDonald, "GM's Cart-Before-the-Horse Car," *True: The Man's Magazine*, November 1959, 65.

5. Ibid., 104.
6. "New Generation."
7. "New Generation."
8. MacDonald, "GM's Cart-Before-the-Horse Car," 104.
9. "New Generation."
10. MacDonald, "GM's Cart-Before-the-Horse Car," 108.
11. Ibid., 61–62.
12. Ibid., 63–64.
13. Chevrolet advertisement, September 27, 1959, NAHC.
14. Ibid. (Emphasis in original.)
15. Ibid.
16. Chevrolet sales brochure, NAHC.
17. Chevrolet sales leaflet, NAHC.
18. Arthur W. Baum, "The Big Three Join the Revolution," *Saturday Evening Post*, October 3, 1959, 141.
19. Ludvigsen, "SCI Analyzes," 30.
20. Ibid., 25.
21. *The 1960 Corvair, Competitive Car Information, Chrysler Corp. Engineering Division*, Chrysler Internal Report, November 1959.
22. Ludvigsen, "SCI Analyzes," 25.
23. Baum, "Big Three," 141.
24. MacDonald, "GM's Cart-Before-the-Horse Car," 109.
25. Chevrolet sales brochure, NAHC.
26. Ibid.
27. "The Chevrolet Corvair" (technical paper presented to the Society of Automotive Engineers convention in Detroit, January 11–15, 1960), NAHC, 6.
28. Ibid., 4.
29. Ibid.
30. "Best Cars for 1964," *Car and Driver*, May 1964, 31.
31. Chevrolet, press release, 1963, NAHC.
32. Ralph Nader, *Unsafe at Any Speed: The Designed-In Dangers of the American Automobile*, 25th Anniversary ed. (New York: Knightsbridge Publishing, 1991), ciii.
33. Ibid., 2.
34. Ibid., 19.
35. Ibid., 11.
36. Ibid., 24.

37. Ibid., 14.
38. Ralph Nader, "Profits vs. Engineering—the Corvair Story," *Nation*, November 1, 1965, 265.
39. Elinor Langer, "Auto Safety: Nader vs. General Motors," *Science*, April 1966, 48.
40. Ibid.
41. Mike Knepper, *Corvair Affair* (Osceola, WI: Motorbooks International, 1982), 82.
42. Langer, "Auto Safety," 48.
43. Charles McCarry, *Citizen Nader* (New York: Saturday Review Press, 1972), 22.
44. "Why Cars Must—and Can—Be Made Safer," *Time*, April 1, 1966, http://www.time.com/time/magazine/article/0,9171,840604,00.html.
45. "The U.S.'s Toughest Customer," *Time*, December 12, 1969, http://www.time.com/time/magazine/article/0,9171,840502-1,00.html.
46. "GM and Nader Settle His Suit Over Snooping," *Wall Street Journal*, August 14, 1970.
47. National Highway Traffic Safety Administration, report PB 211-015, http://www.corvaircorsa.com/handling01.html.
48. Tom McCarthy, *Auto Mania: Cars, Consumers, and the Environment* (New Haven, CT: Yale University Press, 2008), 192.
49. William S. Wells, "TV Bout of '74: Nader vs. Cole," *Detroit Free Press*, October 30, 1974.
50. American Law Institute, Restatement (Second) of Torts, § 402A.
51. Marshall S. Shapo, *Tort Law and Culture* (Durham, NC: Carolina Academic Press, 2003), 10.
52. Robert Marlow, "The Most Important Car Ever," *Old Cars*, November 7, 1996, 21.
53. Author interviews with Corvair collectors.

CHAPTER SIX

1. Author interview with Harold Sperlich, June 2008.
2. Lee Iacocca, text of remarks at press conference, April 13, 1964, NAHC.
3. Chase Morsey Jr., Ford marketing manager, text of remarks at press conference, April 13, 1964, NAHC.
4. Ibid.

5. *TV Guide*, January 1962.

6. President Lyndon B. Johnson, "Great Society" speech (commencement address, University of Michigan, Ann Arbor, MI, May 22, 1964), http://www.americanrhetoric.com/speeches/lbjthegreatsociety.htm.

7. Lee Iacocca with William Novak, *Iacocca* (New York: Bantam Books, 1984), 80.

8. Robert Boyd, "The Mustang—A Planned Miracle," *Detroit Free Press,* December 6, 1965.

9. Sperlich, interview.

10. "Ford's Young One," *Time*, April 17, 1964, http://www.time.com/time/magazine/article/0,9171,875829,00.html.

11. "The Mustang: Newest Breed Out of Detroit," *Newsweek*, April 20, 1964, 98.

12. Ibid.

13. Author interview with Baron Bates, retired Chrysler PR executive, March 2008.

14. Leonard M. Apcar, "Bookshelf: Motor Mouth Speaks Out," *Wall Street Journal*, November 8, 1984.

15. Iacocca, *Iacocca*, 45.

16. Ford Motor Co., Internal Marketing Memo, 1964, NAHC.

17. "Ford's Young One," *Time*.

18. Ford Motor Co., Internal Memo.

19. Author interview with Donald Frey, November 2007.

20. Bradley A. Stertz, "Sperlich, Intense Chrysler Executive, Retires Unexpectedly," *Wall Street Journal*, January 22, 1988.

21. Lee Iacocca with Catherine Whitney, *Where Have All the Leaders Gone?* (New York: Scribner, 2008), 19.

22. "Newest Breed," 98.

23. "Ford's Young One."

24. Iacocca, *Iacocca*, 71.

25. Sperlich, interview.

26. Frey and Sperlich, interviews.

27. Boyd, "The Mustang."

28. "Ford's Young One."

29. "Newest Breed," 100.

30. Boyd, "The Mustang."

31. Sperlich, interview.

32. "Unmasking the Mustang," *Time*, March 13, 1964, http://www.time
    .com/time/magazine/article/0,9171,828277,00.html.
33. "Newest Breed," 97.
34. "Ford's Young One."
35. "Road Research Report: Ford Mustang," *Car and Driver*, May 1964,
    42, 126.
36. Iacocca, *Iacocca*, 77.
37. "Crowds Pack Showrooms for a Look at the Mustang," *Detroit
    News*, May 18, 1964.
38. Ford Motor Co., press release, NAHC.
39. Author interview with John Hitchcock, August 2007.
40. Ford Motor Co. advertisement, 1965, NAHC.
41. Mustang sales brochure, 1965, NAHC.
42. Ford Motor Co. advertisements, 1964 and 1965, NAHC.
43. Author interview with Jack Ready Jr., June 2007.
44. Author interview with Jack Griffith, October 2007.
45. Ford Motor Co. commercial script, NAHC.
46. L. Scott Bailey, "Mustang Rides the Market," *Automobile Quarterly*
    3, no. 3 (Fall 1964), 323.
47. Ford Motor Co., press releases, NAHC.
48. Iacocca, *Iacocca*, 81.

CHAPTER SEVEN

1. Pontiac advertisement, 1964, NAHC.
2. Jim Wangers, *Glory Days: When Horsepower and Passion Ruled Detroit* (Cambridge, MA: Robert Bentley, 1998), 130.
3. Paul Zazarine, *Pontiac's Greatest Decade: 1959–1969: The Wide Track Era* (Hudson, WI: Iconografix, 2006), 102.
4. Pontiac sales brochure, 1967, NAHC.
5. Author interview with Bill Collins, retired Pontiac engineer, June 2007.
6. J. Patrick Wright, *On a Clear Day You Can See General Motors: John Z. DeLorean's Look Inside the Automotive Giant* (Grosse Pointe, MI: Wright Enterprises, 1979), 92.
7. Solon E. Phinney, Pontiac public relations department, to a Pontiac customer, September 25, 1964, NAHC.
8. "Best Performance Sedan: Pontiac Tempest GTO," *Car and Driver*, May 1964, 34.

9. Author interview with Ken Crocie, August 2007.
10. Pontiac sales brochures, 1964, NAHC.
11. "GTO vs. GTO," *Car and Driver*, March 1964, 26.
12. Wangers, *Glory Days*, 115.
13. "Ferocious GTO," *Motor Trend*, February 1965, 31.
14. Wright, *Clear Day*, 96.
15. Wangers, *Glory Days*, 96.
16. Pontiac sales brochure, 1967.
17. Pontiac advertisement, 1967
18. Author's researcher interview with George Poynter, August 2008.
19. Terry Ehrich and Richard A. Lentinello, eds., *The Hemmings Motor News Book of Pontiacs* (Bennington, VT: Hemmings Motor News, 2001), 108.
20. Wangers, *Glory Days*, 157.
21. Pontiac advertisement, 1968, http://www.adclassix.com/ads2/68pontiac gtowoodward.htm.
22. Wangers, *Glory Days*, 184.
23. Ibid., second photo insert.
24. Mitchell, April 8, 1987, ADOH, 40.
25. Wright, *Clear Day*, 97.
26. Pontiac sales brochure, 1971, NAHC.
27. Collins, interview.
28. Author interview with former GM executive, who wished to remain anonymous, August 2007.
29. Jeff Jarvis, "Downfall of an Auto Prince," *People*, November 8, 1982, 41.
30. Ibid., 44.
31. Author interview with John Skwirblies, August 2007.

CHAPTER EIGHT

1. Jerry Knight, "Honda Took Simple Route to Get to No. 1, but Detroit Can't Read the Map," *Washington Post*, January 9, 1990.
2. Edrie J. Marquez, *Amazing AMC Muscle: Complete Development and Racing History of the Cars from American Motors* (Osceola, WI: Motorbooks International, 1988), 128.
3. Author interview with Brad Alty, Honda manufacturing manager, October 10, 2008.

4. Masaaki Sato, *The Honda Myth: The Genius and His Wake*, trans. Hiroko Yoda with Matt Alt (New York: Vertical, 2006), 5.
5. Ibid., 78.
6. Honda Motor Co. advertisement, 1967, http://oldadvertising .blogspot.com/2009_07_01_archive.html.
7. Patrick Neville, "Preview Test: Honda 1300," *Car and Driver*, June 1970, 63.
8. Tetsuo Chino (speech to the College of Engineering, The Ohio State University, Columbus, OH, May 14, 1987), 1.
9. Author interview with Soichiro Irimajiri, October 23, 2007.
10. Neville, "Preview Test," 63.
11. Sato, *Honda Myth*, 150–80.
12. Ibid.
13. "Honda Civic CVCC," *Road & Track*, February 1975, 108.
14. Brock Yates, "Make Way for the Latest in Cult Cars," *Car and Driver*, March 1978, 24
15. Ibid.
16. Author interview with Shige Yoshida, January 8, 2009.
17. Author interview with Chan Cochran, former press secretary to Governor James Rhodes, October 29, 2008.
18. "Sabotage at Lordstown?" *Time,* February 7, 1972, http://www.time .com/time/magazine/article/0,9171,905747,00.html.
19. Chino, speech, 2.
20. Author interview with Toshi Amino, October 9, 2008.
21. Yoshida, interview.
22. Author interview with Brad Alty, September 29, 2010.
23. "Backers See Obstacles to Compensation of Japanese-Americans," *New York Times*, June 18, 1983.
24. Ibid.
25. Cindy Richards, "How Japanese 'Invasion' Fares," *Chicago Sun-Times*, November 22, 1987.
26. Ibid.
27. Author interview with Susan Insley, former Honda executive, January 8, 2009.
28. Don Hensley, ed., *Building on Dreams: The Story of Honda in Ohio* (s.l.: Honda of America Mfg. Inc., 2004), 21.
29. Ito Shuichi, "Interview with Hiroyuki Yoshino, President of Honda

Motor Co.," *Journal of Japanese Trade and Industry*, September 1, 2002.

30. James Risen, "Honda's Accord Drives off with Best-Selling Status," *Los Angeles Times*, January 5, 1990.
31. Reuters, January 9, 1990.
32. Knight, "Honda Took Simple Route."
33. Risen, "Honda's Accord."
34. Ibid.

CHAPTER NINE

1. Craig Shoemaker, Comedy Central's Jokes.com, http://www.jokes .com/funny/craig+shoemaker/craig-shoemaker—never-pulled-over -in-a-minivan.
2. Phil Patton, "A Visionary's Minivan Arrived Decades Too Soon," *New York Times*, January 6, 2008.
3. Al Rothenberg, "Sperlich Speaks Out," *Ward's AutoWorld*, August 1989.
4. Bradley A. Stertz, "Sperlich, Intense Chrysler Executive, Retires Unexpectedly," *Wall Street Journal*, January 22, 1988.
5. Author interview with Harold Sperlich, June 2007.
6. Iacocca, *Iacocca*, 129.
7. Ibid., 134.
8. Ibid., 149.
9. Dow Jones News Service, "Chrysler, Awaiting More Funds, Halts Payments to Suppliers," June 11, 1980.
10. Paul Ingrassia and Joseph B. White, *Comeback: The Fall and Rise of the American Automobile Industry* (New York: Simon & Schuster, 1994), 61.
11. Jim Dunne, "Chrysler's K-car," *Popular Science*, April 1980, 88.
12. Sperlich, interview, June 2008.
13. Ingrassia and White, *Comeback*, 62.
14. Brock Yates, "A Van for All Seasons," *Car and Driver*, May 1983, 39.
15. "Minivan a Hit For Chrysler," *New York Times*, February 7, 1984.
16. Author interview with Baron Bates, 1993.
17. Ingrassia and White, *Comeback*, 80.
18. Alex Taylor III, *Sixty to Zero: An Inside Look at the Collapse of*

General Motors—and the Detroit Auto Industry (New Haven, CT: Yale University Press, 2010), 193.

19. Author interview with R. S. Miller Jr., former Chrysler vice chairman, June 2008.

20. Author interview with Harold Sperlich, May 2010.

21. Cathy Karlin Zahner, "Mama Chauffer," *Kansas City Star*, December 18, 1981.

22. Author interview with Lindy Robinson, April 2010.

23. Ibid.

24. Author interview with Arthur C. Liebler, former vice president of marketing for Chrysler, June 2010.

25. Carey Goldberg, "Suburbs' Soccer Moms, Fleeing the G.O.P., Are Much Sought," *New York Times*, October 6, 1996.

26. Steve Rubenstein, "Political Debate Doesn't Interest Busy Soccer Moms," *San Francisco Chronicle*, October 17, 1996.

27. Author interview with Ben Pearson, the girl's father, June 2010.

28. Author interview with Laurel Smith, founder of MomsMinivan.com, April 2010.

29. Denise Roy, *My Monastery Is a Minivan: Where the Daily Is Divine and the Routine Becomes Prayer; 35 Stories from a Real Life* (Chicago: Loyola Press, 2001), 10.

CHAPTER TEN

1. Jonathan Gold, "What Happens After the Hype?" *Los Angeles Times*, October 1, 1989.

2. Author interview with William Collins, March 2010.

3. Ron Brownstein and Nina J. Easton, "The New Status Seekers in the 1980s," *Los Angeles Times Magazine*, December 27, 1987, http://articles.latimes.com/1987-12-27/magazine/tm-31245_1_status-symbol.

4. Lois Therrien, "Pet Food Moves Upscale," *BusinessWeek*, June 15, 1987, 80.

5. Herb Caen, "Friday Flimflam," *San Francisco Chronicle*, January 11, 1985, 41.

6. David Brooks, *Bobos in Paradise: The New Upper Class and How They Got There* (New York: Simon & Schuster, 2000), 91–92.

7. BMW Group, http://www.bmwgroup.com/e/nav/index.html?http://

www.bmwgroup.com/e/0_0_www_bmwgroup_com/unternehmen/historie/meilensteine/meilensteine.html.

8. Stephen Williams, "BMW Roundel: Not Born From Planes," *New York Times*, January 7, 2010, http://wheels.blogs.nytimes.com/2010/01/07/bmw-roundel-not-born-from-planes/.

9. Goebbels family home movies, http://video.google.com/videoplay?docid=-8973962176504385280.

10. Magda Goebbels to her son, Harald Quandt, Axis History Forum, http://forum.axishistory.com/viewtopic.php?f=45&t=54731.

11. David Kiley, *Driven: Inside BMW, the Most Admired Car Company in the World* (Hoboken, NJ: John Wiley, 2004), 99.

12. Ibid., 103.

13. *Automotive News*, 1967 Almanac Issue, 76.

14. David E. Davis Jr., "Turn Your Hymnals to 2002," *Car and Driver*, April 1968, 66.

15. Richard A. Johnson, *Six Men Who Built the Modern Auto Industry* (St. Paul, MN: Motorbooks, 2005), 67.

16. Paul Ingrassia, "Three for the Road," *SmartMoney*, November 1998.

17. Neal Boudette, "Navigating Curves: BMW's Push to Broaden Line Hits Some Bumps in the Road," *Wall Street Journal*, January 10, 2005.

18. BMW sales brochure, 1982, NAHC.

19. Brooks, *Bobos in Paradise*, 89–90.

20. "Suggested Retail Prices," BMW sales brochures, October 1982, NAHC.

21. *Wards Automotive Yearbook 1986*, 158.

22. Emmett Watson, "Yuppies? What's New About Status Seekers," *The Seattle Times*, Feb. 3, 1985.

23. Johnson, *Six Men*, 62.

24. Author interview with Larry Schultz, April 2010.

25. Ibid.

26. Author interview with Joseph Katz, March 2010.

27. Ibid.

28. Reuters, September 30, 1992.

29. Donna Lee, "3 Greens Find a Home in Status-Conscious Salads," *Chicago Tribune*, November 6, 1986.

30. Robin Hill, "Short Black," *Sydney Morning Herald*, February 6, 1990, 2.

31. Heath Urie, *Daily Camera* (Boulder, CO), March 25, 2009.

32. "Wilmington Horticulturalist Transplants Farming, Teaching Skills Overseas," *Star-News* (Wilmington, DE), December 1, 2009.

CHAPTER ELEVEN

1. Miller, interview.

2. "Shop Talk," *Wall Street Journal*, August 30, 1984.

3. Dale D. Buss, "AMC May Be Denying Itself Oscar for 'Best Supporting Auto Maker,'" *Wall Street Journal*, May 8, 1985.

4. Melinda Grenier Guiles, "AMC Is Granted an Order By Court Against UAW Unit," *Wall Street Journal*, April 25, 1985.

5. Sperlich, interview, June 2007.

6. Herbert R. Rifkind, *The Jeep—Its Development and Procurement under the Quartermaster Corps, 1940–1942,* Historical Section, General Service Branch, General Administrative Services, Office of the Quartermaster General, 1943, 2.

7. Patrick R. Foster, *The Story of Jeep* (Iola, WI: Krause Publications, 2004), 45.

8. Ibid., 41.

9. Ibid.

10. ExplorePAhistory.com, "Science and History," http://explorepahistory.com/hmarker.php?markerId=1-A-2F1.

11. Foster, *Story of Jeep*, 51.

12. Ronald H. Bailey, "The Incredible Jeep," http://www.historynet.com/the-incredible-jeep.htm.

13. Steve Statham, *Jeep Color History* (Osceola, WI: MBI Pub., 1999), 26.

14. Ibid.

15. "Willys-Overland: This Jeep-Riding Independent Is Taking New Leases on Life and Its Own Real Estate," *Fortune*, August 1946, 185.

16. Bailey, "Incredible Jeep."

17. Queen's Film and Media, "The Original Nellybelle," http://www.film.queensu.ca/cj3b/siblings/Nellybelle.html.

18. Frank Zappa with Peter Occhiogrosso, *The Real Frank Zappa Book* (New York: Poseidon Press, 1989), 23.

19. American Motors Corp., *Jeep Corporation: Its Heritage, Its Current*

Products, Its Future, undated American Motors press release, 1971, AMC file, NAHC.

20. U.S. Scouting Service Project, "Boy Scout Oath or Promise," http://usscouts.org/advance/boyscout/bsoathlaw.asp.

21. American Motors Corp., *Jeep Corporation*.

22. T. Rex, "Jeepster," 1971, http://www.metrolyrics.com/jeepster-lyrics-t-rex.html.

23. American Motors Corp., letter to shareholders, 1970 Annual Report.

24. Robert W. Irvin, "Outlook Still Cloudy for Hard-working AMC," *Detroit News*, December 10, 1970.

25. Statham, *Jeep Color History*, 106–7.

26. Laurence G. O'Donnell, quoting industry analyst Arvid Jouppi, "The Little Fourth: Tiny American Motors Struggles to Survive as a Separate Concern," *Wall Street Journal*, July 12, 1971.

27. Mitch McCullough, review of 1997 Jeep Wrangler, http://www.newcartestdrive.com/review-intro.cfm?Vehicle=1997_Jeep_Wrangler&ReviewID=4242.

28. Kevin Klose, "A U.S. Era Ends," *Washington Post*, January 29, 1986.

29. Diane Jennings, "Wagons Ho! Trade in the BMW and Mercedes," *Dallas Morning News*, January 26, 1986.

30. Ibid.

31. Pearson, interview.

32. Ibid.

33. Author interview with John Peterman, July 6, 2010.

34. Pearson, interview.

35. Yvon Chouinard, *Let My People Go Surfing: The Education of a Reluctant Businessman* (New York: Penguin Press, 2005), 48.

36. Ibid., 54.

37. American Motors Corp., *1985 Almanac: U.S. Market Data*.

38. Damon Darlin and Thomas Kamm, "Stalling Out: AMC Is French Now, But Renault's Money Hasn't Put It Right," *Wall Street Journal*, July 30, 1986.

39. *Merriam-Webster Online Dictionary*, s.v. "SUV," http://www.merriam-webster.com/dictionary/suv?show=0&t=1310891469.

40. Statham, *Jeep Color History*, 122.

41. Bradley A. Stertz, "Off-Road Vehicles Do Delicate Duty with the Quiche Set," *Wall Street Journal*, March 5, 1990.

42. Chrysler Corp., Mexico videotape, obtained by the author.

43. Jerry Flint, "Occupied Chrysler," *Ward's AutoWorld*, November 1999, 23.

CHAPTER TWELVE

1. P. J. O'Rourke, *Republican Party Reptile: The Confessions, Adventures, Essays, and (Other) Outrages of . . .* (New York: Atlantic Monthly Press, 1987), 114–15.
2. Don Bunn, "Chevrolet Trucks History: Segment Two: 1929–1936 Early Six Cylinder Pickups," http://www.pickuptrucks.com/html/history/chev_segment2.html.
3. John Steinbeck, *Travels with Charley* (New York: Viking Press, 1962), 6.
4. Ibid., 70.
5. Ibid., 205.
6. *Ward's Automotive Yearbook 1979*, 42.
7. Ford Motor Co., office of sales statistics, 2011.
8. Ibid.
9. Sidney C. Schaer, "Billy Carter Dies: Ex-president's Brother Succumbs to Cancer," *Newsday*, September 26, 1988, accessed on Factiva.com.
10. Songs and artists: Billy Joe Shaver, "Ragged Old Truck," 1981; Jerry Jeff Walker, "Pickup Truck Song," 1989; Toby Keith, "Big Ol' Truck," 1994; Joe Diffie, "Pickup Man," 1994; J. C. Hyke, "That Old Truck," 2002; John Williamson, "This Old Truck," 2002.
11. Keith Bradsher, *High and Mighty: The Dangerous Rise of the SUV* (New York: Public Affairs, 2003), 95.
12. Jack Keebler, "The Ford SVT Lightning," *Motor Trend*, June 1999, 70.
13. Ford Motor Co., *Big, Bad and Bold: The 2005 Harley-Davidson Super Duty*, press release, 2004.
14. Christian Bokich, spokesman for Ford, e-mail message to author, August 9, 2010.
15. Federal Documents Clearing House e-Media, Transcript of Bush-Putin press conference in Slovenia, June 16, 2001, accessed on Factiva.com.
16. Author interview with Douglas Scott, Ford director of truck marketing, July 19, 2010.
17. Boe, e-mail.

18. Author interview with Professor Bryan Jones, July 14, 2010.
19. Author interview with John Williford, August 10, 2010.
20. Ibid.
21. Ibid.
22. Paul Ingrassia, "The Pickup Bar Has Just Been Raised," *Smart-Money*, November 1, 2003, 130.
23. Paul Ingrassia, *Crash Course: The American Automobile Industry's Road from Glory to Disaster* (New York: Random House, 2010), 206.
24. Michael Cooper, "Senate GOP Victory Stuns Democrats," *New York Times*, January 19, 2010.
25. Alexander Burns, "Scott Brown Pulls Off Historic Upset," January 20, 2010, http://www.politico.com/news/stories/0110/31674.html.
26. Kimberley A. Strassel, "The Obama Heyday Is Over," *Wall Street Journal*, September 10, 2010, http://online.wsj.com/article/SB10001424052748704644404575482122517174884.html.
27. Author interview with a Washington automotive lobbyist, who wished to remain anonymous, March 2009.

CHAPTER THIRTEEN

1. Alex Taylor III, "Toyota: The Birth of the Prius," *Fortune*, February 21, 2006.
2. "Pious Prius," posting to "I, Anonymous" column, *Portland Mercury* (Portland, OR), September 4, 2008.
3. Frank Swertlow, "Al Gore's Son Arrested for Speeding, Drugs," *People*, July 4, 2007, http://www.people.com/people/article/0,,20044628,00.html.
4. Horst O. Hardenberg, *The Oldest Precursor of the Automobile: Ferdinand Verbiest's Steam Turbine-Powered Vehicle Model*, (Warrendale, PA: Society of Automotive Engineers, 1995).
5. Dr. Ing. h.c. F. Porsche AG, ed., *Ferdinand Porsche: Hybrid Automobile Pioneer* (Cologne: DuMont Buchverlag, 2010), 16–19.
6. "Present at the Creation" (interview with retired TRW scientist Dr. George Gelb), *Automotive Design and Production*, November 2006, http://www.autofieldguide.com/articles/present-at-the-creation.
7. Ibid.

8. Stuart Lavietes, "Victor Wouk, 86, Dies; Built Early Hybrid Car," *New York Times*, June 12, 2005.

9. Michael Shnayerson, *The Car That Could: The Inside Story of GM's Revolutionary Electric Vehicle* (New York: Random House, 1996), 182.

10. Author interview with Takeshi Uchiyamada, January 2011.

11. Ibid.

12. Hideshi Itazaki, *The Prius That Shook the World*, trans. Albert Yamada and Masako Ishikawa (Tokyo: Nikkan Kogyo Shimbun Ltd., 1999), 48–49.

13. Ibid., 70.

14. Uchiyamada, interview.

15. Ibid.

16. Itazaki, *Prius*, 152, 154.

17. Uchiyamada, interview.

18. Ibid.

19. Itazaki, *Prius*, 270.

20. James B. Treece, "Six from Toyota Die in Crash," *Automotive News*, February 10, 1997.

21. Itazaki, *Prius*, 164.

22. Uchiyamada, interview.

23. Yuri Kageyama, Associated Press, "Toyota Introduces the World's First Gas-Electric Hybrid Car," October 14, 1997.

24. Author interview with Joseph Tetherow, Toyota public relations executive, November 2010.

25. Author interview with Toyota manager, who wished to remain anonymous.

26. Author interview with Csaba Csere, former editor of *Car and Driver*, November 2010.

27. Carole McCluskey, "Squeaky Clean, But Not Much Fun," *Seattle Times*, September 3, 1999.

28. Author interview with Mark Amstock, Toyota marketing executive, November 2010.

29. Ibid.

30. Ibid.

31. Lavietes, "Victor Wouk," *New York Times*, June 12, 2005.

32. "Half Gas, Half Electric, Total California Cool: Hollywood Gets a Charge out of Hybrid Cars," *Washington Post*, June 6, 2002.

33. Ibid.
34. Author interview with Ed LaRocque, Toyota marketing executive, November 2010.
35. "Half Gas, Half Electric."
36. Ned Martel, "Playing a Wife Who's the Other Woman," *New York Times*, February 1, 2004.
37. *Curb Your Enthusiasm*, "Wandering Bear" episode, February 29, 2004.
38. Author interview with Steve Ingrassia, cousin of the author, November 2010.
39. Sharon Waxman, "A Prius-Hummer War Divides Oscarville," *New York Times*, March 7, 2004.
40. Associated Press, October 26, 2004.
41. "Hybrid Hypocrisy; Gas Guzzlers Get Breaks Too," *Sacramento Bee*, August 3, 2005.
42. *South Park*, "Smug Alert!" episode, March 29, 2006, http://www.southparkstudios.com/full-episodes/s10e02-smug-alert.
43. Ed LaRocque, interview.
44. Gary Richards, "Can Prius Top 100 MPH? Ask Wozniak," *San Jose Mercury News*, August 21, 2007.
45. Ibid.
46. Author interview with Scott Jackson, November 2010.
47. Uchiyamada, interview.

AFTERWORD

1. Gayle Pollard-Terry, "Mustang Is His Driving Passion: The '05 Version of the Car Was Designed by a Vietnamese Immigrant," *Los Angeles Times*, January 3, 2005.
2. Paul Ingrassia, "Pony Express Rides Again," *SmartMoney*, April 1, 2005.

SELECTED BIBLIOGRAPHY

Burton, Jerry. *Zora Arkus-Duntov: The Legend Behind the Corvette*. Cambridge, Mass: Bentley, 2002.

Casey, Robert. *The Model T: A Centennial History*. Baltimore: Johns Hopkins University Press, 2008.

Challis, Clive. *Helmut Krone, The Book: Graphic Design and Art Direction (Concept, Form and Meaning) After Advertising's Creative Revolution*. Cambridge: Cambridge Enchorial, 2005.

Chouinard, Yvon. *Let My People Go Surfing: The Education of a Reluctant Businessman*. New York: Penguin Press, 2005.

Ford, Henry, with Samuel Crowther. *My Life and Work*. Garden City, NY: Garden Publishing, 1926.

Foster, Patrick R. *The Story of Jeep*. Iola, Wis.: Krause Publications, 2004.

Genat, Robert. *Woodward Avenue: Cruising the Legendary Strip*. North Branch, Minn.: CarTech, Inc. 2010.

Iacocca, Lee, with William Novak. *Iacocca*. New York: Bantam Dell, 1984.

Itazaki, Hideshi. *The Prius that Shook the World*. Translated by Albert Yamada and Masako Ishikawa. Tokyo: Nikkan Kogyo Shimbun, 1999.

Johnson, Richard A. *Six Men Who Built the Modern Auto Industry*. Minneapolis, Minn: Motorbooks, 2005.

Keats, John. *The Insolent Chariots*. Philadelphia: J. B. Lippencott & Co., 1958.

Kerouac, Jack. *On the Road*. First published 1957; New York: Penguin, 1976.

Kiley, David. *Inside BMW, the Most Admired Car Company in the World*. Hoboken, NJ: John Wiley & Sons, 2004.

Kreipke, Robert C. *The Model T: A Pictorial Chronology of the Most Famous Car in the World*. Evansville, Ind.: M.T. Publishing, 2007.

Lamm, Michael, and David Holls. *A Century of Automotive Style: 100 Years of American Car Design*. Stockton, Calif.: Lamm-Morada Publishing, 1996.

Larson, Matt, and Ron Van Gelderen. *LaSalle: Cadillac's Companion Car*. Paducah, Ky.: Turner Publishing, 2001.

Levenson, Bob. *Bill Bernbach's Book: A History of the Advertising that Changed the History of Advertising*. New York: Villard, 1987.

Lewis, David L., and Laurence Goldstein, eds. *The Automobile and American Culture*. Ann Arbor: University of Michigan Press, 1980.

Lewis, David L. *The Public Image of Henry Ford: An American Folk Hero and His Company*. Detroit: Wayne State University Press, 1976.

McCarthy, Tom. *Auto Mania: Cars, Consumers and the Environment*. New Haven, Conn.: Yale University Press, 2008.

McLaughlin, Paul G. *Ford Pickup Trucks: Buyer's Guide*. Osceola, Wis.: MBI Publishing, 1991.

Muir, John. *How to Keep Your Volkswagen Alive: A Manual of Step by Step Procedures for the Compleat Idiot*. Santa Fe: John Muir Publications, 1969.

Nader, Ralph. *Unsafe at Any Speed*. New York: Pocket Books, 1966.

Nelson, Walter Henry. *Small Wonder: The Amazing Story of the Volkswagen Beetle*. Cambridge, Mass: Bentley Publishers, 1998.

O'Rourke, P.J. *Republican Party Reptile: The Confessions, Adventures, Essays, and (Other) Outrages of P.J. O'Rourke*. New York: Atlantic Monthly Press, 1995.

Packard, Vance. *The Hidden Persuaders*. Updated edition; New York: Pocket Books, 1984.

Railton, Arthur. *The Beetle: A Most Unlikely Story*. Stuttgart: Verlagsgesellschaft, 1985.

Rowsome, Frank. *Think Small: The Story of Those Volkswagen Ads*. Lexington, Mass: S. Greene Press, 1970.

Sato, Masaaki. *The Honda Myth: The Genius and His Wake*. New York: Vertical, 2006.

Shapo, Marshall S. *Tort Law and Culture*. Durham, NC: Carolina Academic Press, 2003.

Steinbeck, John. *Cannery Row*. First published 1945; New York: Penguin, 1993.

——. *Travels with Charley: In Search of America*. New York: Penguin Modern Classics, 1980.

Van Doren Stern, Philip. *Tin Lizzie: The Story of the Fabulous Model T Ford*. New York: Simon and Schuster, 1955.

Wangers, Jim, and Art Fitzpatrick. *Pontiac Pizazz!* Oceanside, Calif.: Jim Wangers Productions, 2007.

Wolfe, Tom. *The Electric Kool-Aid Acid Test*. First published 1968; New York: Bantam, 1999.

Wright, J. Patrick. *On a Clear Day You Can See General Motors: John Z. DeLorean's Look Inside the Automotive Giant*. Brooklyn, NY: Wright Enterprises, 1979.

Yates, Brock. *The Critical Path: Inventing an Automobile and Reinventing a Corporation*. Boston: Little, Brown & Co., 1996.

INDEX

Abbey Road, 102
Academy Awards, 314–15, 331–32, 333
acceleration, 10, 44, 56, 108, 152, 201, 329
accelerator pedals, 152
accidents, xiv, 107, 112, 126–39, 182, 198–200, 266, 278, 282, 318, 345
Adams, Edie, 126
Adventures of Ozzie and Harriet, 65
Advertising Artists Inc., 62
aerodynamic design, 60–61, 72, 76, 85
air-conditioning, 158
aircraft, 25, 60–61, 72, 76, 152
air pollution, 53, 84, 107, 136, 167, 183–84, 191, 199, 200, 252, 313–39
Alfa Romeo, 246
Ali, Muhammad, 144
Allard car company, 40
All in the Family, 17
All Things Must Pass, 183
Altamont music festival, 182–83
Alty, Brad, 191–93, 207–9, 211, 216
Alvord & Alvord, 130
AMC, see American Motors Corp.
AMCAR, 58
American Austin Car Co., 268–69
American Bantam Car Co., 268–72, 273

American Car Club of Norway, 58
American Graffiti, 178
American Law Institute, 137
American Motors Corp. (AMC), 61, 96–97, 112, 177, 192, 265–66, 273, 275–79, 281–83, 285, 317
   AMX, 177
   Gremlin, xviii, 191, 192, 216, 276, 328, 342–43
   Hornet, 276
   Rambler, 96–97, 112, 265
American Trial Lawyers Association, 129
Amerikaner Biltraef Koebenhavn (American Car Club of Copenhagen), 58
AM General, 285–86
   Hummer H1, 285–86, 288
   Hummer H2, 76, 286, 333, 336–37
ANA Hotel, 327
Andrea Doria, 69
Andy Griffith Show, The, 291
Anheuser-Busch, 16
"ankle fatigue," 152
Anna (Ohio) Honda plant, 209–10
antifreeze, 113
anti-Japanese sentiment, 208–9, 214–15

anti-roll bars, 123, 126, 129
anti-Semitism, 14, 49, 86, 131, 132, 246
Arden, Elizabeth, 155
Ardun cylinders, 39
Ardun Mechanical Corp., 38
Arkus, Jacques, 36
Arkus-Duntov, Rachel, 36–37
Arkus-Duntov, Yura, 36, 38
Arkus-Duntov, Zora, 33–56, 120
Arkus sports car, 37
Armed Forces Radio, 344
Army, U.S., 265–71
arugula, 261–62
assembly lines, 2–3, 10, 11, 12–13, 43,
    153, 158–59, 187, 193, 210, 211,
    212–13, 214, 232
"associates," 205, 210, 212–13
Atlantic, 68
Austin, 268–69
Seven, 246
Austin-Healey, 32
Austro-Daimler, 84
autobahns, 250
automatic transmissions, 44, 68–69, 75,
    156, 158, 202, 231, 274, 294, 317
automation, 210
Automobile, 8
Automobile Quarterly, 28, 158
automobiles, automotive industry:
  advertising for, xiv, 9–11, 22, 27, 31,
      52, 73–74, 83, 96, 97–107, 116, 119–
      22, 123, 129, 132, 141, 146, 153–55,
      157, 164–65, 172–73, 174, 175–76,
      179, 180, 196, 232, 236, 255, 270–
      71, 275–76, 277, 292, 301–2, 332
  British, 38, 43, 92, 148–49, 188,
      268–69
  colors of, 11–12, 42, 58, 155, 156, 180,
      201, 271, 275
  compact, 81–109, 112–13, 120–22,
      125, 126, 143, 171–72, 177, 198–99,
      200, 220, 224, 255, 275–79, 318, 322
  competition in, 16–17, 40, 41, 43–44,
      45, 46–49, 64, 67–68, 120–22, 175,
      176–80, 191–202, 214–16, 227,
      283–84, 290, 291–95, 311, 320, 328,
      338–39
  computer technology compared with,
      xiv—xv, 8, 10, 84, 109, 201, 316,
      336–37
corporate culture in, 197–98, 287–88
cost estimates in, 24, 27, 129–30, 151,
    152–53, 157–58, 225, 226–27
cultural impact of, xi—xx, 2–4, 31–33,
    41–42, 50–51, 57–59, 64, 65–67,
    72–74, 82–83, 94–106, 133–34,
    141–45, 147, 148, 150, 154–57,
    160–61, 166–67, 173–84, 189, 192,
    196, 208–9, 210, 220–21, 234–39,
    258–72, 289–90, 313–14, 341–45
dealerships for, 24–25, 93–94, 106,
    154, 171, 175–76, 181, 186, 226,
    232–33, 277, 278
designer, xii—xiii, 3–5, 19–20, 25
design of, xiii, 1–4, 19–26, 32, 43–44,
    49–53, 56, 60–79, 84–88, 106, 108–
    9, 116–18, 119, 126, 148–52, 153,
    154, 221–25, 231, 250–51, 302–3,
    328–29, 341–42
economic conditions and, xiii, 41,
    43–44, 53, 56, 66, 73–76, 90–92, 98,
    102–3, 112, 116, 157–59, 181, 184–
    85, 191–93, 195, 205–9, 220–21,
    292–93, 310–11
electric, 7, 84, 315–19, 323, 335, 339
engineering for, 37, 48, 69, 71–72,
    84–88, 97, 99, 114–15, 116, 118–25,
    126, 129, 143, 149–52, 166–71, 176,
    184, 185–200, 209, 213, 222–23,
    227–30, 250–51, 253, 257, 267, 277,
    283, 308–9, 319–28, 343–45
engines for, *see* engines
European, xiii—xiv, 43–44, 50,
    81–109, 148–49, 152, 157, 224–25,
    227–28, 295–96
exports and imports of, xiii—xiv,
    92–97, 202–3, 212–13, 223, 252,
    253–54, 276–77, 295–98
financing of, 6–7, 8, 12, 14, 16, 35,
    64, 67, 68, 141–42, 188, 197, 223,
    226–27, 228, 233, 247–48, 265–67,
    285, 287–88, 295–98, 308, 310
German, 81–109, 193, 241–63
government oversight of, 48, 112,
    116, 127, 128, 132–39, 147–48,
    226–27, 230, 295–98, 319, 345
hybrid, 84, 313–39
Japanese, 11–12, 107–8, 135, 191–217,
    223, 233, 260, 266, 284, 290, 292,
    297–98, 308–9, 319–38

INDEX

luxury, 19, 46, 74–76, 94, 241–63

management in, 15, 97, 185–86, 204–7, 210–11, 247–48, 319–21

manufacture of, 1–17, 23–24, 27, 28, 43, 64, 69, 87–92, 153, 158–59, 187, 193, 210, 211, 212–13, 214, 232, 322, 326, 327–31

marketing in, 13–14, 16, 21–25, 42, 45–46, 62, 73–75, 95–96, 104, 108, 112, 119–24, 126, 141–42, 143, 148, 153–62, 168, 170–73, 185–86, 189, 197, 232–34, 236, 257, 276–79, 283–84, 301–2, 305, 328–35

media coverage of, xvi–xviii, xx, 12, 17, 18, 21–22, 31–32, 43, 50–51, 87–88, 103–5, 113–14, 119–21, 131–32, 133, 143–47, 153, 154, 158, 172–73, 199, 202, 230–31, 251–52, 272, 314, 327, 328–29, 331–35

mergers of, 272–73, 275, 287–88

military, 28, 269–72, 285

models of, 11, 15–19, 63–64, 69–72, 116–17, 125, 145, 157, 159, 175, 185, 210, 213, 228–29, 246–47, 250–51

names of, 74, 81, 152

optional equipment for, 47, 51–52, 157–58, 165, 170–82, 184, 257

political aspect of, 14, 217, 234–39, 289–90, 295–300, 303–4

prices of, 2, 3, 8, 9, 11, 16, 17, 20–21, 27–28, 44, 50, 69, 75–76, 85, 87, 94, 96, 100, 122, 143, 149, 157–58, 159, 164, 188, 223, 257, 261, 285, 292, 318, 329, 337–38

racing, 6, 7, 10–11, 22, 37, 39, 40–41, 46, 47–48, 84, 147–48, 154, 159, 160, 161, 167–82, 198, 202, 213, 246

repair of, 9–10, 96, 106

research and development (R&D) in, 66, 70–71, 197–200, 213, 319–28

safety of, xiv, 107, 111–39, 169, 176, 192, 198–200, 215, 266, 278, 282, 318, 345

sales of, 10–11, 13, 16, 27–29, 41, 46, 47, 48–49, 54, 63, 64, 67, 68, 75–76, 77, 87, 95–96, 108, 116, 117, 124–25, 141–42, 143, 145, 149, 151, 152–53, 157–58, 161, 171, 176, 180, 181, 184–85, 186, 197, 201, 202–3, 223, 226, 228, 229, 244, 247–48, 250–54, 256, 257, 285, 292–98, 300, 302, 308, 310–11, 327–28, 332

sex appeal of, 64, 66, 67, 73–74, 116, 143–44, 150, 162, 233

size of, 78–79, 83, 112–13, 116–18, 161

sports, xiii, 31–56, 125, 143, 148–49, 157–58, 159, 165, 168–69, 186–89, 197, 342

styles of, 16, 21–28, 64–79, 112, 120–22, 126, 148–52, 233, 234, 328–29, 336

utilitarian, xii, 1–17, 81–109, 148, 149, 232, 234–39, 269–72, 289–311

vintage, 50, 55–59, 159

weight of, 10, 86, 123, 159, 161, 201, 252, 255, 268, 269, 286

youth market for, 41–42, 141–42, 147, 148, 150, 174

Automotive Hall of Fame, 216

axles, 9–10, 124, 269

Babcock Electric Carriage Co., 7

baby boomers, 219–20, 221, 234, 236–37, 254, 343

back seats, 48–49, 150, 152, 342

Back to the Future, 189

Baez, Joan, 147

bank loans, 226–27, 230

"Bata-Batas," 194

batteries, 316, 318, 325, 328

"bat-wing" fins, 77

Bavaria, 245–46, 253–54

Bayerische Flugzeugwerke (BFW), 245–46

Beach Boys, xi, xiv–xv, 173, 196, 299

Bean, Leon Leonwood, 279

Bean Boots, 279, 280–81

Beatles, 102, 144, 183

Beatty, Warren, 166

"Beetle babies," 101

"Beetle stuffing," 95

Begley, Ed, Jr., 335

"belt line," 24, 121

Berlin Automobile Show, 85, 87

Berlin Wall, 215

Bernbach, William, 97

Berry, Jan, 51

"Best Television Commercial of the Century," 100

Big Chill, The, xii, 258
"Big Ol' Truck," 301
"Big Three," 61, 112, 147–48, 161, 191,
    193, 209, 210, 214–15, 229, 272,
    273, 275–76, 297, 298, 319
see also Chrysler Corp.; Ford Motor
    Co.; General Motors Corp.
Bimmers, 244, 256–57, 258, 261
Bismarck Tribune, 17
"Black Day in July," 166
black urban professionals ("buppies"),
    241
Bloomington Gold Corvette exhibition,
    xviii, 55
Bloomington Gold Survivor, 55
"blue-collar blues," 205
BMW (Bayerische Motoren Werke), 78,
    109, 216, 241–63
    3/15 model, 246
    3 Series, 244, 254–55, 256, 260
    5 Series, 255, 256
    7 Series, 255, 256
    318i, 256, 259
    320i, 256, 257
    325i, 260
    325xiT, 256
    328 Berlinetta, 246
    330i, 261
    501 model, 247
    1500 model, 250–51
    1600–2 model, 251, 252
    2002 model, 251–52, 254, 255, 257
    Isetta, 247–48
    M3, 302
    R32, 246
BMW of North America, 254–55, 278,
    315, 331
Bolivia, 83
Bonds for Babies Born in Beetles, 101
Bonnie and Clyde, 166
Bosch fuel injection, 256
bourgeois bohemians ("Bobos"), 243,
    244, 255, 256–57, 260
Bow, Clara, 22
Boy Scouts, 275
brake lights, 76
brakes, 10, 76, 87, 91, 126, 158, 165, 174,
    177, 250, 314, 326, 328
Breech, Ernest, 92
brochures, sales, 124, 125, 176, 256

Brooks, David, 243
Brown, Scott, 309–10
Brush "Everyman's Car" Runabout, 9
"brute utes," 284
"bubbas," 306–7
"bubble car," 247
bucket seats, 125, 148, 157, 180
"Bug," 95
Buick, 21, 24, 74, 112, 114, 118, 168, 245
    Estate, 220
    Skylark, 317
    Y-Job, 25
Bullitt, 160
bumpers, 25, 71, 180, 274
Bush, George W., 111, 139, 303–4, 319,
    345
BusinessWeek, 185–86, 243
Butler, Pa., 267–68
"By the Time I Get to Phoenix," 299

cabins ("cabs"), 64, 231, 293, 296, 300,
    304
"Cadibacks," 115, 117
Cadillac, xiii, xix–xx, 6–7, 15, 18–19,
    20, 22, 24, 26, 28, 50, 57–79, 83,
    104, 115–16, 118, 156, 168, 186,
    242, 245, 262, 271, 320, 337, 342
    Eldorado Biarritz, 57
    Eldorado Brougham, 76
    Series 75, 76
    V12 convertible, 115
Cadillac Club of Denmark, xx, 78
"Cadillac Ranch," 77–78
California Dreamers, 58
Campbell, Bill, 109, 298–99
Campbell, Glen, 291
camper units, 125
Camp Holabird, 269–70
camshafts, 47, 201–2, 250
Canada, 215, 216, 225, 228, 254, 277–78,
    282
Cannery Row (Steinbeck), xix, 1
"captain's chairs," 235
Car and Driver, 49, 81, 103–4, 126, 154,
    170, 172–73, 196, 199, 202, 230–31,
    233, 251–52
Car of the Year award, 122, 180, 228,
    338–39
carpool lanes, 334
Carson, Johnny, 187, 188

Carter, Billy, 300
Carter, Jimmy, 192, 299–300
Cash, Johnny, 291, 299
catalytic converters, 136, 184, 200, 345
Center for Auto Safety, 135
Center for the Study of Responsive
    Law, 135
Center for the Study of Southern
    Culture, 299
Cerberus, 287–88
Challis, Clive, 98
Chamberlain, Wilt, 100
Charlotte Speedway, 160
charts, 124–25
chassis, 9, 10, 11, 15–16, 64, 149–50,
    161, 169–70, 222, 224–25, 228–29,
    233, 234
Chevrolet, 16, 21, 24, 39–46, 50, 59,
    111–12, 116–18, 168, 175, 176, 181,
    192, 219, 291, 292–98, 308, 311
  Astro, 233
  Avalanche, 303
  Bel Air, xv, 116, 342
  Biscayne, 122
  Camaro, 161, 177, 342
  Canyon, 309–10
  Chevelle SS (Super Sport) 396, 177,
    179
  Corvair, xiv, 97, 109, 111–39, 176,
    187–88, 198–99, 215, 278, 344–45
  Corvan, 125
  Corvette, xiii, xviii, 29, 31–56, 164,
    169, 216, 313–14, 342
  Corvette Grand Sport, 48
  Corvette Sting Ray, 49, 51, 58
  Corvette ZR1, 56
  El Camino, 294–95
  Greenbrier Sports Wagon, 125, 222
  Impala, 342
  Lakewood, 125, 126
  Lumina, 233–34
  LUV ("light utility vehicle"), 297, 298
  Monza, 125
  Nova, xv, 201–2
  Rampside, 125
  S-10, 298
  Silverado, 302, 306
  Suburban, 284
  Vega, 205–6
  Volt, 338–39
Chicago Auto Show, 125
"chicken tariff," 295–98
childproof locks, 232
children, 219–20, 232, 234–39, 241–42
Chin, Vincent, 208
Chino, Tetsuo, 197, 206, 213–14
Chouinard, Yvon, 281
chrome, 64, 70, 71, 72, 76, 77, 112, 180,
    274
Chrysler, Walter, 24
Chrysler Airflow, 63, 341
Chrysler Corp., xix, xi, 39, 48, 59, 63–
    64, 74, 77, 96–97, 120, 136, 191–92,
    193, 208, 209, 210, 214–15, 220–21,
    222, 225, 226–39, 266–67, 272,
    275–76, 281–83, 287–88, 290, 297,
    298, 310, 317, 343
  Norseman, 69
Chrysler Corporation Loan Guarantee
    Act (1979), 227
Chrysler Jeep, 281–83, 287–88
  Cherokee Limited, 282–83
  Grand Cherokee, xi, 283–84, 285
  Grand Cherokee Orvis Edition, xi,
    284
Churchill, Beatrice, 43
classified ads, 100
clean-air laws, 107, 167, 183–84
Clear Air Act (1970), 53
Clinton, Bill, 160, 238, 331
Clooney, George, 335
clutches, 10
coaches, 18–19
"Coal Miner's Daughter," 299
Coffey, Shelby, 56
"cold air induction kit," 174–75
Cold War, 36, 43, 75, 193
Cole, Edward N., 36, 40, 42, 45, 46, 48,
    63, 67, 109, 111–18, 122, 134, 136–
    37, 139, 184, 198, 294, 342, 344–45
Cole, Roy, 63
"Cole's turnpikes," 116
Collins, Bill, 168, 169, 185, 186–87
communism, 43, 65, 75, 105, 215–16
compact pickups, 292, 297–98, 309–10
"components," 297
concentration camps, 246, 249
"concept car," 321–28
Concours d'Elegance, xix
Congressional Record, 127

convertibles, 21–22, 44, 57, 58, 115, 159, 160, 172, 254, 256
Copenhagen, xix–xx
Copley Plaza Hotel, 21
"Corvette Night," 55
"counter-measures," 212
country music, 291–92, 298–302, 303, 307–8
coupes, 21–22, 172, 177, 212–13, 254
Couzens, James, 7, 8, 12, 14, 194
crankshafts, 87
Crash Course (Ingrassia), xix
"crazy speed," 206–7, 209–10, 213
"Crew Cab," 300, 308
Crocie, Ken, 171–72
Crosby, Bing, 75, 344
cross-country races, 10–11
"Cruisin' Tigers," 163–65, 189
Cunningham, Briggs, 43, 48
Curb Your Enthusiasm, xx, 314, 332–33
Curtice, Harlow "Red," 35, 66, 113, 117–18
customs duties, 295–98
"cute utes," 284
Czechoslovakia, 86, 88

"Dagmars," 64, 66, 71, 72, 193
Daimler-Benz, 84, 247, 248, 249, 287–88
DaimlerChrysler, 287–88
Dallas Cowboys, 302, 303
Danner, Richard, 130
dashboards, 11, 176, 201, 326, 328
David, Larry, 332–33, 337, 338
David, Laurie, 332, 337
Davis, David E., Jr., 251–52
Daytona International Speedway, 47, 172–73, 335
"Dead Man's Curve," 51
Dearborn Independent, 14
Death of Common Sense, The (Howard), 138
decals, 180
Defense Department, U.S., 145–46
Delage De Villars Roadster, xix
Deliverance, 291
DeLorean, Betty, 181
DeLorean, Cristina Ferrare, 186, 189
DeLorean, John Z., 165–71, 174, 175–76, 179–82, 185–89
DeLorean, Kelly Harmon, 181–82, 185

DeLorean Motor, 186–89
    DMC-12, 187–89
"delta-wing" fins, 77
DeMille, Cecil B., 18
Democratic National Convention (1968), 166
Democratic Party, 14, 166, 237, 309–10
demographics, 141–42
Denmark, xix–xx, 57–59, 78
Denver, John, 299
DeSoto, xv, 64
Detroit, xiii, xvi–xviii, 1–9, 19, 20, 23, 39–40, 43, 45, 61, 69–70, 82, 89, 97, 101, 103, 105, 111, 112, 116, 121, 123, 178–79, 191–93, 202, 205–6, 208, 210, 214–15, 217, 222–23, 226, 247, 295–98, 308, 319, 331
Detroit Athletic Club, 120
Detroit Automobile Co., 6, 284, 310, 338
Detroit Economic Club, 105
Detroit Free Press, 143, 153
Detroit Pistons, 230–31
Detroit Times, 43
DeVito, Danny, 78–79
DiCaprio, Leonardo, 331
"Dick, The" (Ridgeway), 131
Dick Van Dyke Show, 161
Diffie, Joe, 301
"directional stabilizers," 59–60, 72
Dirty Harry, 183
Dirty Jobs, 305
disassembly lines, 11
disc brakes, 250
Disney, Walt, 104
Dodge, 112, 176–77, 292, 302
    Aries, 228–29
    Caravan, 232–33, 239
    Charger, 160, 176–77, 313–14
    Charger R/T Magnum, 160
    Corona Super Bee, 58
    Coronet, 176–77
    Dart, 343
    Omni, 228
    Ram 50, 297, 306, 309
    Stealth, 215
Donahue, Phil, 136
Donna Reed Show, The, 65
door handles, 44
door locks, 232, 283

doors, 44, 165, 187, 235, 274, 283

Double Income, No Kids (DINKs), 241

Dow Jones Industrial Average, 17, 158

Doyle, Dane and Bernbach (DDB), 97–102, 107

Dreyfuss, Richard, 78–79

"Driverized Cab," 293

drive shafts, 86, 113, 201, 224

drive wheels, 221

"drop-fendered" fronts, 302

dual headlights, 174

Duesenberg Automobile & Motors Co., 16–17, 22, 62

Dukes of Hazzard, 313

Dunaway, Faye, 166

Duntov, Josef, 36–37

Duntov, Zora Arkus-, 33–56, 120

"Duntov camshaft," 47

Durant, William "Billy," 20

Dwire, Jeff, 160

Earl, Harley J., 4, 17–29, 35, 40, 44, 60–61, 70–72, 76, 150

Earl, J. W., 18–19

Earl Automobile Works, 18–19

"early adopters," 329

Earth Day, 183–84, 316

"econo-cars," 191–92

Edison Electric Illuminating Co., 5

Ed Sullivan Show, 75, 144

Eisenach Motor Works (EMW), 247

Eisenhower, Dwight D., 32, 43, 66, 112, 274

elections, U.S., 139, 166, 237–38, 296, 309–10, 345

electric automobiles, 7, 84, 315–19, 323, 335, 339

Electric Horseman, 301

Electric Kool-Aid Acid Test, The (Wolfe), 57

electric starters, 15, 16, 118

electro-mechanical transmission (EMT), 317

emission standards, 53, 84, 107, 167, 183–84, 191, 199, 200, 252, 313–39

engines:

air-cooled, 85–86, 94, 97, 107, 109, 111, 113, 121, 198, 199

aluminum, 111, 117–18, 121, 124

assembly of, 11, 27, 91

"big block," 177

"Blue Flame six," 44, 53

carburetors for, 10, 42, 115, 159, 165, 172–74, 176, 256

combustion in, xvii, 8, 9, 10, 39, 42, 47, 173–74, 200–201 315–317

"copper-cooled," 118

CVCC (Compound Vortex Controlled Combustion), 200–201

displacement of, 169–70, 251, 252, 256

eight-cylinder, 157, 176–77

"Flathead" V8, 39

four-cylinder, 9, 10–11, 114, 201, 228, 231, 232, 250, 251, 254, 260, 292, 293

four-stroke, 195–96

front-mounted, 85–86, 116–17, 198–99, 201

fuel efficiency (miles per gallon) of, 98, 107, 113, 115–16, 123–24

fuel injection for, 42, 47, 256, 260

"high-compression," 115–16, 177

high performance, 157–58, 159, 163–89

horsepower of, 8, 22, 39, 47, 51, 52, 54, 76, 94, 105, 113, 115–16, 125, 157, 159, 161, 162–63, 176–77, 182, 184, 193, 201, 210, 216, 247, 252, 260, 361, 270, 285, 286, 292, 294, 302

intake valves for, 174–75, 176, 210

internal pushrods for, 201–2, 250

"knock" in, 136

"mid," 52–54

mounts for, 168–69

multivalve, 210

noise of, 51, 91, 108, 115

overhead cams for, 201–2, 250

overheating of, 115

"pancake" style," 118

pistons of, 118

precombustion chamber in, 200

rear-mounted, 83, 85, 88, 94–95, 97, 98, 107, 109, 113, 115, 116–17, 120, 121, 133, 187–88, 201, 228

revolutions of, 174, 244–45, 252

six-cylinder, xv, 32, 35, 44, 53, 54, 151, 201–2, 254, 260, 292

size of, 94, 111, 154, 156, 161, 168–70, 180, 222

(engines *cont.*)
"small block," 116, 294
"Street Hemi," 176–79
transverse-mounted, 228
turbocharged, 125
two-cylinder, 316
two-stroke, 195–96
V6, 187, 235, 283, 294
V8, xv, 22, 28–29, 37, 39, 46, 52, 67, 76, 115–16, 156, 157–58, 171, 177, 180, 283, 293, 294, 302, 342
V10, 286
ventilation of, 105
water-cooled, 85–86, 107, 108, 113, 115, 198, 199, 200–201
weight of, 111, 115–18, 119, 121, 124, 228
environmentalism, 53, 84, 107, 167, 183–84, 191, 199, 200, 252, 286, 313–39
Environmental Protection Agency, U.S. (EPA), 335–36
Esquire Magazine Fashion Award, 69
Estes, Elliott "Pete," 171, 174
Eugene the Jeep, 269–70
Evans, Dale, 272
EX 122 prototype, 32, 35–36, 39
Executive Suite, 65, 134
exhaust pipes, 165, 183–84, 251
Exner, Iva and George, 61–62
Exner, Virgil, 61–64, 67–69, 77
Expedicion de las Americas (1978), 276
"experiential marketing," 305

factories, 81, 87–93, 97, 88, 193, 194, 202–13, 226–27, 234, 246–47, 260–61, 277, 298, 308
Fallersleben factory, 81, 87–93
fastbacks, 177
Federal Bureau of Investigation (FBI), 43, 130, 188–89
Federal Coal Mine Health and Safety Act (1977), 134
Federation of Danish American Car Clubs, 58
fenders, 60, 274
Ferrari, 43, 148, 170, 331
250 GTO, 170, 172–73
Fiat, 288
field coats, 279, 286

First National City Bank, 135
Fisher, Larry, 19, 20
Fisher Body, 23–24
Fisher brothers, 19, 20, 23–24
Fisker, 339
"Flight-Sweep," 68
Flint, Mich., 43, 65
"flivver," 11, 17, 62
floor carpeting, 157, 274
floor humps, 113
floor-mounted gear shifting, 158
Ford, Edsel, 15, 16, 61, 74
Ford, Henry, 1–17, 20–21, 24, 28, 61, 82, 118, 179, 194, 316
Ford, Henry, II, 61, 74, 92, 105, 122, 149–52, 220, 225, 227, 257–58
Ford, Henry, III, 3
Ford, Walter Buhl, III, 153
Ford Division, 144, 146–47
Ford Manufacturing Co., 8, 40, 41–42, 48
Ford Motor Co., 1–17, 61, 63, 64, 91–92, 96–97, 106, 120–21, 131–32, 135, 136, 141–62, 168, 176, 191–92, 193, 209, 210, 214–15, 219, 220, 222–23, 225, 227, 228, 272, 275–76, 290, 291, 292–98, 308, 311, 343–44
Aerostar, 233
Bronco, 275
Bronco II, 283
Country Squire, 220
Courier, 297, 298
Econoline, 224
Edsel, 141, 143, 151, 294, 319, 342
Edsel Citation, 74
Edsel Corsair, 74
Edsel Pacer, 74
Edsel Ranger, 74
Escape, 338
Excursion, xvii, 286, 287
Expedition, 284
Explorer, 283, 284, 337–38
Explorer Eddie Bauer edition, 284
Explorer Sport Trac, 337–38
F-1, 293
F-2, 293
F-3, 293
F-150, 289–90, 293, 304, 305–6, 307, 308, 337

F-150 King Ranch, 304, 308, 337
F-150 SVT Lightning (Raptor), 302, 309
F-250, 293, 304, 310
F-Series, 293–94, 300, 302–3, 304, 305, 308, 311
F-Series Platinum, 304, 308, 309
Fairlane, 96
Falcon, 97, 120–21, 122, 125, 143, 146, 150, 154, 157, 229
Fiesta, 224–25, 227–28
Fusion, 338
Model A, 7–8, 292–93
Model K, 7–8
Model N, 2, 8, 9
Model R, 2
Model S, 2, 9
Model T, xii, xiii, xviii, xix, 1–17, 28, 62, 82, 108, 111, 114, 128, 138, 194, 201, 274, 303, 341, 343, 345
Model T Runabout, 16
Model T Runabout with Pickup Body, 292
Model T "Torpedo Runabout," 11
Mustang, xii, xiv, xv, 58, 141–62, 166, 175, 177, 185, 216, 220, 221, 222, 223, 229, 233, 234, 282, 290, 342, 343–44
Mustang Fastback, 159, 160
Mustang GT-350, 159, 160
Mustang "'64-1/2s," 157
Pinto, 137–38, 192
Pygmy, 270
Ranchero, 294
Ranger, 290, 298
Taurus, 214, 343
Thunderbird, xi, xvi, 43–49, 149, 150, 342
Thunderbird II, 152
Ford Product Planning Committee, 151
Ford Sociological Department, 12–13
Formula One racing cars, 198, 213
Fortune, 60, 65–66, 271, 313
"Forward Look," 68
four-speed manual transmissions, 47, 125, 252
four-wheel drive, 88, 256, 267, 268, 270, 271, 275, 278, 294, 305–6
Fox, Michael J., 189
France, 36–38, 88, 96, 266, 276–79,

281–82, 283
Franco-American Motors, 277
Frankfurt Motor Show, 250
French Connection, The, 183
"Frenched" headlights, 67
French Grand Prix, 198
Frey, Donald, 148, 149, 151–52
Friedländer, Richard, 249
front-end springs, 168
front-wheel drive, 107, 108, 201, 227–28, 231, 255
fuel efficiency (gas mileage), 84, 184, 200, 201, 202, 210, 228, 231, 255, 260, 286, 313–39, 344
fuel prices, 53, 54, 56, 136, 167, 184–85, 191, 228, 260, 286, 287, 296, 297, 298, 310, 319, 321
fuel pumps, 10, 220
FUH2.com, 286
Fujisawa, Takeo, 194–95, 197, 199, 206
"Fun, Fun, Fun," xi, 342
"Further," 103

Garcia, Jerry, 82–83
gas-electric hybrid autos, 84, 313–39
gas gauges, 101
gasoline, 10, 42, 53, 54, 56, 84, 107, 136, 137, 138, 167, 173–74, 184–85, 191, 192, 195–96, 200, 201, 202, 210, 228, 231, 255, 260, 286, 287, 296, 297, 298, 310, 313–39, 344
gas tanks, 138, 192
gay urban professionals ("guppies"), 241
Gee, Russ, 168
GeeTO Tiger, 175
Geneen, Harold, 158
General Motors Corp. (GMC), xix, 18, 19–21, 24–25, 27, 31, 32–36, 39–46, 48, 52, 54, 60–61, 64, 66, 74–76, 96–97, 106, 111–12, 113, 116–18, 128, 129, 131–37, 150, 161, 165–68, 171, 175–76, 179–80, 185–87, 191–92, 193, 209, 210, 214–16, 228, 268, 272, 275–76, 286, 288, 297, 310, 316, 317–19, 344–45
EV1, 317–19, 323, 328, 329
Firebird, 77, 251
Hummer H2, 76, 286, 333, 336–37
Impact, 317

(General Motors Corp *cont.*)
  see also Buick; Cadillac; Chevrolet;
    Oldsmobile; Pontiac
General Motors Engineering Policy
  Committee, 117–18, 169–70
General Motors Institute, 39, 114–15,
  320
General Motors Technical Center, 66,
  70–71
General Purpose (GP) vehicle ("jeep"),
  269–72, 285–86, 288
see also Jeep
Gentlemen Prefer Blondes, 33
"Gentle on My Mind," 299
Georgia, Republic of, 261–62
German Automobile Manufacturers
  Association, 86
Germany, 37, 81–109, 193, 215–16,
  241–63
Germany, East, 215–16, 247
Germany, Nazi, 81–89, 97, 248–49
Germany, West, 215, 253
Getrag transmissions, 256
Gillen, Vincent, 130, 132
"Glass House," 145
Glen Campbell Goodtime Hour, 298–99
globalization, 183, 215–16
"Goat," 170, 179
"Go-Devil" engine, 270
Goebbels, Joseph, 248–49
Goebbels, Magda, 248–49
"Goes Like Schnell," 255
golf carts, 316
"gonads," 76, 193
Google, 109
Gordon, John, 49
Gore, Al, 139, 315
Gore, Al, III, 315, 336
Gran Turismo Omologato (GTO), 170
Grapes of Wrath, The (Steinbeck), xix, 293
Great Britain, 38, 43, 88, 92, 148–49,
  188, 268–69
Great Depression, xiii, 17, 27, 32, 66,
  269, 292–93
"Great One" campaign, 175, 179, 181
Great Society, 143
Griffith, Jack, 156–57
Griffith, Mildred, 156–57
grilles, 20, 22, 25, 27, 71, 76, 116, 121,
  154, 255, 286

Grossman, Richard L., 128
Grossman Publishers, 128
"G.T.O.," 173–74, 196
GTO Association of America, 163–65
GTO package, 170–82
gull-wing doors, 165, 187
Gung Ho, 212, 266

Hafstad, Lawrence R., 66
Hahn, Carl, 96, 97, 101–2
hand cranks, 15, 16
Harley-Davidson, 291, 302–3
Harrison, George, 183
Harvard Business School, xiii, 60, 72,
  73, 213
headlights, 25, 67, 68, 76, 174, 273, 277
heaters, 44, 108, 151
Hefner, Hugh, 33, 66
Hells Angels, 183, 196
Helsingør, Denmark, 57–59
Hemingway, Ernest, 50
Hendrix, Jimmy, 238, 259
Henry Ford Co., 6–7
Henry Ford Museum, 216
Hidden Persuaders, The (Packard), xx,
  73–74
Highland Arms, 127, 130
High Mobility Multi-purpose Wheeled
  Vehicles (HMMWV) ("Humvees"),
  285–86
highways, 43–44, 114, 277, 295, 320,
  327, 334
Hinsley couple, 100–101
"hippies," 82–83, 103–4, 181, 222, 243,
  258–60
Hirohito, Emperor of Japan, 270
Hirst, Ivan, 89–91, 92, 108
Hispano-Suiza ("Hisso"), xix, 20
Hitchcock, Janette and John, 155
Hitler, Adolf, xiii–xiv, 17, 18, 81–82,
  83, 84, 85–88, 89, 93, 94, 95, 245,
  246, 248, 249, 268, 270
"Hitler's Flivver Now Sold in the U.S.,"
  94
Hoffman, Maximilian E., 94
Hoffman Motor Car, 94
Holden, William, 65, 134
"Holden car," 119
Holls, David, 26, 70
Honda, 54, 107, 191–217, 343

Accord, 193, 201–2, 209, 210, 212–17, 219, 220, 224, 298
Civic, 107–8, 200–201, 210
CR-Z, 338
Insight, 329, 338
N360, 198–99
S360, 196–97
S500, 197
Super Cub 50 motorcycle, 195–96
Honda, Soichiro, 194–95, 196, 197, 198, 199, 203, 204, 206, 214, 216
Honda Giken Kogyo, 194–95
Honda Research and Development (Honda R&D), 197–200, 213
Hondells, 196
hood latches, 275
hood ornaments, 255
hoods, 20, 63, 64, 154, 231, 255, 274, 275, 326–27
Hootenanny, 147
Hot Rod, 34–35, 52
hot rods, xiv–xv, 34–35, 47, 52, 115, 167–82, 197, 216, 228, 291, 302
House Un-American Activities Committee (HUAC), 43
Howard, Philip, 138
Howard, Ron, 266
Howdy Doody, 73
How to Keep Your Volkswagen Alive: A Manual of Step by Step Procedures for the Compleat Idiot (Muir et al.), 106
hubcaps, 76
Hudson Motor Car, xv, 273
Hurricane Katrina, 287, 310, 336
hybrid cars, 84, 313–39
hydrogen-fuel cells, 316

Iacocca, Lee, 141–42, 144–53, 157, 167, 220–21, 223, 224–30, 234, 265, 267, 282, 283, 287, 343–44
ichiban ("number one"), 214
Indianapolis 500, 22
"Indian Uprising" convention, 163–65, 189
Industrial Designers Institute, 72
Ingrassia, Charlie, xvii
Insolent Chariots, The (Keats), xx, 73–74
installment plans, 87, 145

insurance, auto, 171–72, 182
International Harvester, 103
International Harvester Scout, 275
International Yoga Convention, 335
Intuit, 109
Irimajiri, Shoichiro, 197–98, 213
Isle of Man races, 195
Isuzu, 297
"I Was Country When Country Wasn't Cool," 291

Jackson, Alan, 303
Jackson, Katie and Scott, 337–38
Jaguar, 32, 43, 94, 278, 331
    XK120, 35
    XKE, 51
Jan & Dean, xi, 51, 156
"Japan black enamel," 11–12
J. D. Power and Associates, 209
Jeep, xi, xviii, 88, 265–88, 294–95, 313–14, 317
    see also Chrysler Jeep; Kaiser Jeep; Willys Motors; Willys-Overland
Jeepers Jamboree, 273, 288
Jeep Wrangler YJ, 277–78
Jenkins, Holman, Sr., 95–96
Jews, 41, 37, 40, 49, 86, 97, 131, 132, 246
"Job One" ceremony, 232
Jobs, Steve, 84
"Joe Six-packs," 232–33
John Deer lawn tractors, 94
Johnny Cash Show, The, 299
Johnson, Lyndon B., 133, 143, 296
Jones, Bryan, 305–6
Joplin, Janis, 183
Jordan, Chuck, xiii, 69–70, 71, 76–77
Justice Department, U.S., 135

Kaiser, Edgar, 275
Kaiser, Henry J., 272–73, 274, 275
Kaiser-Frazer, 272–73
Kaiser-Frazer Henry J., 272, 274
Kaiser Jeep, 274
    Cherokee, 274, 282–83
    Jeepster series, 275
    Tuxedo Park, 274–75
    Wagoneer, 274, 275
Kangxi, Emperor of China, 315
Kansas City Star, 234–35
Katz, Joseph, 259–60

K-cars, 228–29, 230, 233
Keating, Thomas, 116
Keats, John, 73–74, 98
Keith, Toby, 303
Keller, K. T., 63, 64
Kennedy, Edward M., 106–7, 309
Kennedy, John F., 32, 141, 144, 145, 146, 147, 324
Kennedy, Robert F., 132, 166
Kennedy, Robert F., Jr., 331
Kent State University, 183
Kerouac, Jack, xx, 50, 98, 295
Kesey, Ken, 103
Kettering, Charles "Boss," 118
Khrushchev, Nikita, 75
Kindergarten Cop, 285
King, Martin Luther, Jr., 166
Kongso, Leif, 78
Kopechne, Mary Jo, 107
Korean War, xiii, 32, 271
Kovacs, Ernie, 126
Kraft durch Freude Wagen (KdF), 81–91
Krone, Helmut, 99
Kübelwagen, 88–89, 95
Kuenheim, Eberhard von, 252–54, 257, 260–61
Kuperman, Bob, 100
Kyushu J7W1 Shinden, 324

Labor Department, U.S., 127–28
labor unions, 43, 193, 205–6, 210–11, 226–27, 266–67, 296, 298
Lamsteed Kampkar, 16
Land Rover, 289
Range Rover, 285, 289
LaSalle, xii–xiii, 3–5, 17–29, 341
V-8, 28–39
Laugh-In, 180–81
lawsuits, product-liability, xiv, 107, 112, 126–39, 198–200, 266, 278, 282, 318, 345
leaded gas, 53, 136, 137, 184, 200, 345
leather seats, 283, 284
Leave It to Beaver, 65
Lee, Harper, xviii
Leland, Henry, 6–7
LeMans auto race, 35, 40–41, 46, 47
"lemons," 74–76
Lennon, John, 102

"Leroy the Redneck Reindeer," 301
Let It Be, 183
Life, 98
Lightfoot, Gordon, 166
limousines, 278, 331
Lincoln, 74, 75, 245
    Continental, 61, 104
Lindbergh, Charles, 17
Ling, Jimmy, 158
Liston, Sonny, 144
"Little G.T.O.," 173–74, 196
"Little Honda," 196
"Little Old Lady from Pasadena, The," xi, 156
L. L. Bean, 267, 279–81, 286
Lockheed P-38 Lightning fighter, 60
Loewy, Raymond, 62–63, 68
Look, 67
Lordstown (Ohio) Honda plant, 205, 206
Los Angeles Times, 215, 242
louvered back windows, 187
louvered vents, 20, 159
Love Bug, The, 104
Lutz, Bob, 284
Lynn, Loretta, 291, 299

M5 tanks, 115
McCahill, Tom, 32
McCarthy, Joseph, 34
McDonald's, 138, 295
Mack trucks, 302
McLellan, David, 53
McNamara, Robert, 145–47, 149, 342
McQueen, Steve, 160
Maddox, Lester, 300
Maharis, George, 31–32
mainstreaming, 279–81, 283, 286
Mandrell, Barbara, 291
Man in the Gray Flannel Suit, The (Wilson), 83
manuals, car, 106
manual transmissions, xv, 10, 44, 47, 68–69, 158, 201, 231, 252, 259, 274
Margaret, Princess, 102
Maris, Roger, 157
marketing research, 26, 112, 141–42, 143, 159–60
market share, 16, 64, 67, 68, 141–42, 223, 228

Marlow, Robert, 138–39
marques, 77, 245–46, 260, 266, 271–72, 282
Marsh, Stanley, 77–78
Marshak, Seymour, 143–44
Marysville (Ohio) Honda plant, 202–14, 216, 343
Mary Tyler Moore Show, 160–61
Maserati, 47
mass production, 1–17, 23–24, 27, 28, 43, 64, 69, 87–92, 153, 158–59, 187, 193, 210, 211, 212–13, 214, 232, 322, 326, 327–31
Mauldin, Bill, 270
Mazda, 216, 297
Mead, Margaret, 59
meat packing, 11, 133
Mechanix Illustrated, 32, 122
Mercedes-Benz, 82, 84, 216, 243, 245, 248, 251, 253, 257, 260, 287, 306
Mercury News, 336–37
Meredith, James, 300
Merry Pranksters, 103, 104
"metallic brake package," 165
Mexico, 108, 215, 285
MG, 38, 43
   MGB, 148–49
Mid America Motorworks, 56
midmarket models, 157
Milford Proving Ground, 22, 40–41, 168
military vehicles, 88–89, 115, 194, 265–71
Mille Miglia road race, 246
Miller, R. S., Jr., 265
Milner, Martin, 31–32
minicars, 198–99
"Mini-Max" design, 224–25
Ministry of Transportation, German, 86
minivans, xvi, 219–39, 265, 267, 282, 284, 292, 343
mirrors, 274
Miss America pageant, 266
Mitchell, Bill, 26–27, 40, 49–50, 70–72, 77, 182
Mitsubishi Motors, 215, 216, 297
Mix, Tom, 4
Mobilgas Economy Run, 123
model years, 157, 159, 175
"mommy-mobiles," xvi, 234–39

MomsMinivan.com, 238–39
Monroe, Marilyn, 33
Monster Jam, 303, 305
Moore, Marianne, 74
Motorama, 36, 39, 77
motorcycles, 37, 54, 86, 192, 193, 194–96, 197, 198, 204–5, 206, 245, 246, 247, 291, 302–3
Motor Life, 68, 76
Motor Trend, 34, 122, 174, 180, 228, 251, 302, 338–39
Mound Road, 69–70
movies, xii, xv, 65, 78–79, 104, 134, 160, 183, 189, 212, 299
mules, 268
"multi-car," 161
"muscle cars," 58, 105, 162, 163–89, 191, 251
Museum of Modern Art, 288
music, xi, xiv–xv, 51, 147, 156, 160, 173–74, 183, 196, 291–92, 298–302, 303, 307–8, 342
Mussolini, Benito, 18
"Mustang Sally," 160
My Monastery Is a Minivan (Roy), 239
"Mystery Tiger," 175, 176
"My Sweet Lord," 183

Nader, Nathra and Rose, 127
Nader, Ralph, xiv, 111, 127–39, 175, 278, 345
Nader's Raiders, 135
Nantucket prototype, 224
Nash-Kelvinator, 273
Nation, 129–30, 131
National Alliance of Businessmen, 186
National Corvette Museum, xviii, 55
National Guard, U.S., 183
National Lampoon, 107
National Museum of American History, 108
National Traffic and Motor Vehicle Safety Act (1966), 133
Natural Gas Pipeline Safety Act (1968), 133
Nawrot, Bill, 165
Nazi Germany, 81–89, 97, 248–49
"Negligent Automobile Design and the Law" (Nader), 127

Nellybelle (Jeep), 272, 294–95, 313
Newman, Paul, 39–40, 102
New Republic, 127–28, 131
Newsweek, 104–5, 146, 154
New Yorker, 21, 103, 104, 280
New York Times, 88, 131–32, 237, 243, 309, 333
New York World's Fair (1964), 153
Nicklaus, Jack, 204
Night of the Long Knives, 86
"19th Nervous Breakdown," 166
Ninomiya, Masahito, 326
"nipples," 71, 72
Nissan, 202, 216, 297, 298, 308
 LEAF, 339
 Titan, xvii, 308
Nixon, Richard M., 133–34, 136, 202
"noise, vibration, and harshness" (NVH), 320
Nordhoff, Heinz, 91–95, 105
Northern Ireland, 187, 188
Norway, 14, 58
nuclear power, 65–66

Obama, Barack, 310, 313
off-road vehicles, 278, 285, 288
oil crisis, 53, 54, 107, 184–85, 192, 225, 255, 296, 316
oil industry, 136, 319, 321
Okuda, Hiroshi, 323–24, 327, 328
Oldsmobile, 21, 24, 27, 74, 112, 118, 168, 176
 442 model, 58, 177, 179
 Cutless Supreme, 300
Olley, Maurice, 39, 45, 120, 122, 124
"one-box" design, 231
102nd Infantry Division, U.S., 89
One Recent Child, Hideously in Debt ("ORCHIDs"), 241–42
On the Road (Kerouac), xx, 98
Opel, 215–16
Operation Americas, 124
Orbison, Roy, 291
Organization of Petroleum Exporting Countries (OPEC), 316
Original Ford Joke Book, 13–14
"Original 64," 206, 213
Oros, Joseph, 150
O'Rourke, P. J., xx, 289, 295
Orvis, xi, 280, 281, 284

outdoors retailers, 279–81
oversteering, 123

Packard, Vance, 73–74, 98
Packard Motor, 19, 167
"parallel" hybrids, 322
Paris Auto Show, 20
Parks, Bert, 258–59
Partnership for a New Generation of Vehicles, 319
Paske, Raymond, 28
passenger space, 121, 150, 176, 216–17, 224, 228–29, 250, 283, 318
Patagonia, 267, 281, 291
Patton, George S., 230
Paul, Alexandra, 318
"Peace Ship," 14
Pearlstine, Norm, xvi
Peggy Sue club, 58
People, 331
Perkins, Jim, 54
Peterman, John, 280, 291
Peterson, Robert "Pete," 34–35
phaetons, 271
Phinney, Solon E., 170
Picasso, Pablo, 40, 159, 182
Pickett, Wilson, 160
"Pickup Man," 301
pickup trucks, xiv, xvii, xix, 125, 289–311
Pierini, Rose, 128
Pikes Peak, 123–24
Piquette Street factory, 1–9
piston rings, 194
pistons, 194, 197–98
"Plain Mac," 146
Playboy, 33, 66, 103, 259
Plymouth, 64, 112, 120–21, 192
 Barracuda, 161, 177, 179
 Horizon, 228
 Reliant, 228–29
 Road Runner, 180, 181
 Street Hemi GTX, 178–79
 Valiant, 97, 120–21, 122
 Voyager, 232–33
"Polo White," 42
Pon, Ben, 93
Pontiac, 21, 62, 112, 118, 162, 167–68, 178
 Bonneville, 168

GTO, xv, xvi, 58, 162, 163–89, 196, 208, 216, 251, 252, 315
GTO Judge, 180–81
Tempest, 168, 169–72, 185, 317
Tempest LeMans, 170, 171–75, 184, 185
Pontiac PIZAZZ! (Wangers), 164–65
"pony cars," 162
Popeye, 269–70
Popular Science, 94, 228
Porsche, 46, 285
    Cayenne, 285
    Cayenne GTS, 285
Porsche, Ferdinand, 84–88, 90, 93, 106, 108, 198, 316–17
Porsche-Konstruktionsbüro, 84, 86–87
Portland Mercury, 313
Power, Aloysius F., 130
power brakes, 158
PowerFlite, 68–69
"Powerglide" transmission, 44
power locks, 283
power windows, 69, 283
"Pregnant Buick," 24
Presley, Elvis, 32–33, 59, 66
pressure, tire, 119, 120, 122, 129
Probst, Karl, 269
product development, 197–98, 213, 222–23, 253, 283, 308–9
productivity, 11, 12–13, 92–93, 205, 210, 212–13
Professional Bull Riders (PBR), 304–5
profit-sharing, 12
"Profits vs. Engineering—The Corvair Story" (Nader), 129–30
program reviews, 151, 222–23
Prohibition, 16
Project 12, 84–85
Project G21, 321–28
prototypes, 32, 35, 49, 69, 84–87, 115, 116–20, 125, 149–53, 185–89, 224, 230, 268–70, 316–28
Public Interest Research Groups, 135
public relations (PR) campaigns, 101, 104–5, 119–20, 123, 126, 132, 141, 146, 153–54, 172–73, 232, 277
Puebla, Mexico, 108
push-button automatic transmission, 68–69, 75

Putin, Vladimir, 303–4
Pyle, Ernie, 270

"Quad Cab," 300, 304
Quadricycle, 5–6
quality control, 99, 188, 191–92, 193, 195, 209, 213, 308
Quandt, Antonie, 248
Quandt, Günther, 248, 249
Quandt, Harald, 248, 249–50, 252
Quandt, Herbert, 248, 249–50, 251, 252, 253, 257, 258
Quandt, Johanna, 257–58

"Racing Spirit, The" (Irimajiri), 213
racing stripes, 177
Radiation Control for Health and Safety Act (1968), 133
radiators, 113, 118
radios, 44, 158
"Ragged Old Truck," 301
Railton, Arthur, 105
rally wheels, 177
Randy & the Rainbows, 164
"range anxiety," 319
Range Rover, 285, 289
Rapaille, Clotaire, 302
Ready, Jack, Sr., 155–56
Reagan, Ronald, 274
rear-end stability, 119, 121–25, 126, 128–30, 135–36, 187–88, 198–200, 278
rear seats, 48–49, 150, 152, 342
rear suspension, 49, 250
rear-wheel transmissions, 201, 224, 233, 255, 271
rear windows, 40–50, 69, 187
recessions, 75–76, 208–9, 287–88, 310–11
Redford, Robert, 301
"Red Scare," 43, 65
"regenerative" energy, 326
Reiner, Rob, 331, 335
Renault, 96, 266, 276–79, 281–82, 283
    Alliance, 266, 277, 282
Republican Party, 14, 237
Republican Party Reptile (O'Rourke), xx, 289
Rhodes, James A., 203–4, 213
Ribicoff, Abraham, 127, 128, 132–33, 134
Richards, Gary, 336–37

Ridgeway, James, 131
Road & Track, 201, 251
roadsters, xix, 35, 143, 148–49, 187, 271
"Robin Hood" features, 157
Robinson, Edrie, 235–36
Roche, James M., 132, 134, 175–76, 181
"Rocinante" Chevy El Camino, 294–95
Rogers, Roy, 272, 294–95, 313
"rolling model changes," 210, 213
Rolling Stone, 82–83
Rolling Stones, 166, 182–83, 258
Rolls-Royce, 126
roll stability, 124–25, 266, 278, 282
Romney, George, 112, 177, 265
Ronny and the Daytonas, 173–74
Roosevelt, Theodore, 10
Rootes, William, 90
Rosenberger, Adolf, 87
"roundels," 245–46
Route 66, xx, 31–32, 50–51, 313
Royal Electrical and Mechanical
    Engineers, British, 89
Royal Pontiac dealership, 178
Roy Rogers Show, xx
RPO 684 performance package, 47
RPO L88 performance package,
    51–52
rubber-composite front bumpers, 180
Rugaber, Walter, 131–32
rumble seats, 21
Russert, Tim, 309

Saab, 243
Sachtleben, Dale, xvi
Sacramento Bee, 334
"sales bank," 226
San Francisco Chronicle, 238
Sarnoff, David, 65–66
Saturday Evening Post, 32, 122
Saxon, 114
Schmidt, Eric, 109
Schultz, Larry, 258–59
Schwarzenegger, Arnold, 285, 333
Schwimmwagen ("swimming car"), 89
"scoops," 150, 174–75
scout cars, 270
seat cushions, 274
seats, 21, 48–49, 125, 148–49, 150, 152,
    157, 176, 180, 231–32, 274, 283,
    284, 342

sedans, 2, 21, 28, 44, 93–94, 95, 112, 125,
    126, 177, 216–17, 229, 236, 254–55,
    266, 305
Seeger, Pete, 147
Seger, Bob, 301–2, 303
Selfridge Air Base, 60
Senate Subcommittee on Executive
    Reorganization, 127, 128, 132–33,
    134
"series" hybrids, 322
"Seven Dwarfs," 24
Shapo, Marshall, 137
shareholders, 14, 20–21, 225, 233, 248,
    249–50
Shelby, Carroll, 159, 160
shock absorbers, 47
Shoemaker, Craig, 219
Shore, Dinah, 35–36, 42
"Shut Down," xiv–xv
"side curtains," 44
"side splitters," 165
Sierra Club, 318, 333
Simca, 227–28
Sinatra, Frank, 75, 147, 275
Single Income, Two Children,
    Oppressive Marriage
    ("SITCOMs"), 242
Skwirblies, John, 189
slave labor, 89, 246–47, 249
sliding side doors, 235
Sloan, Alfred P., Jr., 19, 20–21, 23, 118
"Slug Bug," 95
Small World, 101
SmartMoney, xvii
Smith, Laurel, 238–39
Smothers Brothers Comedy Hour, 298
"Snobs' Guide to Status Cars," 103
"soccer moms," 234–39, 241, 284
Society of Automotive Engineers, 41,
    120, 124, 320
solar panels, 329
Sorensen, Theodore, 133
sound systems, 257
South Park, 334
Soviet Union, 36–37, 60, 75, 89, 90, 247,
    249, 253
space race, 60, 69, 324
Spartanburg (South Carolina) BMW
    plant, 260–61
"specialty" trucks, 291

speed, 10, 47, 56, 113, 172–75, 185, 277, 315, 327
speed limits, 185
speedsters, 2
Speed Week, 48
Sperlich, Harold K. "Hal," 141, 144, 148–49, 151–53, 157, 220–30, 233, 234, 239, 267
spinouts, 126, 129, 187–88
split windows, 49–50
spoilers, 180
"sport hybrid," 338
Sportone trim, 72
Sports Car Illustrated, 113–14, 123, 126
Sports Illustrated, 39, 47, 52, 230
sport-utility vehicles (SUVs), xvii, 216, 239, 265–88, 289, 303, 314, 321, 332, 334, 337–38
Sputnik, 75
"stability enhancement" kits, 129
Stanley Steamer, 316
starters, 15, 16, 118
"static prototype," 322
station wagons, xv, 112, 125, 126, 156, 220, 222, 224, 235, 254, 256, 271, 274
steering, xv, 10, 47, 123, 176, 328
steering columns, xv, 10
steering wheels, 176, 328
Steinbeck, John, xix, 1, 3, 293, 294–95
Stevenson, Adlai, 103
Stewart, Jon, xviii, 342
Steyr, 84
stick shifts, xv, 68–69, 158, 259, 274
stock cars, 154, 167–68
Stout, William, 221
Stout Scarab, 221–22
Studebaker, 39, 62–63, 68, 167
  Dictator, 18
Subaru, 216
  Brat, 297
subcompacts, 200, 224, 227–28
suburbia, 65, 121
Sukarno, 66
Summer of Love, 166
Sunbeam Alpine, 148–49
Sun Microsystems, 109
sunroofs, 283
"Super Cab," 300

Supreme Court, U.S., 56, 139
suspensions, 10, 47, 49, 107, 126, 129, 168, 169, 177, 187–88, 200, 210, 257, 266
Suzuki Grand Vitara, 284
swing-axle rear suspensions, 124
Sydney Morning Herald, 261

T115 prototype, 230
tachometers, 174, 176, 251
tail fins, xiii, 56, 57–79, 83, 96, 134, 193, 274, 286, 288, 320, 337, 342
tailgates, 271
taillights, 76
"Take Me Home, Country Roads," 299
tanks, 88, 115
tariffs, 295–98
television, xx, 31–32, 50–51, 65, 73, 99–100, 153, 157, 160–61, 180, 236, 272, 291, 298–99, 305, 309, 313–14, 332–33
Television Quarterly, 50
Territory Ahead, 280
Tesla, 339
test tracks, 118, 152, 168, 172–73, 324–25
"Texas Mustang," 290
Thai-Tang, Hau, 344
"Thank God I'm a Country Boy," 299
"thermal incidents," 318
"This Old Truck," 301
Thomas, Clarence, 56
"Those Were the Days," 17
"Thoughts Pertaining to Youth, Hot Rodders and Chevrolet" (Duntov), 41–42, 55
three-box design, 64, 67, 231, 250
"three deuces," 173–74, 176
"three joys," 212
"three-point" chassis, 10
Tiffany Design Award, 153
Time, 43, 44, 63, 66, 75, 94, 103, 111, 122, 133, 134, 145, 154, 166, 230, 319
"Tin Lizzie," 2
Tin Men, 78–79
tires, 119, 120, 122, 129, 159, 169, 315
To Kill a Mockingbird (Lee), xviii
Tokyo Motor Show, 196–97, 206, 322
Tokyo Stock Exchange, 195

Toledo, Ohio, 266–67, 273
touring cars, 157
Tourist Trophy, 195
Toyoda, Eiji, 321, 328
Toyota, 202, 216, 319–38
  Camry, 343
  City, 319, 320, 328
  Corolla, 322, 327, 329
  4Runner, 284
  Land Cruiser, 284
  Previa, 234
  Prius, xviii, 121, 313, 319–39
  RAV4 (Recreational Activity
    Vehicle), 284
  Tundra, 308
Toyota Vehicle Development Center
  2, 321
Trabant, 215–16
trailer hitches, 304, 307
transmissions, 44, 47, 75, 113, 125, 151,
  156, 158, 201, 224, 233, 255, 256,
  270, 271, 322
"Trap at Cordova," 51
Travels with Charley (Steinbeck), xix,
  294–95
Travolta, John, 301
Treasury Department, U.S., 288
T. Rex, 275
trial lawyers, 129–30, 137–39
trim, 231–32
Tri-Power carburetors, 165, 176
Triumph TR4, 148–49
Truck Room, 304–5
trucks, 229, 291, 294, 302
  see also pickup trucks
True: The Man's Magazine, 114, 118
trunks, 64, 150, 229, 231
TRW, 316–17, 326
Tucker, Preston, 115
Tucker, Stanley, 154, 159
"Turn Your Hymnals to 2002" (Davis),
  251–52
TV Guide, 31, 142
two-car families, 112–13, 114, 142,
  219–20, 228
two-seaters, 148–49
two-speed automatic transmissions,
  44
two-wheel drive, 88, 278
"Type 1 sedan," 95

"Type 2 station wagon," 95

Uchiyamada, Takeshi, 319–28, 338,
  339
"Ultimate Driving Machine," 255
"unibody" construction, 283
Uniroyal, 118
United Auto Workers (UAW), 206,
  210–11, 296, 298
unleaded gas, 53, 136, 137, 184, 200, 345
Unsafe at Any Speed (Nader), xiv,
  128–30, 133
Urban Cowboy, 290, 301
U.S. Electric Vehicle Association, 335
U.S. Rubber Co., 119

V-1 rockets, 89
vanadium steel, 9–10
vans, 125, 221–25
  see also minivans
Verbiest, Ferdinand, 315
"vertically stacked" dual headlights, 174
Vietnam War, 53, 134, 150, 183, 192,
  258, 344
Vogue, 22
Volkswagen, 78, 81–109, 166
  Beetle, xiii–xiv, 79, 81–109, 112, 113,
    119, 121, 123, 133, 150, 155, 166,
    199, 202, 206, 221, 247, 258, 260,
    316, 317, 337, 344
  Dasher, 108
  EuroVan, 108
  Microbus, 2, 82–83, 84, 93, 95, 96,
    102–4, 108, 125, 222, 258–59, 260,
    343
  New Beetle, 108–9
  Rabbit, 206
  Super Beetle, 107
volkswagen ("People's car"), 86–87
Volkswagen of America, 95–102,
  166, 203, 206, 213, 216, 254, 255,
  276
"Volkswagen Sedan," 81
Volkswagenwerk, 93, 247, 257
Volvo, 243, 257

wages, 2, 3, 12–13
Walker, George William, 61, 67, 75
Wall Street Journal, xvi–xvii, 12, 135,
  146, 280, 285

Wangers, Jim, 164–65, 172–73, 174, 178–79
Ward's Automotive Yearbook, 297
Warhol, Andy, 99
Washington Post, 191, 215, 331
Watergate scandal, 53, 192
water pumps, 113
Watkins Glen race track, 35
Wayne, John, 42
weight distribution, 124–25
welding robots, 210
wheelbase, 22, 277
wheels, 9, 22, 177, 277
see also tires
"White Christmas," 344
Whitman, Charles, 166
Whitney, Clay, xvi
Who Killed the Electric Car?, 318–19
Whole Foods, 313, 335
Wholesale Poultry Products Act (1968), 133
"Wide Track" marketing campaign, 168
Wilder, Billy, 126
Wilhelm II, Emperor of Germany, 245
Wilkin, John "Bucky," 173, 174
Will, George, 55
"Willie and Joe," 270, 285, 288
Williford, John, 306–7
Willys Motors, 273–74
    CJ-5 (Civilian Jeep), 273–75, 277, 278, 288
    Jeep Maverick, 274
    Surrey, 274
Willys-Overland, 269–73
    CJ-2A (Civilian Jeep), 271
    Jeepster, 271
    Jeep Truck, 271
Wilson, Charles "Engine Charlie," 116
Wilson, Sloan, 83
windows, 44, 49–50, 187, 201, 202, 221–22, 283

Winton, Alexander, 6
Wisconsin, University of, (Madison), 183
Wolfe, Tom, 57
Wolff, Elfi, 37, 38, 39
Wolfsburg, Germany, 97, 99
Woman's Angle, The, 155
women, 12, 22, 67, 102, 142, 143–44, 153, 155, 160–61, 195, 232, 234–39, 284–85
Woods Interurban, 316
Woodward Avenue, 178–79, 208
workers, auto, 12–13, 43, 193, 205–11, 226–27, 232–33, 266–67, 287, 296, 298
World War I, 14, 245, 246, 268
World War II, xiii, 28, 88–91, 95–96, 115, 152, 230, 246–47, 266, 267–71, 272, 274, 278, 286, 288, 324
Wouk, Victor, 317, 331
Wozniak, Steve, 336–37

X-rays, 133

"Yak Pack," 238
Yates, Brock, 202
Yeats, William Butler, 262
Yoshida, Shige, 203–4, 206, 207
Yoshino, Hiroyuki, 214
"You Can't Always Get What You Want," 258
young urban professionals ("yuppies"), 4, 241–63, 267, 278, 302
YouTube, 109
"Yuppie cotillion," 243

Zappa, Frank, 272
Zarichny, James, 43
ZF Friedrichshafen ("Zed F") steering systems, 256